# delivery of protein and peptide drugs in cancer

# delivery of protein and peptide drugs in cancer

vladimir p torchilin
Northeastern University, USA

ICP Imperial College Press

*Published by*

Imperial College Press
57 Shelton Street
Covent Garden
London WC2H 9HE

*Distributed by*

World Scientific Publishing Co. Pte. Ltd.
5 Toh Tuck Link, Singapore 596224
*USA office:* 27 Warren Street, Suite 401-402, Hackensack, NJ 07601
*UK office:* 57 Shelton Street, Covent Garden, London WC2H 9HE

**British Library Cataloguing-in-Publication Data**
A catalogue record for this book is available from the British Library.

**DELIVERY OF PROTEIN AND PEPTIDE DRUGS IN CANCER**



ISBN 1-86094-627-5

Printed in Singapore by Mainland Press

# Contents

# Contents

# Contributors

**Robert B Campbell**
Department of Pharmaceutical Sciences,
Northeastern University,
360 Huntington Ave
Bouvé College of Health Sciences,
110 Mugar Hall
Boston, MA 02115,
USA

**Sunil Chada**
Introgen Therapeutics, Inc.,
2250 Holcombe Blvd.,
Houston, TX 77030,
USA

**Daan J A Crommelin**
Department of Pharmaceutics,
Utrecht Institute for Pharmaceutical Sciences,
Utrecht University,
Utrecht,
The Netherlands

**Steven F Dowdy**
Howard Hughes Medical Institute,
and Department of Cellular and Molecular Medicine,
George Palade Laboratories,
Room 231B,
USCD School of Medicine,
9500 Gilman Drive,
Dept. 0686,
La Jolla, CA 92093-0686

**James F Eliason**
Asterand, Inc.,
TechOne Suite 501,
440 Burroughs,
Detroit, MI 48202-3420
USA

**Julia Fahrmeir**
Pharmaceutical
Biology-Biotechnology,
Center of Drug Research
Department of Pharmacy,
Butenandtstrasse 5-13,
81377 München,
Germany

**Jun Fang**
Faculty of Pharmaceutical
Sciences, Sojo University,
4-22-1, Ikeda Kumamoto,
860-0082, Japan

**Khaled Greish**
Faculty of Pharmaceutical
Sciences, Sojo University,
4-22-1, Ikeda, Kumamoto,
860-0082, Japan

**Arun K Iyer**
Faculty of Pharmaceutical
Sciences, Sojo University,
4-22-1, Ikeda, Kumamoto,
860-0082, Japan

**Michio Kawasuji**
Department of Cardiovascular
Surgery, Kumamoto
University Graduate
School of Medical Sciences,
Honjo 1-1-1, Kumamoto, Japan

**Christopher P Leamon**
Endocyte, Inc.,
3000 Kent Avenue,
Suite A1-100
West Lafayette,
IN 47906, USA

**Philip S Low**
Department of Chemistry,
Purdue University,
560 Oval Drive
West Lafayette,
IN 47906, USA

**Hiroshi Maeda**
Faculty of Pharmaceutical
Sciences, Sojo University,
4-22-1, Ikeda, Kumamoto,
860-0082, Japan

**Margherita Morpurgo**
Department of
Pharmaceutical Science,
University of Padova,
via F. Marzolo 5,
35131-Padova,
Italy

**Herbert B Newton**
Dardinger Neuro-Oncology
Center and Division
of Neuro-Oncology,
Departments of Neurology
and Oncology,
Ohio State
University Medical
Center and James
Cancer Hospital
465 Means Hall
1654 Upham Drive,
Columbus,
Ohio 43210, USA

**Manfred Ogris**
Pharmaceutical
Biology-Biotechnology,
Center of Drug Research
Department of Pharmacy,
Butenandtstrasse 5-13 (Haus D)
81377 München
Germany

**Gianfranco Pasut**
Department of
Pharmaceutical Science,
University of Padova,
via F. Marzolo 5,
35131-Padova,
Italy

**Rajagopal Ramesh**
Department of Thoracic
and Cardiovascular
Surgery, Unit 445
M.D. Anderson Cancer Center,
P.O. Box 301402
1515 Holcombe Blvd
Houston, TX 77230-1402, USA

**Joseph A Reddy**
Endocyte, Inc.,
3000 Kent Avenue, Suite A1-100,
West Lafayette,
IN 47906, USA

**Curzio Rüegg**
Lausanne Cancer Center (LCC)
Swiss Institute for
Experimental Cancer
Research (ISREC)
Chemin des Boveresses 155
CH-1066 Epalinges
Switzerland

**Cheryl C Saenz**
Department of
Reproductive Medicine,
USCD School of Medicine,
La Jolla, California
92093-8433, USA

**Raymond M Schiffelers**
Department of Pharmaceutics,
Utrecht Institute for
Pharmaceutical Sciences,
PO Box 80.082,
3508 TB Utrecht,
The Netherlands

**Gert Storm**
Department of Pharmaceutics,
Utrecht Institute for
Pharmaceutical Sciences,
Utrecht University,
Utrecht,
The Netherlands

**Vladimir Torchilin**
Department of Pharmaceutical
Science, Bouvé College
of Health Sciences,
Northeastern University
312 Mugar Hall,
Boston,
MA 02115,
USA

**Francesco M Veronese**
Department of
Pharmaceutical Science,
University of Padova,
via F. Marzolo 5,
35131-Padova
Italy

**Ellen K Wasan**
Advanced Therapeutics,
Faculty of Pharmaceutical Sciences,
University of British Columbia,
BC Cancer Agency
675 West 10th Ave
Vancouver, BC V5Z 1L3
Canada

**Kishor M Wasan**
Division of Pharmaceutics
and Biopharmaceutics,
Faculty of Pharmaceutical Sciences,
University of British Columbia,
2146 East Mall
Vancouver, British Columbia
Canada V6T 1Z3

# 1

# Introduction

*Vladimir Torchilin*

To effectively treat cancer, we have to be able to selectively attack the tumor and individual cancer cells, while effectively protect normal tissue from possible toxicity and other side effects of anti cancer drugs. It is not an easy task, since systemically administered drugs may rapidly metabolize in the blood or be cleared from the body before reaching the tumor cells. In addition, on its way to the target, it has to overcome multiple physiological barriers, such as irregularities in the tumor blood flow, the high interstitial pressure, and the absence of a lymphatic drainage in tumors (Campbell, Chapter 2). Many drugs have also been found to perform their action inside the cells which requires their intracellular delivery through low permeable cell membranes. All these obstacles are especially pronounced in the case of protein and peptide drugs, whose successful application needs effective means of drug delivery into tumors. This book will consider various problems associated with tumor delivery of protein and peptide drugs and some of the current strategies to solve these problems.

It is well known that any proteins and peptides possess biological activity that makes them therapeutically potent, in particular, anticancer agents. Advances in solid-phase peptide synthesis and recombinant DNA and hybridoma technology allow for production of unlimited quantities of clinical grade protein and peptides. The use of proteins and peptides as therapeutic agents is hampered, however, by their fast elimination from circulation mostly because of renal filtration, fast enzymatic degradation, uptake by the reticuloendothelial system (RES) and accumulation in non-targeted organs and tissues. Fast elimination and distribution into non-targeted organs and tissues cause the need to administer a drug in large quantities, which is often uneconomical and sometimes impossible due to non-specific toxicity. Low permeability of cell membranes for macromolecules often represents an additional obstacle for the development of protein and peptide based anticancer formulations. Numerous approaches to overcome fast elimination and non-specific biodistribution of conventional drugs have been developed and can be adapted for the delivery of anticancer protein and peptides. This book focuses on injectable microscopic systems for the delivery of protein and peptide anticancer agents to and into tumors. Main advantages of these systems over macroscopic devices include greater convenience and less invasive administration, the ability to reach delocalized targets and a lower manufacturing cost.

One of the reasons for fast clearance from systemic circulation of proteins and peptides with molecular weight of 40 kDa or lower is renal filtration. This issue may be addressed by conjugation of the biomolecules with water-soluble polymers, which results in a complex with high enough molecular weight.[1] Additional benefits of protein (peptide)–polymer conjugation are increased resistance against enzyme degradation and lowered immunogenicity. Both enzymatic degradation and immune response against a protein cause its fast elimination from the systemic circulation. The developing of the immune response, in addition, is potentially dangerous because of the possibility of allergic reactions and anaphylactic shock upon repetitive administrations. Polymer molecules attached to the protein globule create steric hindrances, which interfere with active sites of proteases,[1] opsonins or antigen-processing cell.[2] Currently, poly(ethylene glycol) (PEG) is the most popular polymer for modification of proteins with therapeutic potential[3–5] (Veronese, Chapter 4; Eliason, Chapter 6). PEG modified L-asparaginase has been proposed as an anticancer agent as early as in 1984.[5] This formulation (Oncospar® from Enzon) was approved as an orphan drug in the US for use in lymphoma and leukemia treatments.[6] It has a longer circulation time than the original enzyme and does not induce hypersensitivity reaction in patients with such reaction to the non-modified enzyme.[7,8] In some cases, drugs are conjugated with polymers that can attach themselves and conjugated drug to natural long-circulating blood plasma components, like serum albumin or lipoproteins. Thus, the conjugation of proteins and peptides with poly(styrene-co-maleic acid/anhydride) (SMA)[9] increases the circulation time of anticancer proteins and peptides *via* the binding of the conjugates to plasma albumin.[10] The conjugation with SMA also protects proteins from enzymatic degradation, and decreases immunogenicity of modified proteins.[9] SMA-modified neocarzinostatin is currently approved in Japan for hepatoma treatment.[11]

High molecular weight (40 kDa or higher), long-circulating macromolecules, including proteins and peptides conjugated with water-soluble polymers, are capable of spontaneous accumulations in solid tumors via the enhanced permeability and retention effect (EPR).[9,11] This effect is based on the fact that tumor vasculature, unlike vasculature of healthy tissues, is "leaky", i.e. penetrable for macromolecules and nanoparticulates, which allows macromolecules to accumulate in the interstitial tumor space (see Maeda, Chapter 3). Such accumulation is also facilitated by the fact that lymphatic system, responsible for the drainage of macromolecules from normal tissues, is virtually not working in case of many tumors as a result of the disease.[11]

Nanoparticulate drug delivery systems may represent a valid alternative to soluble polymeric carriers. This type of systems includes liposomes, micelles, polymer microparticles, etc. The use of this type of carriers allows achieving much higher active moiety/carrier material ratio compared with "direct" molecular conjugates. They also provide better protection of protein and peptide drugs against enzymatic degradation and other destructive factors upon parenteral administration because the carrier wall completely isolates drug molecules from the environment. All nanooparticulate carriers have the size, which excludes a possibility of renal filtration. The main disadvantage of microreservoir carriers is their tendencies to be taken up by the RES cells primary in liver and spleen.[12] Among particulate drug carriers, liposomes are the most extensively studied and poses the most suitable characteristics for protein (peptide) encapsulation (Torchilin, Chapter 8). Similar to macromolecules, liposomes are capable of accumulating in tumors of various origins *via* the EPR effect.[13,14] In some cases, however, the liposome size is too large to provide an efficient accumulation via the EPR effect presumably due to relatively small tumor vasculature cut off size.[15,16] In such cases, alternative delivery systems with smaller sizes such as micelles (prepared, for example, from PEG-phospholipid conjugates) can be used. These particles lack the internal aqueous space and are smaller than liposomes. Protein or peptide pharmaceutical agent can be covalently attached to the surface of these particles or incorporated into them via chemically attached hydrophobic group ("anchor").[15]

The use of vector molecules can further enhance tumor targeting of protein/peptide drugs or protein/peptide-loaded nanocarries or make them EPR effect independent. The latter is especially important for the cases of tumors with immature vasculature, such as tumors on the earlier stages of their development, and delocalized tumors. Vector molecules (those having affinity toward ligands characteristic for target tissues) capable of recognizing tumors were found among antibodies, peptides, lectines, saccharides, hormones, and some low molecular weight compounds[17] (Reddy, Chapter 9; Ogris, Chapter 10). From this list, antibodies and their fragments provide the most universal opportunity to target various targets and have the highest potential specificity. Antibodies capable of recognizing specific antigens were derived for the majority of known tumors.[18] Recent advances in recombinant engineering make it possible to produce anticancer antibodies on industrial scale at relatively low cost. Humanized versions of antibodies and their fragments in which rodent-derived binding sites and human conservative regions are combined using recombinant technology became available.[19,20]

The successful delivery of anticancer drugs, proteins and peptides among them, into tumors does, however, solve only a part of a general efficiency problem. The following task is to achieve their intracellular delivery, since many targets for anticancer drugs are located inside cells (e.g. the surface of mitochondria may serve as a promising target for apoptosis-inducing drugs). A huge body of available information about cellular metabolic and signaling pathways essential for tumorogenesis and tumor cell development allows for identifying protein targets for interference with the tumor growth. Quite a few molecular targets have already been identified.[21] The creation of a working draft of the human genome sequence[22,23] in combination with high-throughput methods of molecular biology promises continued rapid growth in identifying such targets.[24,25] Sometimes tumor results from the malfunctions of tumor suppressor genes, as well as the lack of activity of the proteins they encode.[25,26] In this case, the delivery into tumor cells of working copies of proteins obtained by recombinant methods would provide indispensable tools for the validation of gene functions and potential development of protein or gene therapy-based methods of treatment (Chada, Chapter 15; Crommelin, Chapter 7).

Generally, the use of peptides and proteins for molecular target validation and eventual development of anticancer drugs is hampered by the low permeability of cell membranes. The very nature of cell membranes prevents protein/peptide entering unless there is an active transport mechanism, which is usually the case for very short peptides.[27] As mentioned above, vector molecules promote the delivery of associated drug-carriers inside the cells via receptor-mediated endocytosis.[28] An efficient cellular uptake via endocytosis is generally observed, but the delivery of intact proteins and peptides is compromised by an insufficient endosomal escape and lysosomal degradation. An enhanced endosomal escape can be achieved through the use of, for example, lytic peptides,[29,30] pH-sensitive polymers[31] or swellable dendritic polymers.[32] Although these agents have provided encouraging results in overcoming limitations of endocytosis-based cytoplasmic delivery, there is still a need for further improvements or alternative delivery strategies. An approach recently emerged, which provides for a much more straightforward and efficient way for delivery of proteins and peptides to the cytoplasm. This approach is based on the phenomenon called transduction (Dowdy, Chapter 11), and uses the ability of certain peptides to ferry conjugated macromolecules, such as proteins[33] and DNA, and even particles as large as 40 nm iron oxide colloidal particles[34,35] and 200 nm liposomes,[36,37] across cell membranes directly into cytoplasm. Peptides that cause transduction (PTDs, protein transduction domains, or CPPs, cell-penetrating peptides) can be as short as 10-to-16-mer.[33,38,39] Several proteins including those involved in oncogenesis, cancer-related signal transductuction and cell proliferation pathways have been delivered in active form into various human cells *in vitro* using fused PTD peptides.[40–43] It has also been shown that TAT PTD allows delivery of biologically active proteins into various cells *in vivo*.[44] These results open new avenues in the development of protein and peptide-based anticancer therapeutics with intracellular molecular targets.

Thus, current knowledge provides some promising approaches how to deliver protein and/or peptide-based anticancer drugs into tumors (see also Wasan, Chapter 13) and further inside tumor cells. This opens new opportunities for improved therapy of various cancers (see Ruegg,

Chapter 12, Newton, Chapter 14). This book certainly covers only a fraction of issues related to the use of protein and peptide drugs in cancer therapy. Still, we hope that the information it contains will be useful for academics and clinicians involved in related research.

## References


1. Torchilin VP (1991) Immobilized enzymes in medicine. *Progress in Clinical Biochemistry and Medicine.* **11**:206, Heildelberg Springer-Verlag: Berlin.
2. Harris JM, Martin NE and Modi M (2001) Pegylation: A novel process for modifying pharmacokinetics. *Clin Pharmacokinet* **40**:539.
3. Veronese FM and Harris JM (2002) Introduction and overview of peptide and protein pegylation. *Adv Drug Deliv Rev* **54**:453.
4. Roberts MJ, Bentley MD and Harris JM (2002) Chemistry for peptide and protein PEGylation. *Adv Drug Deliv Rev* **54**:459.
5. Abuchowski A, Kazo GM, Jr, Verhoest CR, Van Es T, Kafkewitz D, Nucci ML, Viau AT and Davis FF (1984) Cancer therapy with chemically modified enzymes. I. Antitumor properties of polyethylene glycol-asparaginase conjugates. *Cancer Biochem Biophys* **7**:175.
6. Ettinger AR (1995) Pegaspargase (Oncaspar), *J Pediatr Oncol Nurs* **12**:46.
7. Asselin BL (1999) The three asparaginases. Comparative pharmacology and optimal use in childhood leukemia. *Adv Exp Med Biol* **457**:621.
8. Holle LM (1997) Pegaspargase: An alternative? *Ann Pharmacother* **31**:616.
9. Maeda H (2001) SMANCS and polymer-conjugated macromolecular drugs: Advantages in cancer chemotherapy. *Adv Drug Deliv Rev* **46**:169.
10. Mu Y, Kamada H, Kaneda Y, Yamamoto Y, Kodaira H, Tsunoda S, Tsutsumi Y, Maeda M, Kawasaki K, Nomizu M, Yamada Y and Mayumi T (1999) Bioconjugation of laminin peptide YIGSR with poly(styrene co-maleic acid) increases its antimetastatic effect on lung metastasis of B16-BL6 melanoma cells. *Biochem Biophys Res Commun* **255**:75.
11. Maeda H, Sawa T and Konno T (2001) Mechanism of tumor-targeted delivery of macromolecular drugs, including the EPR effect in solid tumor and clinical overview of the prototype polymeric drug SMANCS. *J Control Rel* **74**:47.
12. Senior H (1987) Fate and behavior of liposomes in vivo: A review of controlling factors. *Crit Rev Ther Drug Carrier Syst* **3**:123.
13. Yuan F, Leunig M, Huang SK, Berk DA, Papahadjopoulos D and Jain RK (1994) Microvascular permeability and interstitial penetration of sterically stabilized (stealth) liposomes in a human tumor xenograft. *Cancer Res* **54**:3352.

14. Gabizon AA (2001) Pegylated liposomal doxorubicin: Metamorphosis of an old drug into a new form of chemotherapy. *Cancer Invest* **19**:424.
15. Weissig V, Whiteman KR and Torchilin VP (1998) Accumulation of protein-loaded long-circulating micelles and liposomes in subcutaneous Lewis lung carcinoma in mice. *Pharm Res* **15**:1552.
16. Hobbs SK, Monsky WL, Yuan F, Roberts WG, Griffith L, Torchilin VP and Jain RJ (1998) Regulation of transport pathways in tumor vessels: Role of tumor type and microenvironment. *Proc Natl Acad Sci USA* **95**:4607.
17. Gregoriadis G (1977) Targeting of drugs. *Nature* **265**:407.
18. Torchilin VP, Ed. (1995) *Handbook of Targeted Delivery of Imaging Agents.* CRS Press, Boca Raton.
19. Jurcic JD, Scheinberg DA and Houghton AN (1997) Monoclonal antibody therapy of cancer. *Cancer Chemother Biol Response Modif* **17**:195.
20. Dillman RO (2001) Monoclonal antibodies in the treatment of malignancy: Basic concepts and recent developments. *Cancer Invest* **19**:833.
21. Gibbs JB (2001) Mechanism-based target identification and drug discovery in cancer research. *Science* **287**:1969.
22. Venter JC, *et al.* (2001) The sequence of the human genome. *Science* **291**:1304.
23. Lander ES, *et al.* (2001) Initial sequencing and analysis of the human genome. *Nature* **409**:860.
24. Workman P (2001) New drug targets for genomic cancer therapy: Successes, limitations, opportunities and future challenges. *Curr Cancer Drug Targets* **1**:33.
25. Balmain A (2001) Cancer genetics: From Boveri and Mendel to microarrays. *Nat Rev Cancer* **1**:77.
26. Hussain SP, Hofseth LJ and Harris CC (2001) Tumor suppressor genes: At the crossroads of molecular carcinogenesis, molecular epidemiology and human risk assessment. *Lung Cancer* **34**(Suppl 2):S7.
27. Egleton RD and Davis TP (1997) Bioavailability and transport of peptides and peptide drugs into the brain. *Peptides* **18**:1431.
28. Park JW, Kirpotin DB, Hong K, Shalaby R, Shao Y, Nielsen UB, Marks JD, Papahadjopoulos D and Benz CC (2001) Tumor targeting using anti-her2 immunoliposomes. *J Control Rel* **74**:95.
29. Kamata H, Yagisawa H, Takahashi S and Hirata H (1994) Amphiphilic peptides enhance the efficiency of liposome-mediated DNA transfection. *Nucleic Acids Res* **22**:536.
30. Midoux P, Kichler A, Boutin V, Maurizot JC and Monsigny M (1998) Membrane permeabilization and efficient gene transfer by a peptide containing several histidines. *Bioconjug Chem* **9**:260.

31. Lackey CA, Press OW, Hoffman AS and Stayton PS (2002) A biomimetic pH-responsive polymer directs endosomal release and intracellular delivery of an endocytosed antibody complex. *Bioconjug Chem* **13**:996.
32. Padilla De Jesus OL, Ihre HR, Gagne L, Frechet JM and Szoka FC Jr (2002) Polyester dendritic systems for drug delivery applications: In vitro and in vivo evaluation. *Bioconjug Chem* **13**:453.
33. Vives EP, Brodin P and Lebleu B (1997) A truncated HIV-1 Tat protein basic domain rapidly translocates through the plasma membrane and accumulates in the cell nucleus. *J Biol Chem* **272**:16010.
34. Lewin M (2000) Tat peptide-derivatized magnetic nanoparticles allow in vivo tracking and recovery of progenitor cells. *Nat Biotechnol* **18**:410.
35. Bhorade R (2000) Macrocyclic chelators with paramagnetic cations are internalized into mammalian cells via a HIV-tat derived membrane translocation peptide. *Bioconjug Chem* **11**:301.
36. Torchilin VP, Rammohan R, Weissig V and Levchenko TS (2001) TAT peptide on the surface of liposomes affords their efficient intracellular delivery even at low temperature and in the presence of metabolic inhibitors. *Proc Natl Acad Sci USA* **98**:8786.
37. Tseng YL, Liu JJ and Hong RL (2002) Translocation of liposomes into cancer cells by cell-penetrating peptides penetratin and tat: A kinetic and efficacy study. *Mol Pharmacol* **62**:864.
38. Derossi D, Joliot AH, Chassaing G and Prochiantz A (1994) The third helix of the Antennapedia homeodomain translocates through biological membranes. *J Biol Chem* **269**:10444.
39. Elliott G and O'Hare P (1997) Intercellular trafficking and protein delivery by a herpesvirus structural protein. *Cell* **88**:223.
40. Vocero-Akbani A, Lissy NA and Dowdy SF (2000). Transduction of full-length Tat fusion proteins directly into mammalian cells: Analysis of T cell receptor activation-induced cell death. *Methods Enzymol* **322**:508.
41. Soga N, Namba N, McAllister S, Cornelius L, Teitelbaum SL, Dowdy SF, Kawamura J and Hruska KA (2001) Rho family GTPases regulate VEGF-stimulated endothelial cell motility. *Exp Cell Res* **269**:73.
42. Zezula J, Casaccia-Bonnefil P, Ezhevsky SA, Osterhout DJ, Levine JM, Dowdy SF, Chao MV and Koff A (2001) p21cip1 is required for the differentiation of oligodendrocytes independently of cell cycle withdrawal. *EMBO Rep* **2**:27.
43. Hsia CY, Cheng S, Owyang AM, Dowdy SF and Liou HC (2002) c-Rel regulation of the cell cycle in primary mouse B lymphocytes. *Int Immunol* **14**:905.
44. Schwarze SR, Ho A, Vocero-Akbani A and Dowdy SF (1999) *In vivo* protein transduction: Delivery of a biologically active protein into the mouse. *Science* **285**:1569.

# 2

# Influence of Tumor Physiology on Delivery of Therapeutics

*Robert B. Campbell*

## 1. Introduction

The concept of delivering therapeutic peptides and larger sized proteins to tumors has developed rapidly over the last few decades. The goal is to maximize delivery of therapeutics to tumor targets while minimizing effects on healthy organ tissues. Current approaches aim to selectively target cancer cells that have invaded host tissues, or to attack tumor vessels in order to arrest neovascularization or abolish mature vascular function. In view of spectacular advances, it is no surprise that drug delivery has achieved such prominence and has now emerged at the forefront of biomedical research and in many clinical environments.

It is important that investigators developing new treatment approaches against cancer both understand and safeguard against the many barriers impeding the optimal delivery of peptides, proteins and other therapeutics to solid tumors. In this chapter, we highlight the obstacles confronted today by drug delivery experts in their efforts to streamline global research in the fight against cancer and progression of disease. More specifically, we discuss the physiology of tumors in terms of the structure and function of vessels in normal and tumor tissues and in exploitable tumor targets to improve drug-target recognition. The role of physiological factors will be evaluated, including ways to exploit specialized features of solid tumors and reduce the influence they have on drug delivery and transport.

## 2. Blood Vessels: Modulation of Normal and Pathologic Function

### 2.1. *General features of blood vessels in biological systems*

The endothelium is a structural barrier separating the intravascular compartment from the interstitial environment. Because of wide variability in anatomical structure, the vascular compartment is further grouped into three different subcategories. The subcategories are continuous, fenestrated or discontinuous endothelia.

*Continuous endothelia* are the most common type and found most frequently in blood vessels lining chambers of the heart, walls of capillaries and arterioles in skeletal, skin, cardiac muscle, and connective tissue and are well known for their relatively tight cellular junctions.[1,2] This particular category of vessels is also critical in the regulation and rapid exchange of ions and solutes.[1,2] Plasmalemmal vesicles involved in endothelial transport are abundant in myocardial endothelia but are far less frequently observed in capillaries of the brain.[3] The actual number of plasmalemmal vesicles existing along the continuous endothelium is thus heterogeneous, varying as a function of organ and tissue environment. These vesicular structures are also highly sensitive to charge characteristics, favoring associations with anionic over cationic proteins and other small circulating molecules.[4]

*Fenestrated vessels* are normally found in vessels of organs that secrete (or excrete) biological fluids as in the gastrointestinal mucosa and in the glomerular capillaries of the kidneys. Fenestrae are usually between 50 and 80 nm in size and appear either as individual gap openings in the wall of functional vessels or as clusters. Similar to plasmalemmal vesicles, their frequency of occurrence along vessels depends on organ type and microenvironment.

Often two or more capillaries may join to form post-capillary venules. These newly formed networks are composed of a single lining of endothelial cells with a basement membrane with no smooth muscle cell attachment. These vessels are heavily involved in exchange of molecules and are preferential sites of plasma extravasation as a result of the actions of vasoactive and humoral factors. Fenestrae possess a negatively charged surface density due to high heparin sulfate proteoglycan content, and unlike plasmalemmal vesicles of continuous endothelia, they favor interaction with cationic over anionic molecules.[5]

*Discontinuous endothelia* are found primarily in the liver, spleen and bone marrow organs.[1] In the liver sinusoids, the endothelia are not continuous and possess an average fenestrae size between 100 and 150 nm in diameter, with the size of the fenestrate often changing in response to local mediators. These changes include, but are not limited to, response to luminal pressures and potent vasodilators such as histamine and bradykinin.[6–9] An investigation into the size of vascular pore openings of tumors revealed gap openings that are significantly larger than those observed along vessels in normal tissues, around 4 microns (4000 nm) in at least one tumor type, but normally falling within the range of 0.4 to 0.6 microns (400 to 600 nm) in others studied.[10,11] Nonetheless, the evidence is overwhelmingly in favor of the development of tumor targeted delivery of therapeutic carrier molecules that are small enough to enter through tumor vascular pores without passing through openings in normal healthy organ tissues.

The endothelium is responsible for synthesizing a variety of molecules regulating endothelial cell migration, proliferation, blood vessel maturation and function. It has been shown to synthesize vascular growth factors, nitric oxide, collagen IV, laminin, glycosaminoglycans and proteoglycans to highlight several proven functions.[9,12–18] The physical barrier organizes very rapidly to form monolayers, reassembles to form vascular tubes,[19] and can change specific inter- and intracellular signaling patterns to meet highly specialized needs of the host. Additionally, endogenous and exogenous mediators of immune and inflammatory response regulate specialized functions at the surface of the endothelium. The endothelium can thus be considered an effective mediator of organ homeostasis.[9,20]

In many ways, the vascular networks found in solid tumors poorly resemble the more regular, well-defined vascular structure observed in disease-free tissues. Tumors, for example, have a highly chaotic arrangement of vessels compared with vessels in normal tissues. Tumor vessels also have an overabundance of anionic phospholipids in addition to a number of other negatively charged functional groups.[21–24] In view of the negatively charged molecules, glycosaminoglycans carry out important functions in the metastatic disease process, and much like phospholipids, can serve as useful targets of peptide and protein therapeutics. Evaluation of altered proteoglycan expression in human breast tissue revealed a total proteoglycan content that was significantly increased in comparison with that in healthy tissues.[25] Proteoglycans isolated from malignant breast tissue have been shown to stimulate endothelial cell proliferation, and the total glycoprotein content in tumors is produced by many cell types, including cancer and tumor endothelial cells alike.

The vascular networks of tumors have an increased permeability for macromolecules and a higher proliferation rate of endothelial cells compared with vessels in quiescent tissues.[26,27] An estimated 30- to 40-fold increase in the growth rate of endothelial cells lining vessels in tumors compared with that in normal tissues has been demonstrated.[26]

Direct access to intravenously administered agents, rapid proliferation rate of endothelial cells and over-expression of negatively charged functional groups along vessels are potentially exploitable features of tumors. Peptide-free or endothelium-specific drug carrier molecules conjugated to potent peptide therapeutics can impede tumor growth on molecular and pharmacological levels.

## 2.2. *The tumor vasculature: Specialized features and angiogenesis*

The basic structure of a solid tumor, including the existence of its blood supply and some other important structural-related features, were first discovered by Rudolph Virchow during the 1860s.[28] During the early 1900s, Goldman[29] investigated the increased vascular supply in malignant diseases and the disorganized growth patterns of tumor vessels. Roughly 40 years later, it was confirmed that the growth of a transplanted tumor was connected to its ability to induce continuous endothelial cell growth.[30] The most relevant contribution to the study of *tumor vasculature* was made around the early 1970s when Gimbrone and Folkman first discovered that solid tumors require the development of new blood vessels to reach maturity, and that when tumor vascular growth was prevented, tumor dormancy was observed.[31,32] Today, nearly 35 years later the vast majority of new treatments are developed with the understanding that neovascularization is absolutely essential for malignant transformation.

Under normal circumstances the process of forming new blood vessels from pre-existing vessel networks (aka ~ angiogenesis) is observed during embryonic development and wound healing.[31] Angiogenesis is more commonly associated with pathological diseases involved in tissue regeneration; some other conditions include diabetic retinopathy, rheumatoid arthritis, chronic inflammatory diseases and cancer.[33] Microvascular networks are a vital component in the development of solid tumors. Efficient gas exchange, waste removal and delivery of nutrients to tissue-invading cancer cells depend on angiogenesis, without which the maximum size a tumor can reach is ~1–2 mm.[31] Until a new blood supply is recruited, tumors obtain the oxygen and nutrients they need for survival through passive diffusion. Once angiogenesis has begun, it remains as an active part of a tumor's life. Angiogenesis is not very active near the center of the tumor, but is a routine and highly efficient process near the tumor periphery. Given that angiogenesis does not occur with the same efficiency in all tumor regions, peptide and protein based therapeutics should be applied accordingly.

As tumors develop, a region deprived of oxygen and nutrients near the center of the tumor is formed. Many cells then die due to the severe hypoxic conditions. Tumor ischemic necrosis is therefore apparent in many solid tumors. Hypoxic conditions are probably linked to an insufficient number of blood vessels that undoubtedly influence the cells belonging to this hostile environment. For this reason, differential growth kinetics exists between cancer cells in well-oxygenated tumor regions, and neoplastic cells in regions that possess an inadequate blood supply. Assuming both cellular populations have found a way to adapt to their respective environments, all cells of a particular tumor region must thus share region-specific cell survival mechanisms that ensure adaptation of cells to a particular environment. Regardless of the tumor regions selecting for particular cellular characteristics (or expressed features), angiogenesis is absolutely essential for sustaining and maintaining the life of all solid tumors. Furthermore, the endothelial cells recruited by a developing tumor mass during angiogenesis are useful targets of peptide and protein therapeutics.[34,35]

As a tumor grows, it soon develops a nutrient-deprived tumor center. The poor-nutrient environment leads to dead cells due to hypoxia. Tumor vessels are either recruited from pre-existing vessels of the host or develop as a result of neovascularization.[31] Small venules and capillaries are involved in these processes. The formation of arteries and arterioles in tumors and the invasion of cancer cells in these vascular types are not unprecedented, but rarely observed. Furthermore, vessels possessing layers of smooth muscle do not respond to the instructional call of proangiogenic stimuli, i.e. vascular endothelial growth factor (VEGF), to form new vessels. Regardless of the angiogenic stimulus and the mechanism(s) corresponding to the initiation signaling event, some important steps are commonly associated with the development of a new blood vessel. First, once an angiogenic stimulus has been recognized by the recruited venule or capillary, the basement membrane surrounding this vessel begins to degrade. Ausprunk and Folkman confirmed that this process is controlled by endothelial cells, and that the migration of endothelial cells approaches from the direction of the host vessel to the tumor.[36] A new capillary sprout next forms through openings created in the basement membrane.[36,37] The cytoskeleton of endothelial cells (lining the sprout) begins to curve and a lumen is formed. This event was first witnessed *in vivo* and later longitudinal vacuole formation was demonstrated in cultures of endothelial cells derived from bovine and human tissues.[19] The exact location and orientation of each endothelial cell with respect to the developing sprout will determine its overall role in this process. In general, the endothelial cells located at the tip and middle sections of the sprout perform highly specialized roles related to cell migration and mitosis, respectively. Once an individual sprout has formed, the same process continues to occur elsewhere in the vicinity of the new sprout, and two or more sprouts will frequently join together. Blood flows through the lumen; pericytes (or mural cells) arrange along the sprout; and a new basement membrane is constructed soon after the lumen forms.[38,39] The more often this process occurs, the more likely the tumor expands its vascular bed and increases in overall size.

## 3. Transport of Peptide and Protein Molecules across Tumor Capillary Networks

### 3.1. *Barriers limiting drug transport*

Suboptimal expression of tumor-associated antigens, multi-drug resistance, and insufficient binding of therapeutics to intended cellular targets cannot explain all the problems associated with delivery of therapeutics to solid tumors. It is generally understood that several physiological barriers also contribute to this problem. These barriers collectively represent some of the most serious issues facing formulation experts today.

The first barrier is the structural arrangement of tumor vessels. Vascular networks of tumors are structurally and functionally unique. The formation of loops and trifurcations clearly distinguish them from vessels in normal tissues.[29,38,40,41] There is insufficient evidence to relate the average number of vessels in a tumor to its size. Some reports suggest a constant vascular fraction over the lifespan of a tumor, while others report an actual reduction in vascular volume over a similar period.[42] Other factors include tumor type and microenvironment. This barrier is important given that the total vascular volume, the organization of tumor vessels, and irregular blood flow velocities all contribute as one unit to limit optimal delivery of macromolecules.

The second barrier involves issues related to tumor vascular permeability. Permeability to therapeutics is also a function of tumor type and anatomical location, and tumor vessels are therefore heterogeneous in terms of their permeability to circulating therapies.[10,11,22] In this regard, it is not uncommon to observe two tumor vessels in different microenvironments (or two vessels in a similar environment within the same tumor) exhibiting markedly different levels of vascular leakiness. Interstitial drug delivery is often unpredictable due to this structural feature.

Once a therapeutic agent has entered the tissue compartment, it must fight against other physiological factors. The third barrier is the actual journey therapeutic molecules must take to selectively target and eradicate cancer cells. This process is otherwise known as "interstitial transport". The ability of the tumor interstitial matrix to limit transport of a desired agent possessing single or multiple physiochemical features (such as charge, size and shape of molecule) is a significant problem.[41,43–45] These factors must be taken into account whenever possible. To date, cationic liposomes (positively charged drug carrier molecules) have been used to selectively deliver therapeutic molecules to target, e.g. the p53 and interferon-beta genes, and to deliver antisense oligonucleotides and other therapies to tumors.[46–48] An issue of particular concern is when therapies are delivered by the intravenous route of administration and the intended target is located within the tumor interstitial matrix. Studies have shown that cationic liposomes (approx. ~150 nm in size) preferentially target the intravascular compartment of tumors and are not as likely to penetrate the interstitial matrix as their similarly sized anionic and electroneutral counterparts.[49–51] The same issue holds true for micelles, nanoparticles and other potential carriers of peptide and protein therapeutics of a similar size and charge.

The next barrier limits the extent to which drugs can penetrate the tumor interstitium and thus reach their intended cellular targets. Barrier #4 is elevated, interstitial fluid pressure (IFP) also known as *interstitial hypertension*. An increase in tumor IFP reduces the transcapillary pressure gradient and drives an outward pressure gradient (or fluid flux) over the capillary wall.[52] An IFP gradient from the tumor center to the periphery is responsible for the characteristic outward convection tissue gradient commonly linking elevated IFP with limited delivery and transport.[41,52] *In vivo* investigations involving the use of isolated tumor preparations revealed significant fluid in areas surrounding the tumor periphery as a result of elevated tumor pressures. In these studies, mammary carcinomas MTW9 and Walker 256 were transplanted in rats, and the net fluid loss was estimated at around 0.14 to 0.22 ml/hr per gram of tissue. A highly reproducible measure of increased hydrostatic pressure was the main reason for the significant loss of fluid.[53,54]

In areas near the tumor periphery bordering the interface of normal host tissues, the IFP was estimated at near 0 mm Hg.[40,55] Tumor interstitial pressure is, however, quite variable and region-dependent; IFP is closer to zero near the tumor periphery but near the center of the tumor, it is significantly higher and more uniform.[55] A "wick-in-needle" (WIN) technique *(see Ref. 56 for a full description of the technique)* was used to evaluate IFP in skin (melanoma) and cervical carcinomas and values were estimated around 45 and 36 mm Hg, respectively. In another study, the average IFP values ranged between 5.8 to 22.8 mm Hg.[52,56] A strong correlation exists between IFP and tumor size: the larger the tumor the higher IFP values in human and animal tumors.[56] Another report showed that the mean IFP for human breast and liver tumors derived from a primary colorectal tumor was estimated at around 33 and 21 mm Hg, respectively.[56,57] No matter the experimental tumor model used to investigate IFP, it is clearly evident that the interstitial fluid pressure is significantly higher in tumors compared with normal tissues and is associated with poor prognosis.

Due to significant vascular permeability and insufficient lymphatic drainage, the sum accumulation of fluid pressure in the vascular compartment (~aka microvascular pressure (MVP)) directly influences IFP.[58,59] MVP is dependent on differences in both arteriovenous pressure and "geometeric and viscous resistance" to blood flow; experimental evidence supports higher geometric and viscous resistance in tumors, estimated to be 1–2 magnitudes higher than in healthy tissues.[41,56,57,60] To date, a mechanism-based understanding of elevated IFP is not completely mapped out. A combination of factors such as increased vessel permeability, dysfunctional lymph vessels and structural alternations in the basic design and composition of the interstitial matrix compared with normal tissues are likely contributing factors.

Clinical investigations support a relationship between IFP and patient survival that is seemingly independent of prognostic factors. Cervical cancer patients treated with radiotherapy with relatively high IFP tumors were more likely to present within the pelvis and at untreated anatomical locations, compared with those with tumors with lower IFP.[61] Moreover, disease-free survival for patients with low and relatively high IFP (>19 mm Hg) was 68% and 34%, respectively.[61] The relationship between IFP and angiogenesis was investigated in studies involving the use of intravital microscopy. Tumor IFP was found to be highly dependent on neovascularization. Immature tumors without a developed vascular supply (stage 1 of development) had reportedly lower IFP values compared with the same tumors at a more established stage of physiological development (stages 2 and 3).[55] One cannot rule out the possibility that other inactive tumor stage-dependent factors might turn on later during stages 2 and 3 of the tumor development, contributing at least in part to interstitial hypertension. Elevated IFP is thus an important factor impeding effective penetration and distribution of therapeutics, including peptides and proteins.

How efficiently peptide and protein therapeutics travel through leaky blood vessels to desired target locations is regulated by the extent to which these barriers influence the process. In order to improve delivery, we seek to understand how experimental agents affect the density and diameter of tumor vessels, volume surface area, and blood flow. Highly sophisticated *in vivo* imaging techniques now offer ways to investigate effects of therapeutics on the structure and function of tumor-associated blood vessels.[62,63] An intravascular compartment and a formidable interstitial matrix may represent common features of all solid tumors, but tumors are unique in that they all possess a different microvessel structure and organization.

Comprehensive perfusion rate studies revealed the existence of at least three separate tumor zones. Starting from the outermost region of the tumor and working towards the central core, the periphery is the most vascularized region (zone 1) and is often the intended target of most peptide therapeutics. Zone 2 represents the semi-necrotic region and is characterized by relatively lower perfusion rates compared with zone 1. The poorly perfused, highly necrotic zone 3 is devoid of blood vessels (avascular).

Tumor cells within approximately 110 $\mu$m of the vasculature are viable, and necrosis due to prolonged hypoxia is observed in regions exceeding this critical limit of nutritional support.[64] In this way, geometric organization and location of neoplastic cells in relation to the blood supply is critical for tumor progression.[31,65] It would also stand to reason that an increase in necrotic tissue mass results in limited perfusion of oxygen, nutrients, and therapies to these tumor areas. Several lines of evidence suggest that hypoxic conditions give rise to the potent upregulation of VEGF (vascular endothelial growth factor). Upregulation results in higher expression levels in areas deprived of oxygen compared with more highly vascularized, oxygen-rich regions. In this regard, it is not difficult to understand why tumors are difficult to treat. The extent to which a particular agent can exert a desired therapeutic effect is determined by the extent to which the intended target zone is affected by treatment.

Three main parameters are generally used to define dynamics involved in the transport of all circulating therapeutics. These are blood flow rate, transport across the vascular wall, and transport within the interstitial matrix. The rate of blood flow is proportional to the drop in blood pressure across a vascular bed.[41,57] The drop in pressure is inversely proportional to "geometric and viscous resistance". Rate of blood flow, therefore, depends on these specialized features of blood vessels, including the number of blood vessels, patterns of their branching, vessel length, and diameter. The most effective approaches to date take one or more of these parameters into account.

Unlike normal tissues, tumor vessels do not respond as well to vasoactive agents (i.e. histamine, bradykinin, and serotonin) normally used to regulate blood flow resulting from injuries or inflammatory stimuli. How then do blood vessels become leaky? Tumors secrete a multifunctional cytokine called VPF/VEGF (vascular permeability factor/vascular endothelial growth factor).[66] VEGF induces rapid and reversible increases in extravasation of proteins (a function discovered by Dvorak's group with underlying mechanisms by Ferrara and colleagues).[66–68] Relatively wide inter-endothelial spaces are created, thus allowing for extravasation of proteins. It is important to note that not all tumor vessels are leaky; therefore, tumors exhibit spatial heterogeneity. Upon intravenous administration of stealth liposomes, perivascular localization of liposomes in tumor tissue was observed. This further supports the notion of heterogeneous distribution of hyperpermeable regions along the length of a single vessel.[11,69]

VEGF can exert an effect on tumor vessels that is 50 000 times more potent than histamine, and a number of significant physiological effects can result from the influence of VEGF on vascular permeability.[70] Some changes include extravasation of plasma proteins in tissue and elevated levels of cytoplasmic calcium. Specific changes to endothelial cells are alterations in cell morphology, patterns of migration, and gene expression. All of these changes are essential to the functional development of tumors. Since the lining of tumor vessels possesses relatively high affinity VEGF receptors (VEGF1 and VEGF2), over-expression of VEGF in tumors has formed the basis of many rational peptide and protein therapies against cancer.[67,70]

## 3.2. *The interstitial matrix: MMPs, collagen, invasion and metastasis*

In order for a primary tumor to expand beyond its local environment, cancer cells must first detach and migrate to (and establish growth at) a secondary tumor site. A successful journey involves degrading the basement membrane and invading the surrounding region to gain access to the intravascular compartment for the purpose of traveling to a favorable distant location. This process has been described elsewhere as the "three-step theory of invasion".[71] The first phase describes cellular attachment to the interstitial matrix through interactions with extracellular glycoproteins (i.e. laminin and fibronectin). The second step involves local proteolysis resulting from the release of specific hydrolytic enzymes synthesized by cancer cells, or generated by host cells which have been instructed by cancer cells to synthesize them. The third step involves the actual migration (or locomotion) of cells into areas structurally rearranged by hydrolytic enzymes.[71]

The tumor interstitial matrix is composed of multiple proteins involved in intercellular communication and in interactions of cells with components of the interstitial matrix. Cadherins, integrins, laminin, fibronectin, and matrix metalloproteinases (MMPs) are but a few of the proteins shown to play a role in the functional regulation of invasion and metastasis.[71–76] These specialized components of the interstitial matrix perform important roles in both health and disease. It is, however, beyond the scope of this chapter to discuss each in detail. Due to the steady increase in the number of publications relating MMP and collagen function to tumor invasion and metastasis, some attention will be focused on these components.

The field of cancer research has benefited enormously from investigations into the structural and functional relationships of MMPs and cancer. MMPs are the interstitial enzymes that degrade collagen, but require $Ca^{++}$ and $Zn^{++}$ to exert function.[77,78] Of note is the regulation of MMP synthesis and functional action in tissues by the specific local action of MMP-inhibitors, without which the functional activity of MMPs might go unregulated in host tissues.

MMPs are involved in physiological processes, including bone remodeling and embryogenesis, and in pathological conditions such as tissue destruction, arthritis, cancer and other diseases.[73,74] MMPs degrade collagen in pathological tissues. The turnover rate of interstitial collagen in normal tissues is relatively slow compared with that of tumors, with an estimated half-life in years. Two types of MMPs involved in metastasis and in the rapid breakdown of collagen in tumors are MMP-2 and MMP-9; digestion of collagen type IV and type V have been reported for each, respectively.[74] Type IV collagen is the main component of the basement membrane, whereas type V is found in areas located between the basement membrane and interstitial stroma.[37,77–79]

The role of MMP-2 in angiogenesis and cancer has been investigated.[76] In one study, MMP-2 knockout mice demonstrated a reduced response to B16-BL6 and Lewis lung carcinoma cells when implanted intradermally. MMP-2 deficient mice exhibited significantly lower tumor growth, demonstrating 39 and 24% reduced growth in comparison with MMP-2 competent mice, respectively.[76] Subsequent studies later linked MMP-9 activity with invasion of high grade gliomas, and to effective therapeutic action of Interferon $\beta$-1b.[80,81]

A fragment of collagen IV $\alpha 3$ chain generated by MMP-9 proteolysis (aka ~ tumstatin) inhibited angiogenesis associated with tumor progression, without exerting effects on the physiology of normal tissues. This is possible due to the over-expression of $\beta 3$ integrin in tumors compared with normal tissues, and because tumstatin requires $\beta 3$ integrin to exert its therapeutic effect.[82] Mice deficient in MMP-9 were shown to have lower circulating levels of tumstatin linking MMP-9 with integrin-mediated tumor suppression. The product of MMP-9 proteolysis was later demonstrated to inhibit invasive properties of metastatic melanoma *in vivo* by triggering an intracellular signaling cascade.[83] Altogether, these data support an endogenous role of MMP-9 in tumor biology, including regulation, invasion and metastasis.

A number of additional components of the interstitial matrix offer benefits in terms of tissue organization and structure, many of which limit optimal delivery of larger protein therapeutics to tumors due to unfavorable physicochemical characteristics. These are glycosaminoglycans, proteoglycans, and high collagen content in stroma; several studies have investigated their structural and functional significance in normal and tumor tissues.[12,15,25,43,84] It was recently shown (with the use of noninvasive techniques) that successful modification of the tumor interstitial matrix can result in improved diffusive transport. In this study, the pregnancy hormone relaxin was administered to HST26T containing SCID mice and a significant increase in diffusion coefficients of IgG and dextran was observed.[85] The effect of relaxin on diffusive properties of IgG and dextran is likely due to upregulated functional effects of MMPs.[86] Elevated levels of relaxin associated with tissue remodeling in breast cancer patients have since been discovered.[85] Important findings resulted from a study of the effects of extracellular matrix composition, structure, and distribution of molecules in tumors. Investigations confirmed that diffusion of small proteins was not affected by tumor location; however, tumor location was an important consideration for diffusion of significantly larger protein molecules of a similar chemical composition. The diffusional hindrance of larger molecules correlated with relatively high collagen type I and fibrillar collagen content in the diffusion limiting site.[43] The design of the interstitial matrix, including the role of various protein components, is thus critical in the regulation of cancer and progression of disease.

### 3.3. *Overcoming barriers: Exploiting tumor physiology for therapeutic gain*

Depending on tumor type and the tumor-associated microenvironment, optimizing delivery of peptides and protein therapeutics to tumors will require success with at least one of two different approaches. The first approach is to circumvent one or more tumor physiological barriers. The second is to temporarily impair physiological barrier function(s) prior to administering interstitial targeting agents. The realization of completely eliminating or suppressing any one of these barriers is no easy task.

Tumor vascular targeting might represent a rational way to circumvent elevated tumor interstitial pressures and other impeding factors of the interstitial matrix. However, for tumors to respond optimally to treatment, sufficient access of therapeutics to the tumor blood supply is essential. How can this be accomplished? In one study, Hong *et al.*[87] successfully used adenosomes (adenovirus proteins combined with cationic liposomes) to deliver AAV/CMV-LacZ to human endothelial cells. Peptide-mediated therapy involving specific ligands has been used to deliver genes to endothelial cells. In one study, two different derivatized RGD peptides delivered cationic lipid-plasmid DNA to human umbilical derived endothelial cells (HEVEC), and a 4-fold increase in transfection efficiency was reported.[88] Several studies have reported successful efforts to deliver peptide-based therapeutics to endothelial cells *in vitro* and *in vivo*. By way of example, a lipid-mediated *peptide nucleic acid* (PNA) agent was used to deliver peptide therapeutics to pulmonary endothelial cells *in vivo*.[89] The peptide derived from the LDL receptor (LDLr) binding domain of apolipoprotein E (apo E) improved the uptake of liposomes by endothelial cells lining the brain. The authors note that non-protein coated liposomes were not successfully taken up by the brain endothelial cells.[90] Also noteworthy is the opportunity to target therapeutics to the brain while avoiding the highly elevated IFP normally associated with tumors in this anatomical location.

Natural and synthetic sources of angiostatic proteins and peptides have been evaluated and the majority of them demonstrate the ability to inhibit neovascularization *in vivo*.[35] Macromolecule-assisted delivery of these agents to tumor vessels could improve vascular recognition and outcomes associated with treatment.

Earlier in this chapter, the multicytokine function of VPF/VEGF in tumors was discussed. It is reasonable to discuss using VEGF to induce tumor vascular permeability to circulating therapies. The delivery of VEGF in sufficient levels to endothelia could result in venules that are more permeable to therapeutics. From this perspective both delivering VEGF directly to tumor vessels, and delivering plasmids encoding for VEGF to this site as well, might represent equally effective approaches. To illustrate, when plasmid encoding for VEGF was injected as a VEGF-cationic liposome complex, gene expression was detected after 1 to 3 weeks, with DNA detected up to several months following the initial injection.[91] An *in vivo* quantitative measure of the effects of various growth factors (VEGF, PIGF-1 and PIGF-2 (platelet growth factors), and bFGF (basic fibroblast growth factor)) on the permeability of tumor vessels to circulating macromolecules (dextran, liposomes, and albumin) demonstrated that only VEGF significantly increased vascular permeability.[92] Such studies have clinical implications for tumor and non-tumor vascular related diseases.

In some situations a more practical approach could involve delivering therapeutic peptides or proteins directly to populations of neoplastic cells invading the tumor matrix. In this approach some specialized feature of neoplastic cells is usually exploited for therapeutic gain. These studies are usually investigated with the use of human tumor xenograft models in the presence of fully functioning physiological barriers; suboptimal to adequate levels of success is commonly associated with treatment outcome. Upon closer evaluation, better treatment outcomes would probably result if more clinically relevant therapeutic concentrations were delivered to the tumor interstitial compartment. Simply increasing the injected dose may improve delivery of therapeutics to the intended target over normal tissues, but the final dose should be optimized in relation to tumor and non-tumor targets to maximize therapeutic effect. The aim, however, is to improve drug location and duration of drug exposure at the intended site(s) of drug action; increasing the ratio of drug to tumor as opposed to normal tissue is a critical first step.

Over the last decade, several groups have investigated the use of agents to lower tumor IFP with the aim of reducing the pressure long enough to allow for better penetration of therapeutics into interstitial tumor areas. The eventual hope is to treat regions that would otherwise go unaffected by more conventional approaches. To summarize a few of these studies, the following agents have demonstrated the ability to lower tumor IFP: tumor necrosis factor-alpha (TNF-$\alpha$); tumor necrosis factor-beta (Fc:T$\beta$RII); dexamethasone; pentoxifylline (PTX); and taxol.[93–97] In experimental animal models, several agents have been shown to lower tumor IFP.[93,94,96–103] Table 1 shows a list of 10 agents that have been shown to lower tumor interstitial fluid pressure, with a summary of the injected dose, route of

**Table 1** Tumor models used to investigate effect of agents on tumor IFP and related properties. U87 (human glioblastoma); HSTS-26T (human soft tissue sarcoma); MCaIV (murine breast carcinoma); S-MEL (human melanoma); MeWo (human malignant melanoma derived from lymph node); KAT-4 (human thyroid carcinoma); LS174T (human colon adenocarcinoma); FSaII (fourth generation, fibrosarcoma); MATB-111 (rat mammary adenocarcinoma); NT [aka ~ CaNT] murine breast adenocarcinoma; C-IMC (chemically-induced mammary carcinoma).

| The following agents have been shown to lower tumor interstitial fluid pressure | | | | | |
|---|---|---|---|---|---|
| Agent name | Injected dose | Injection route | Tumor type | Species/ strain | Result summary/Conclusions |
| | | | | | *(Griffon-E et al., 1999)* |
| (1) Taxol | 40 mg/kg | tail vein | *U87mg<br>*HSTS-26T<br>*Mca-IV | Nude<br>NCr/Sed<br>C3H/Kam | *48 hrs post-injection:*<br>↑ tumor vascular diameter<br>HSTS-26T & Mca-IV showed↓ IFP ($P < 0.05$)<br>IFP of U87mg (taxol resistant) was not affected by taxol |
| | | | | | *(Kristensen et al., 1996)* |
| (2) TNF-$\alpha$ | 500 $\mu$g/kg | tail vein | *S-MEL<br>*P-MEL<br>*MeWo | Male nude | *5 and 24 hrs post-injection:*<br>All 3 tumors = ↓IFP by 50–70% ($P < 0.05$)<br>" = ↓ MABP by 30% ($P < 0.01$)<br>no pressure lowering effect observed at 24 hrs |
| | | | | | *(Salinikov et al., 2005)* |
| (3) Fc:T$\beta$RII | *Fc:TβRII;* 1 mg/kg<br>*Dox*; 3 mg/kg, every 2 days for 2 wks | i.p. | *KAT-4 | athymic<br>C57 bl/6 | Pretreatment with Fc:Tbeta RII ↑ doxorubicin efficacy<br>FcT$\beta$RII normalized blood vessels<br>FcT$\beta$RII and rh-IL-1 ↓ IFP |

**Table 1** *(Continued)*

| The following agents have been shown to lower tumor interstitial fluid pressure | | | | | |
|---|---|---|---|---|---|
| Agent name | Injected dose | Injection route | Tumor type | Species/strain | Result summary/Conclusions |
| | | | | | *(Kristjansen et al., 1993)* |
| (4) Dexamethasone | 0.3–30 mg/kg | i.p. | *LS174T | SCID | 0.3 mg/kg = no significant effect on IFP observed<br>1.0 mg/kg = marginal ↓ IFP between *day 1 versus day 4*<br>3–10 mg/kg = significant ↓ IFP between *day 1 versus day 4* |
| | | | | | *(Lee et al., 1994)* |
| (5) Pentoxifyllin | 25 and 100 mg/kg | i.p. | *FSaII (fourth generation) | female C3Hf/Sed | *2 hrs post injection:*<br>25 mg/kg = no effect on PO2<br>100 mg/kg = ↑PO2; ↑RBC flux; ↓IFP by 40%<br>100 mg/kg = no effect on MABP was observed |
| | | | | | *(Ansiaux et al., 2005)* |
| (6) Thalidomide | 200 mg/kg | i.p. | *FSA II | male C3H/HeCuJIco | 2 days post injection = ↑ in tumor reoxygenation ↓ IFP; effect due to tumor vascular remodeling |
| | | | | | *(Emerich et al., 2001)* |
| (7) Cereport (Bradykinin agonist) | 0.1, 1.0, and 0.15 $\mu$g/kg/min | I.V. infusion | *MATE-III | male Fisher/rats | *I.V. infusion 5–10 min:*<br>0.1 mg/kg/min = ↓ IFP significantly ($P < 0.01$)<br>1.0 mg/kg/min = IFP ($P > 0.1$); 66% ↓ in perfusion ($P < 0.001$)<br>0.15 mg/kg/min = ↑ vascular pore size; ↑ vol. surface area |

**Table 1** (*Continued*)

| The following agents have been shown to lower tumor interstitial fluid pressure | | | | | |
|---|---|---|---|---|---|
| Agent name | Injected dose | Injection route | Tumor type | Species/ strain | Result summary/Conclusions |
| | | | | | *(Peters et al., 1997)* |
| (8) Nicotinamide | 100, 500, 1000 mg/kg | i.p. | *NT(CaNT) | mice CBA | *20 minutes post-injection:*<br>100 mg/kg = no effect on IFP ($P = 0.05$)<br>500–1000 mg/kg = ↓ IFP ($P = 0.0001$)<br>1000 mg/kg = ↓ IFP ($P < 0.0001$)<br>Radiosensitizing effect after 80 min but not after 10 min. |
| | | | | | *(Pietras et al., 2002)* |
| (9) ST1571 (PDGF-$\beta$ inhibitor) | 12 mg/kg/day | i.p. | Kat-4 (in mice)/ PROb (in rats) | SCID mice/ BDIX rats | *40 minutes post-injection:*<br>ST1571 blocked PDGF signaling; ↓ IFP ; ↑ 51Cr-EDTA uptake<br>" ↑ antitumor effect of taxol against Kat-4<br>" ↑ antitumor effect of 5FU against PROb |
| | | | | | *(Rubin et al., 2000)* |
| (10) PGE(1) | 15 $\mu$g | s.c. (area around tumor) | PROb *C-IMC | Rats | *8 and 24 hrs post-injection:*<br>PEG(1) ↓ IFP by 30%<br>↑ transcapillary transport by 39.6%-(by microdialysis)($P < 0.05$)<br>↑ uptake of 51Cr-EDTA by 86.9% ($P < 0.05$)<br>Well developed collagen and hyaluronan stromal content |

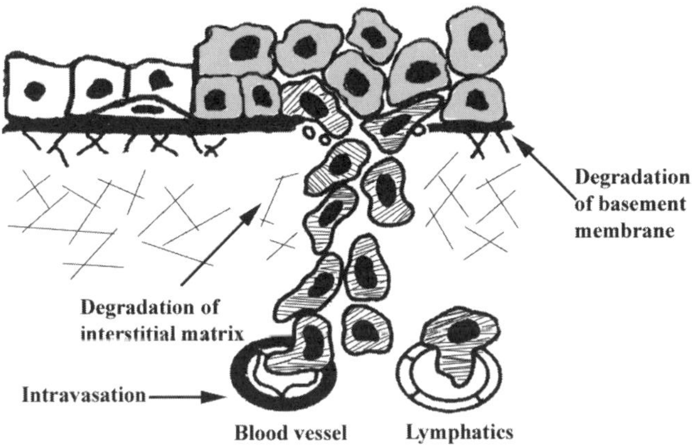


**Fig. 1** This figure models the process of invasion. The continuous basement membrane is penetrated by neoplastic cells. These cells gain access to components of the interstitial matrix. After degradation of the interstitial matrix tumor cells invade blood vessels and lymphatics. In order for the cells to enter the blood stream the cells must penetrate the continuous endothelial basement membrane. Blood vessels are surrounded by a continuous endothelial cell membrane but lymphatics lack this form of structural support.

administration, investigated tumor model(s), species and strain of tumor bearing rodents, and conclusions as they relate to IFP and other interesting related findings. The tumor IFP lowering agents (among other therapeutic functions) were also evaluated under different experimental conditions and should be applied with the understanding that what works under one set of conditions, may not work in another situation. The effects of a few tumor IFP lowering agents will be discussed here; for more information see published findings in respective journals.[93,94,96–103]

TNF-$\alpha$ was shown to reduce IFP and MABP by approximately 50–70% ($P < 0.05$) and 30% ($P < 0.01$), respectively. In this study, human melanoma tumors were implanted in nude mice and a reversible effect of TNF-$\alpha$ on IFP was observed 5 hours post-injection ($P < 0.05$), with no observable influence on IFP after 24 hours.[94]

TGF-$\beta$ lowered IFP in a KAT-4 anaplastic thyroid carcinoma model. Affymetrix microarray studies and other protein expression assays confirmed that the chimeric protein (FC:T$\beta$RII) exerted its effect in part through modulating macrophage activity. When KAT-4 was pretreated with Fc:T$\beta$RII, a significant effect of doxorubicin on tumor growth was observed.[93]

Concentration dependent effects of dexamethasone on IFP in LS174T models were also investigated. Injected doses ranged from 0.3 to 30 mg/kg with i.p. injections administered over 4 consecutive days. Relatively high doses resulted in lower IFP values, reduced permeability and vascular hydraulic conductivity.[96]

As observed with the effects of other tumor IFP lowering agents, PTX was found to lower tumor IFP without altering MABP.[97] Based on concentrations per/kg of body weight, dexamethasone exerted a more potent effect on IFP than PTX. PTX lowered IFP at concentrations around 100 mg/kg but failed to effectively lower IFP at the 25 mg/kg dose. Prior to a decrease in tumor IFP, a dose-specific effect of PTX on RBC flux near the tumor center was observed.[97] Inducing RBC velocity in tumor areas near zone 3 could potentially create new opportunities to deliver peptide and protein therapeutics to the deeper regions of the tumor. Studies of this type could eventually contribute to resolving issues related to radioresistant hypoxic tumor cells.

Taxol (aka ~ paclitaxel), a popular chemotherapeutic agent used in the treatment of a variety of solid malignancies, has also been shown to lower tumor IFP, decompress blood vessels and improve oxygenation ($PO_2$) in patients.[95,99] Clinical data now supports the role of taxol as a tumor IFP lowering agent over other chemotherapeutic agents. In a study involving breast cancer patients, taxol significantly lowered IFP by 36% and increased the $PO_2$ by approximately 100%, whereas doxorubicin did not exert a similar or significant effect on IFP. The ability of taxol to lower IFP is a bonus effect of this chemotherapeutic agent.[95,99] It is possible that other drugs belonging to its class may exert similar or additional anti-barrier functions. It is also entirely possible that some of the conventional and experimental therapeutics that have made their way into the clinic may lower tumor IFP in addition to exerting other widely accepted antitumor effects. Although taxol was never initially used in a clinical setting to limit a major physiological barrier such as interstitial hypertension, it has the potential to improve penetration of small therapeutic peptides and proteins into the interior tumor mass when used in combination with other agents proven to lower tumor IFP.[93,94,96–103] Subsequent clinical investigations are warranted.

## 4. Concluding Remarks

Given the urgency for development of novel therapeutics against cancer, an understanding of the major barriers hindering optimal delivery and transport is necessary. In view of the time and energy invested in the discovery, clinical testing, and marketing of a new agent, the physiology of tumors should be taken into account as soon as possible to ensure the greatest level of clinical success in a time-sensitive manner.

Although a number of researchers have managed to achieve some success with interstitial tumor targeting methods, tumor vascular targeting still represents a viable alternative. Improving methods to selectively deliver therapeutic peptides and proteins to tumor vessels is certainly time well invested, but other methods of circumventing barriers are also needed. Another approach is to reduce the burdens of tumor physiology on drug delivery and transport. Experiments and technologies developed to optimize the use of agents already shown to limit tumor IFP and/or modify the tumor interstitial matrix, will likely assist the next generation of treatments against cancer and progression of disease.

## References

1. Simionescu M, Simionescu N and Palade GE (1974) Morphometric data on the endothelium of blood capillaries. *J Cell Biol* **60**:128–152.
2. Simionescu M, Simionescu N and Palade GE (1978) Structural basis of permeability in sequential segments of the microvasculature. I. Bipolar microvascular fields in the diaphragm. *Microvasc Res* **15**:1–16.
3. Simionescu M, *et al.* (1985) Differentiated microdomains of the luminal plasmalemma of murine muscle capillaries: Segmental variations in young and old animals. *J Cell Biol* **100**:1396–1407.
4. Ghinea N and Simionescu N (1985) Anionized and cationized hemeundecapeptides as probes for cell surface charge and permeability studies: Differentiated labeling of endothelial plasmalemmal vesicles. *J Cell Biol* **100**:606–612.
5. Simionescu M, Simionescu N and Palade GE (1981) Differentiated microdomains on the luminal surface of the capillary endothelium. I. Partial characterization of their anionic sites. *J Cell Biol* **90**:614–621.
6. Movat HZ (1987) The role of histamine and other mediators in microvascular changes in acute inflammation. *Can J Physiol Pharmacol* **65**(3):451–457.

7. Mylecharane EJ (1990) Mechanisms involved in serotonin-induced vasodilation. *Blood Vessels* **27**(2–5):116–126.
8. Vercellotti GM and Tolins JP (1993) Endothelial activation and the kidney: Vasomediator modulation and antioxidant strategies. *Am J Kidney Dis* **21**(3):331–343.
9. van Hinsbergh VW (2001) The endothelium: Vascular control of haemostasis. *Atherosclerosis* **95**(2):198–201.
10. Hobbs SK, *et al.* (1998) Regulation of transport pathways in tumor vessels: Role of tumor type and microenvironment. *Proc Natl Acad Sci USA* **95**:4607–4612.
11. Yuan F, *et al.* (1995) Vascular permeability in a human tumor xenograft: Molecular size dependence and cutoff size. *Cancer Res* **55**:3752–3756.
12. Hudson BG, Reeders ST and Trygvason K (1993) Type IV collagen: Structure, gene organization and role in human diseases. *J Biol Chem* **268**:26033–26036.
13. Gorog P and Pearson JD (1985) Sialic acid moieties on surface glycoproteins protect endothelial cells from proteolytic damage. *J Pathol* **146**:205–212.
14. Hallmann R, *et al.* (2005) Expression of function of laminins in the embryonic and mature vasculature. *Physiol Rev* **85**:979–1000.
15. Sund M, Xie L and Kalluri R (2004) The contribution of vascular basement membranes and extracellular matrix to the mechanics of tumor angiogenesis. *APMIS* **112**(7–8):450–462.
16. Ruoslahti E (1988) Structure and biology of proteoglycans. *Annu Rev Cell Biol* **4**:229–255.
17. Ruoslahti E (2002) Specialization of tumor vasculature. *Nat Rev Cancer* **2**:83–90.
18. Satoh A, *et al.* (2000) New role of glycosaminoglycans on the plasma membrane proposed by their interaction with phosphatidylcholine. *FEBS Lett* **477**:249–252.
19. Folkman J and Haudenschild C (1980) Angiogenesis by capillary endothelial cells in culture. *Trans Ophthalmol Soc UK* **100**(3):346–353.
20. Michiels C, Arnould T and Remacle J (2000) Endothelial cell responses to hypoxia: Initiation of a cascade of cellular interactions. *Biochim Biophys Acta* **1497**(1):1–10.
21. Charonis AS and Wissig SL (1983) Anionic sites in basement membranes. Differences in their electrostatic properties in continuous and fenestrated capillaries. *Microvasc Res* **25**:265–285.
22. Roberts WG and Palade GE (1997) Neovasculature induced by vascular endothelial growth factor is fenestrated. *Cancer Res* **57**(4):765–772.
23. Ran S, Downes A and Thorpe PE (2002) Increased exposure of anionic phospholipids on the surface of tumor blood vessels. *Cancer Res* **62**(21): 6132–6140.

24. Vincent S, DePace D and Finkelstein S (1988) Distribution of anionic sites on the capillary endothelium in an experimental brain tumor model. *Microcirc Endoth Lym* **1988**(4):45–67.
25. Vijayagopal P, Figueoroa JE and Levine EA (1998) Altered composition and increased endothelial cell proliferative activity of proteoglycans isolated from breast carcinoma. *J Surg Oncol* **68**(4):250–254.
26. Denekamp J and Hobson B (1982) Endothelial-cell proliferation in experimental tumours. *Br J Cancer* **46**:711–720.
27. Denekamp J (1984) Vasculature as a target for tumor therapy. *Prog Appl Microcirc* **4**:28–38.
28. David H (1988) Rudolf Virchow and modern aspects of tumor pathology. *Pathol Res Pract* **183**(3):356–364.
29. Goldman E (1907) The growth of malignant disease in man and lower animals with special reference to the vascular system. *Lancet* **2**:1236.
30. Algire GH and Chalkley HW (1945) Vascular reaction of normal and malignant tissues *in vitro*. I. Vascular reactions fo mice to wounds and to normal and neoplastic implants. *J Natl Cancer Inst* **6**:73–85.
31. Folkman J (1971) Tumor angiogenesis: Therapeutic implications. *N Engl J Med* **285**:1182–1186.
32. Gimbrone MA, *et al.* (1972) Tumor dormancy *in vivo* by prevention of neovascularization. *J Exp Med* **136**:261–276.
33. Jain RK and Carmeliet PF (2001) Vessels of death or life. *Sci Am* **285**(6):38–45.
34. Brower V (1999) Tumor angiogenesis-new drugs on the block. *Nat Biotechnol* **17**:963–968.
35. Steege B-t JC, Mayo KH and Griffioen AW (2001) Angiostatic proteins and peptides. *Crit Rev Eukar Gene* **11**(4):319–334.
36. Ausprunk DH and Folkman J (1977) Migration and proliferation of endothelial cells in preformed and newly formed blood vessels during tumor angiogenesis. *Microvasc Res* **14**(1):53–65.
37. Madri JA and William SK (1983) Capillary endothelial cell cultures: Phenotypic modulation by matrix components. *J Cell Biol* **97**:153–165.
38. Morikawa S, *et al.* (2002) Abnormalities in pericytes on blood vessels and endothelial sprouts in tumors. *Am J Pathol* **160**(3):985–1000.
39. Hirschi KK and D'Amore PA (1996) Pericytes in the microvasculature. *Cardiovasc Res* **32**(4):687–698.
40. Jain RK (1987) Transport of molecules across tumor vasculature. *Cancer Metastasis Rev* **6**:559–593.
41. Jain RK (1990) Vascular and interstitial barriers to delivery of therapeutic agents in tumors. *Cancer Metastasis Rev* **9**:253–266.
42. Jain RK (1988) Determinants of tumor blood flow: A review. *Cancer Res* **6**:559–594.

43. Pluen A, *et al.* (2001) Role of tumor-host interactions in interstitial diffusion of macromolecules: Cranial vs. subcutaneous tumors. *PNAS* **98**(8):4628–4633.
44. Renkin EM (1985) Transport of macromolecules: Pores and other endothelial pathways. *Appl Physiol* **58**:315–325.
45. Dellian M, *et al.* (2000) Vascular permeability in a human tumor xenograft: Molecular charge dependence. *Br J Cancer* **82**(9):1513–1518.
46. Zhao XB and Lee RJ (2004) Tumor-selective targeted delivery of genes and antisense oligodeoxyribonucleotides via the folate receptor. *Adv Drug Deliv Rev* **56**(8):1193–204.
47. Yoshida J, Mizuno M and Wakabayashi T (2004) Interferon-beta gene therapy for cancer: Basic research to clinical application. *Cancer Sci* **95**(11):858–865.
48. Xu L, *et al.* (2001) Systemic p53 gene therapy of cancer with immunolipoplexes targeted by anti-transferrin receptor scFv. *Mol Med* **7**(10):723–734.
49. McLean JW, *et al.* (1997) Organ-specific endothelial cell uptake of cationic liposome-DNA complexes in mice. *Am J Physiol* **273(1 pt 2)**:387–404.
50. Thurston G, *et al.* (1998) Cationic liposomes target angiogenic endothelial cells in tumors and chronic inflammation in mice. *J Clin Inves* **101**(7):1401–1413.
51. Campbell RB, *et al.* (2002) Cationic charge determines the distribution of liposomes between the vascular and extravascular compartments of tumors. *Cancer Res* **62**:6831–6836.
52. Aukland K and Reed RK (1993) Interstitial-lymphatic mechanisms in the control of extracellular fluid volume. *Physiol Rev* **73**(1):1–78.
53. Young JS, Lumsden CE and Stalker AL (1950) The significance of the tissue pressure of normal testicular and of neoplastic (Brown-Pearce carcinoma) tissue on the rabbit. *J Pathol Bacteriol* **62**(3):313–333.
54. Butler TP, Grantham FH and Gullino PM (1975) Bulk transfer of fluid in the interstitial compartment of mammary tumors. *Cancer Res* **35(11 pt 1)**:3084–3088.
55. Boucher Y, Leunig M and Jain RK (1996) Tumor angiogenesis and interstitial hypertension. *Cancer Res* **56**(18):4264–4266.
56. Boucher Y, *et al.* (1991) Interstitial hypertension in superficial metastatic melanomas in humans. *Cancer Res* **51**(24):6691–6694.
57. Sevick EM and Jain RK (1989) Geometric resistance to blood flow in solid tumors prefused *ex vivo*: Effects of tumor size and perfusion pressure. *Cancer Res* **49**(13):3506–3512.
58. Zlotecki RA, *et al.* (1995) Pharmacologic modification of tumor blood flow and interstitial fluid pressure in a human tumor xenograft: Network analysis and mechanistic interpretation. *Microvasc Res* **50**:429–443.
59. Zlotecki RA, *et al.* (1993) Effect of angiotensin II induced hypertension on tumor blood flow and interstitial fluid pressure. *Cancer Res* **53**:2466–2468.

60. Less JR, *et al.* (1997) Geometric resistance and microvascular network architecture of human colorectal carcinoma. *Microcirculation* **4**(1):25–33.
61. Milosevic M, *et al.* (2001) Interstitial fluid pressure predicts survival in patients with cervix cancer independent of clinical prognostic factors and tumor oxygenation measurements. *Cancer Res* **61**:6400–6405.
62. Jain RK, Munn LL and Fukumura D (2002) Dissecting tumour pathophysiology using intravital microscopy. *Nat Rev Cancer* **2**(4):266–276.
63. Brown EB, *et al.* (2001) *In vivo* measurement of gene expression, angiogenesis, and physiological function in tumors using multiphoton laser scanning microscopy. *Nat Med* **7**(7):864–868.
64. Hlatky L, Hahnfeldt P and Folkman J (2002) Clinical application of antiangiogenic therapy: Microvessel density, what it does and doesn't tell us. *J Nat Cancer Inst* **94**(12):883–893.
65. Folkman J (2002) Looking for a good endothelial address. *Cancer Cell* **2**:113–115.
66. Ferrara N and Davis-Smyth T (1997) The biology of vascular endothelial growth factor. *Endocr Rev* **18**:4–25.
67. Dvorak HF, *et al.* (1995) Vascular permeability factor/vascular endothelial growth factor, microvascular hyperpermeability and angiogenesis. *Am J Pathol* **146**:1029–1039.
68. Senger DR, *et al.* (1993) Vascular permeability factor (VPF, VEGF) in tumor biology. *Metastasis Rev* **12**:303–324.
69. Yuan F, *et al.* (1994) Microvascular permeability and interstitial penetration of sterically stabilized (Stealth) liposomes in a human tumor xenograft. *Cancer Res* **54**:3352–3356.
70. Brown LF, *et al.* (1997) Vascular permeability factor/vascular endothelial growth factor: A multifunctional angiogenic cytokine, in Rosen IDGEM (ed.), *Regulation of Angiogenesis*, Birkhauser Verlag Basel/Switzerland.
71. Liotta LA, Rao CN and Barsky SH (1983) Tumor invasion and the extracellular matrix. *Lab Invest* **49**(6):636–649.
72. Marhava R and Zoller M (2004) CD44 in cancer progression: Adhesion, migration and growth regulation. *J Mol Histol* **35**(3):211–231.
73. Woessner JF, Jr (1968) Biological membranes of tissue resorption, in BSG (ed.), *Treatise on Collagen*, Academic Press: New York, pp. 253–330.
74. Woessner JF, Jr (1998) The matrix metalloproteinase family, in WCPaRPM (eds.), *Matrix Metalloproteinases*, Academic Press, San Diego: pp. 1–14.
75. Agnantis NJ, *et al.* (2004) Tumor markers in cancer patients. An update of their prognostic significance. Part II. *In Vivo* **18**(4):481–488.
76. Itoh T, *et al.* (1998) Reduced angiogenesis and tumor progression in gelatinase A-deficient mice. *Cancer Res* **58**(5):1048–1051.

77. Timpl R, *et al.* (1978) Nature of the collagenous protein in a tumor basement membrane. *Biochem Biophys Res Comm* **84**(43).
78. Timpl R, *et al.* (1985) Structure and biology of the globular domain of basement membrane type IV collagen. *Ann NY Acad Sci* **460**:58–72.
79. Hernandez-MA and Amenta PS (1979) The basement membrane in pathology. *Lab Invest* **48**:656–680.
80. Uhm JH, *et al.* (1997) Mechanisms of glioma invasion: Role of matrix metalloproteinases. *Can J Neurol Sci* **24**(1):3–15.
81. Stuve O, *et al.* (1996) Interferon beta-1b decreases the migration of T lymphocytes *in vitro*: Effects on matrix metalloproteinase-9. *Ann Neurol* **40**(6): 853–863.
82. Hamano Y, *et al.* (2003) Physiological levels of tumstatin, a fragment of collagen IV alphaVbeta3 integrin. *Cancer Cell* **3**:589–601.
83. Pasco S, *et al.* (2004) *In vivo* overexpression of tumstatin domains by tumor cells inhibits their invasive properties in a mouse melanoma model. *Exp Cell Res* **301**:251–265.
84. Gribbon PM, *et al.* (1998) in Reed RK and Rubin k (eds.), *Connective Tissue Biology: Integration and Reductionism*, Portland, London, pp. 95–124.
85. Brown E, *et al.* (2003) Dynamic imaging of collagen and its modulation in tumors in vivo using second-harmonic generation. *Nat Med* **9**(6):796–800.
86. Binder C, *et al.* (2004) Elevated concentrations of serum relaxin are associated with metastatic disease in breast cancer patients. *Breast Cancer Res Treat* **2004**(87):157–166.
87. Hong Z, *et al.* (1995) Enhanced adeno-associated virus vector expression by adenovirus protein-cationic liposome complex. *Chinese Med J* **108**(5): 332–337.
88. Anwer K, *et al.* (2004) Peptide-mediated gene transfer of cationic lipid/plasmid DNA complexes to endothelial cells. *J Drug Target* **12**(4):215–221.
89. Yuan X, *et al.* (2003) Lipid-mediated delivery of peptide nucleic acids to pulmonary endothelium. *Biochem Biophys Res Comm* **302**(1):6–11.
90. Sauer I, *et al.* (2005) An apolipoprotein E-derived peptide mediates uptake of sterically stabilized liposomes into brain capillary endothelial cells. *Biochem* **44**(6):2021–2029.
91. Nikol S, *et al.* (2002) Local perivascular application of low amounts of a plasmid encoding for vascular endothelial growth factor (VEGF165) is efficient for therapeutic angiogenesis in pigs. *Acta Physiol Scand* **176**(2):151–159.
92. Monsky WL, *et al.* (1999) Augmentation of transvascular transport of macromolecules and nanoparticles in tumors using vascular endothelial growth factor. *Cancer Res* **59**(16):4129–4135.

93. Salnikov AV, *et al.* (2005) Inhibition of TGF-beta modulates macrophages and vessel maturation in parallel to a lowering of interstitial fluid pressure in experimental carcinoma. *Lab Invest* **5**(4):512–521.
94. Kristensen CA, *et al.* (1996) Reduction of interstitial fluid pressure after TNF-alpha treatment of three human melanoma xenografts. *Br J Cancer* **74**(4): 533–536.
95. Taghian A, *et al.* (2005) Paclitaxel decreases the interstitial fluid pressure and improves oxygenation in breast cancers in patients treated with neoadjuvant chemotherapy: Clinical implications. *J Clin Oncol* **23**(9):1951–1961.
96. Kristjansen PE, Boucher Y and Jain RK (1993) Dexamethasone reduces the interstitial fluid pressure in a human colon adenocarcinoma xenograft. *Cancer Res* **53**(20):4764–4766.
97. Lee I, *et al.* (1994) Changes in tumour blood flow, oxygenation and interstitial fluid pressure induced by pentoxifylline. *Br J Cancer* **69**(3):492–496.
98. Emerich DF, *et al.* (2001) Bradykinin modulation of tumor vasculature: II. Activation of nitric oxide and phopholipase $A_2$/prostaglandin signaling pathways synergistically modifies vascular physiology and morphology to enhance delivery of chemotherapeutic agents to tumors. *J Pharmacol Exp Ther* **296**(2):632–641.
99. Griffon-EG, *et al.* (1999) Taxane-induced apoptosis decompresses blood vessels and lowers interstitial fluid pressure in solid tumors. *Cancer Res* **59**:3776–3782.
100. Pietras K, *et al.* (2002) Inhibition of PDGF receptor signaling in tumor stroma enhances antitumor effect of chemotherapy. *Cancer Res* **62**:5476–5484.
101. Peters CE, Chaplin DJ and Hirst DG (1997) Nicotinamide reduces tumour interstitial fluid pressure in a dose- and time-dependent manner. *Br J Radiol* **70**:160–167.
102. Rubin K, *et al.* (2000) Lowering of tumoral interstitial fluid pressure by prostaglandin E(1) is paralled by and increase uptake of (51) Cr-EDTA. *Int J Cancer* **86**(5):636–643.
103. Ansiaux R, *et al.* (2005) Thalidomide radiosensitizes tumor through early changes in the tumor microenvironment. *Clin Cancer Res* **11**:743–750.

# 3

# Enhanced Permeability and Retention (EPR) Effect and Tumor-Selective Delivery of Anticancer Drugs

*Khaled Greish, Arun K. Iyer, Jun Fang, Michio Kawasuji and Hiroshi Maeda*

## 1. Introduction

Research aimed at achieving an effective cure for cancer — a major killer of humans in many nations — is a great challenge. This effort is reflected in the immense number of scientific research articles, which numbered more than 1 million in 2002. In that same year, cancer took the lives of more than 6.7 million patients throughout the world.[1] A major focus in cancer research concerns the unique characteristics of tumor cells or tumor tissues. Understanding of these characteristics will aid development of strategies for selective destruction of abnormal cancer cells without any harm to patients. Tumor vasculature is an ideal and attractive target for such strategies because it demonstrates more extensive abnormalities than vessels in normal tissues or organs. For example, it exhibits uniquely different fluid and molecular transport dynamics to meet an ever-increasing demand for nutrients and oxygen of the cancer cells.

Among these hallmark characteristics is the enhanced permeability and retention (EPR) effect of macromolecular agents in solid tumors, or the EPR effect, which was described about 20 years ago.[2–4] By means of the EPR effects, accumulation of macromolecules at the interstitium of tumor tissues is facilitated. One can take advantage of this macromolecular accumulation, via the EPR effect, for the delivery of polymeric or macromolecular drugs. EPR effect-based drug design is thus becoming more important for tumor-selective drug delivery, and the development of anticancer agents of polymeric or nanosize particles, which are believed to be more effective and safer than conventional chemotherapeutic agents, is of great interest. In this chapter, we briefly discuss the principles and factors involved in this mechanism, as well as the contribution of various fluid dynamic forces working across blood vessels in and near tumor tissue.

## 2. Theory Underlying the EPR Effect

When a tumor reaches a diameter of about 2–3 mm,[5–6] or under hypoxic state, it starts to induce formation of its own vasculature, or neovasculature. In response to an increasing need for a supply of nutrients and oxygen, cancer cells influence the host tissue to produce various factors such as vascular endothelial growth factor (VEGF) [which was initially named as vascular permeability factor (VPF)], bradykinin, nitric oxide (NO) and hypoxia-responsive element, and others.[7–17] Also, this induced vasculature in tumor tissues differs greatly from its counterpart in normal tissues in terms of microscopic morphology.[18–19] The defective anatomy, alone or together with functional abnormalities, results in considerable extravasation of blood plasma contents. These findings prompted us to investigate the possibility of delivering macromolecular anticancer drugs to tumor tissues in a selective fashion. For example, 20 years ago, we found that Evans blue dye, which bound with plasma albumin, concentrated selectively in tumor tissues[2] (Fig. 1). Other plasma proteins, including transferrin (90 kDa) and IgG (160 kDa), that were labeled with radioisotopes behaved similarly, whereas smaller proteins such as neocarzinostatin (12 kDa) and ovomucoid (29 kDa) did not.[2] We named the phenomenon of macromolecule or microparticle accumulation in solid tumor tissue the EPR effect. Since then, we have focused our efforts on analyzing the vascular mediators that facilitate the EPR effect, as well as on influencing the extravasation (a consequence of the EPR effect), so as to enhance this phenomenon for

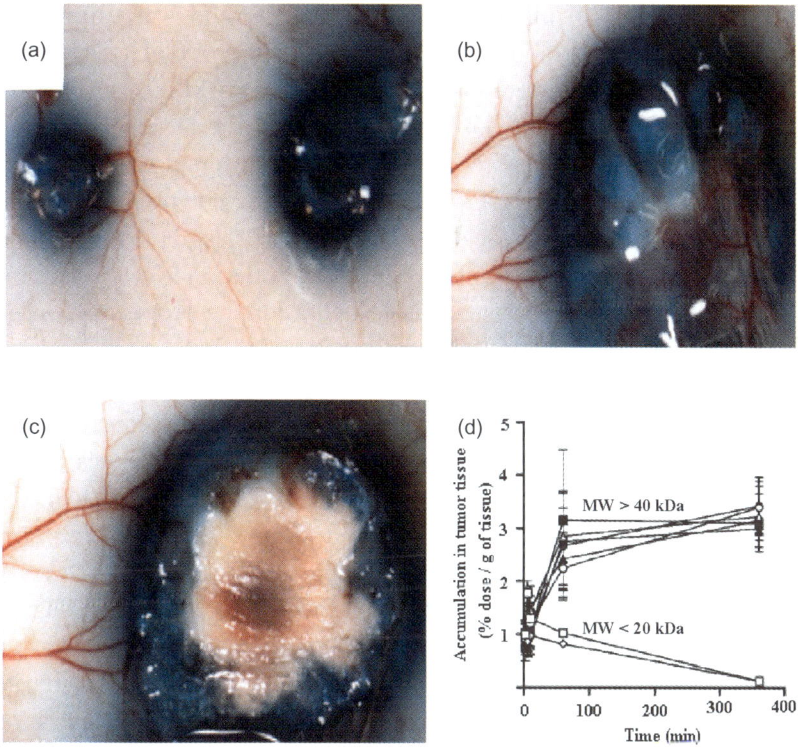


**Fig. 1** Selective accumulation of Evans blue dye bound to albumin (70 kDa) in a small tumor (a), large tumor (b), and cross-section of the tumor in (b) (c). The depth of penetration of the dye was showed. The central region of the tumor is necrotic and avascular, and hence does not facilitate the uptake of macromolecules, because it is not a growing area (from Ref. 7) (d) shows the accumulation of $^{125}$I-labeled N-(2-hydroxypropyl)methacryalmide (HPMA) copolymer in the solid tumor tissues (from Ref. 3). Note that large macromolecules, but not smaller ones, manifest progressive accumulation.

improved delivery of anticancer drugs, especially macromolecular agents, as discussed later. During this period, investigations of the EPR effect continued by use of various biocompatible polymers.[3,20,21]

## 3. Anatomical and Pathophysiological Abnormalities Related to EPR Effect

The density of the vasculatures in many tumors is often higher than that in normal tissues. The newly developed tumor vessels have abnormally wide pores between endothelial cells lining the lumina. Davies *et al.*[22]

measured the rate of diffusion of IgG (160 kDa) and was estimated to be 100 μm/hr. A macromolecular anticancer drug, copoly(styrene-maleic acid)-conjugated neocarzinostatin, or SMANCS (16 kDa, also bound to albumin) diffuses at about 1 mm/hr. In view of the fact that functional cells can be located as far as 20–30 μm from tumor blood vessels, one can understand the easy access of macromolecular drugs to tumor tissue.

Furthermore, tumor vessels were found to lack a smooth muscle layer, which usually surrounds capillary endothelial cells.[18–19] In normal blood vessels, the smooth muscle layer is essential for the proper response to vascular mediators such as acetylcholine, NO, bradykinin, and calcium, and hence for maintaining a constant blood flow volume. Vascular mediators cause relaxation of the smooth muscle layer or vasoconstriction, as a function of smooth muscle constriction. Normal blood vessels have an amazingly sophisticated system for monitoring the blood supply to healthy organs such as the heart, brain, liver, and kidney. The vascular smooth muscle tone is controlled by the autonomic nervous system, as well as by various vascular mediators, via receptors on smooth muscle cells. The tonus of the muscles is involved in maintaining a constant blood flow volume by means of blood pressure and blood flow rate in normal tissues. Thus, angiotensin II (AT-II)-induced hypertension will cause high blood pressure (generated by constriction of smooth muscle that leads to a narrowing of vascular diameter) and faster blood flow, but blood flow volume in the capillaries in normal tissues remains constant.[23] This blood flow control mechanism does not operate in tumor vessels because of the lack of smooth muscles. Therefore, in the presence of high blood pressure, capillaries of tumor tissue show much higher blood flow volume than do those in normal tissue[24] (Fig. 2).

In addition to the enhanced permeability, other important relevant abnormalities include a lack of receptors for vascular mediators, and a lack of functional lymphatic drainage which is needed for clearing lipidic or macromolecular particles.[2–4,23,25,26] Also, as tumors grow, hypoxia may result, causing activation of genes responsive to this state, and thus generation of factors such as hypoxia-induced factor,[16,17] VEGF, NO and others.[7–15] Table 1 summarizes the factors that contribute to the EPR effect in solid tumors.

Tumor cells and tissues, as well as surrounding normal cells and especially leukocytes that have infiltrated the tumor, produce large amounts of

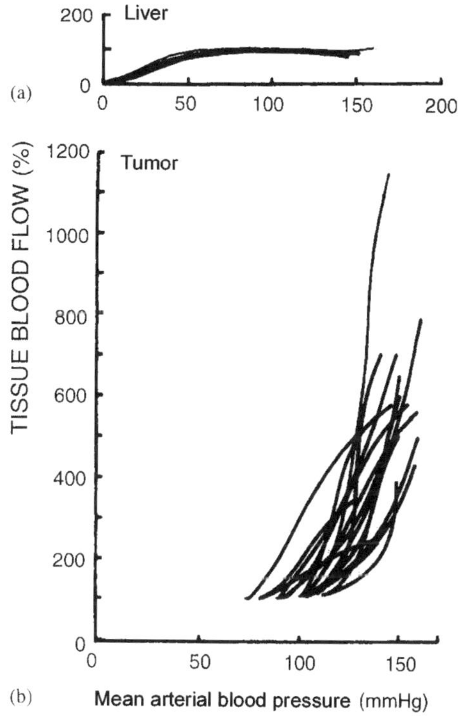


**Fig. 2** Effect of AT-II-induced hypertension on blood flow volume in (a) normal organs as represented by the liver and (b) tumor. Other normal organs, including the brain, bone marrow, and kidney, were also valuated; the tumor used was AH109B hepatoma in Donryu rats (from Ref. [24]).

angiogenic factors that lead to neoangiogenesis. Most of these angiogenic factors function as both permeability factors and vasodilators. For instance, VEGF is linked to the generation of NO. We have demonstrated excessive production of bradykinin, NO, and prostaglandin in tumors.[7,8,10,27] We found that matrix metallo proteinase (MMP) (collagenase), which is known to facilitate metastasis and angiogenesis to support tumor growth, also enhances the vascular permeability of solid tumor in mice.[7,8,14,25] This effect is inhibited by many MMP inhibitors.[14] NO and superoxide are simultaneously generated at sites of inflammation and cancer, and they react very rapidly with each other to form peroxynitrite, which can activate proMMP and affect the EPR effect and tumor metastasis.[28]

Macrophages play an important role in host defense in infection and cancer. However, they secrete VEGF and NO, which potentiate

**Table 1** Factors contributing to the EPR effect: anatomical aspects and overproduction of vascular permeability mediators.

Anatomical Factors

(1) Extensive angiogenesis: High vascular density, irregular formations
(2) Lack of smooth muscle layer: In a hypertensive state, passive opening of gaps induced by vasoconstrictors, e.g. AT-II
(3) Microanatomical defect: Wide endothelial gap junctions
(4) Lack of functional lymphatic system: Prolonged retention of macromolecules after extravasation
(5) Slow venous return

Permeability Mediators

(1) VEGF/(VPF)
(2) Low-molecular-weight mediators: NO, bradykinin, prostaglandins and others
(3) Matrix metalloproteinase (MMP)/collagenase: Facilitates metastasis; peroxynitrite and proteinases also activated (proMMP)
(4) Lack of response to vasoconstrictors (lack of smooth muscle layer, AT-II receptor)

angiogenesis and promote tumor growth. Moreover, the thrombin clotting system is also involved in angiogenesis and tumor progression via multiple mechanisms, including interaction with and enhancement of VEGF.[29]

## 4. Augmentation of the EPR Effect in Solid Tumor by Influencing Vascular Mediators

As discussed earlier, tumor vasculature has irregularities that make it highly leaky, even for macromolecular plasma components that are not found in the vasculature of normal tissues or organs. The absence of a smooth muscle layer means that tumor vessels cannot respond to vasoconstrictors. Therefore, administration of a vasoconstrictor such as AT-II that affects normal vessels and increases blood pressure would be expected to have no effect on tumor vessels. However, hypertension would have mechanical effects and cause dilation of tumor vasculature in a passive manner. Enhanced extravasation would result as discussed below. Indeed, Hori *et al.* showed clearly in a window model of solid tumor that some tumor vessels cannot be seen under normotensive conditions but can when an AT-II-induced hypertensive state is generated.[30] This finding means that apparently avascular tumor tissue actually does have vessels but that they are visible or functional sporadically, e.g. once every 15 min or 25 min, and that the blood flows in unpredictable directions.[30] However, the absence of smooth muscle in tumor vessels accounted for a three- to five-fold increase in blood flow volume under conditions of induced hypertension, when systolic pressure increased from 100 to 160 mmHg by infusion of AT-II.[23,30,31]

With regard to macromolecular agents, deposition of radiolabeled albumin and SMANCS in the tumor tissues increased two- to three-fold compared with that in normal tissues.[30] This greater delivery of macromolecules to tumor tissue was also observed with many solid cancers, including hepatoma, cholangiocarcinoma, metastatic liver cancer, pancreatic cancer, and others, after arterial injection of SMANCS in Lipiodol (SMANCS/Lipiodol) under hypertensive conditions induced by AT-II.[23,31,32] In a converse approach, we also utilized vasodilators, such as the NO-releasing agent isosorbide dinitrate (ISDN; Nitrol), to enhance the EPR effect via widening the tumor feeding artery. This result was accomplished by infusing ISDN by catheter (31, 32 and unpublished results, Maeda H, Greish K, *et al.*).

It is well known that bradykinin induces intense pain and also increases vascular permeability and that NO promotes angiogenesis.[33] Interplay between bradykinin and other mediators including NO, prostaglandins (PGs), and VEGF will also lead to angiogenesis. Biosynthesis of PGs particularly prostaglandin $E_2$ via cyclooxygenase isozymes (COX-1 and -2) is markedly elevated in inflammation and cancer. These increased levels of PGs can also enhance vascular permeability in solid tumor, as evidenced by significant suppression of vascular permeability in sarcoma 180 and other solid tumor models by the COX inhibitor indomethacin and salicylic acid.[7,8,25] We showed that a prostaglandin $I_2$ analogue (beraprost sodium) with a much longer *in vivo* half-life (about 30 min vs 3 sec for prostaglandin $I_2$) was useful for the delivery of macromolecules,[34] although a therapeutic advantage of beraprost sodium needs to be demonstrated.

We had also reported earlier a significant activation of the bradykinin-generating cascade in the tumor compartment, and bradykinin was shown to be involved in the accumulation of malignant ascitic and pleural fluid.[9,10,27] So-called angiotensin-converting enzyme (ACE) inhibitors such as enalapril and other similar agents can inhibit degradation of bradykinin *in vivo* and lead to higher bradykinin concentrations at sites of tumor and infection, because of an amino acid sequence homology near the C-termini. Consequently, ACE inhibitors did enhance the EPR effect[9,27,35] mediated by either bradykinin or NO. It should be noted that the ACE inhibitors or prostaglandin $I_2$ analogue beraprost, in addition to enhancing vascular permeability, significantly suppressed blood flow only in tumor tissue.[27,34,35] It should be remembered that bradykinin is actively generated only at sites of infection, inflammation and cancer, and not in normal tissues[8,25]; ACE inhibitors and beraprost have only a slight, if not negligible effect on systemic blood pressure in the normotensive population. Therefore, increasing the local concentration of bradykinin by means of an ACE inhibitor, and thereby improving tumor-selective delivery of macromolecular drugs should be possible. It should also be mentioned that when [$^{14}$C]methylglucose, a representative low-molecular-weight drug mimic, was studied, accumulation of this agent in tumor was much less than that of polymeric drugs and lasted no longer than 10 min.[36] Such low-molecular-weight drugs seem to be washed out rapidly into the general circulation and are excreted via the urine.

## 5. Relation of Fluid Dynamics in Cancer Tissue to the EPR Effect

Physiologists have studied fluid dynamics in normal tissue for centuries.[23] E.H. Starling referred to the constant blood volume between the arterial end and the venous end of a capillary under normal conditions. In both normal and tumor vessels, the difference between the hydrostatic and colloid osmotic pressures is known to affect the movement of fluid and solutes through the capillary vessel wall.[23,37] In normal human tissue blood vessels, the arterial end of a capillary has an average hydrostatic positive pressure of about 25 mmHg.[23,37] This value drops to about 10 mmHg in the venous end of the capillary. The interstitial colloid osmotic and hydrostatic pressures remain constant at both arterial and venous ends of the capillary. This pressure difference facilitates leakage of fluid and nutrients into the interstitial space at the arterial end of the capillary, and then reabsorption at the venous end. This continuous translocation of fluid from the arterial end to the venous end of capillary through the interstitial space, ensures a continuous supply of oxygen and nutrients for cells, as well as an efficient removal of metabolic waste products.

Tumor vessels, however, demonstrate two main differences compared with normal vessels. Firstly, significantly enhanced or almost unrestricted leakage of plasma proteins occurs because of the wide endothelial gap openings with large pore sizes estimated to be 0.2∼0.5 μm.[22,38] Secondly, the lack of functional lymphatics in tumor tissue,[2,4,25,39] as described in an earlier section, would lead to higher interstitial accumulation of macromolecules or nanoparticles than in normal tissue because these components cannot be cleared. As can readily be perceived, the raised interstitial colloid osmotic pressure would facilitate transfer of low-molecular-weight components as well as macromolecules from tumor vesseles into the interstitial space of the tumor tissues, due to the higher solute level there. Also, the increased interstitial hydrostatic pressure at the arterial end of the capillaries will drive low-molecular-weight components in fluid from the interstitial space into the venous end of the capillaries and back to the luminal circulation (Fig. 3).

Only the lymphatic system can clear macromolecules, nanoparticles, and lipids in normal tissues, whereas small molecules can move into and out of blood vessels in both normal and tumor tissues more freely.[2,3,25,26,38,40–42] However, transfer of macromolecules to the luminal side of blood capillaries does not occur effectively.[30,40] Actively dividing tumor cells with accelerated metabolic activity can eliminate low-molecular-weight waste products via venous return, but the venous return in tumor tissues was found to be an order of magnitude lower than that in normal tissue.[26,34] Thus, the impaired lymphatic clearance of macromolecules and lipidic components can also account, in part, for the EPR effect in solid tumors.

## 6. Implications for Delivery of Drugs to Tumors

In cancer treatment, the concentration of a drug in tumor tissue is of utmost importance. However, it is not usually possible to achieve a drug concentration in tumor tissue that is higher than that in plasma by using conventional anticancer drugs, which are often low-molecular-weight agents without carriers. However, the EPR effect allowed achievement of anticancer drug concentrations that were several to 10 times higher in tumor tissue than in normal tissues or organs. For example, Lipiodol is a lipid contrast medium and becomes minute particles after arterial injection. In our experience,

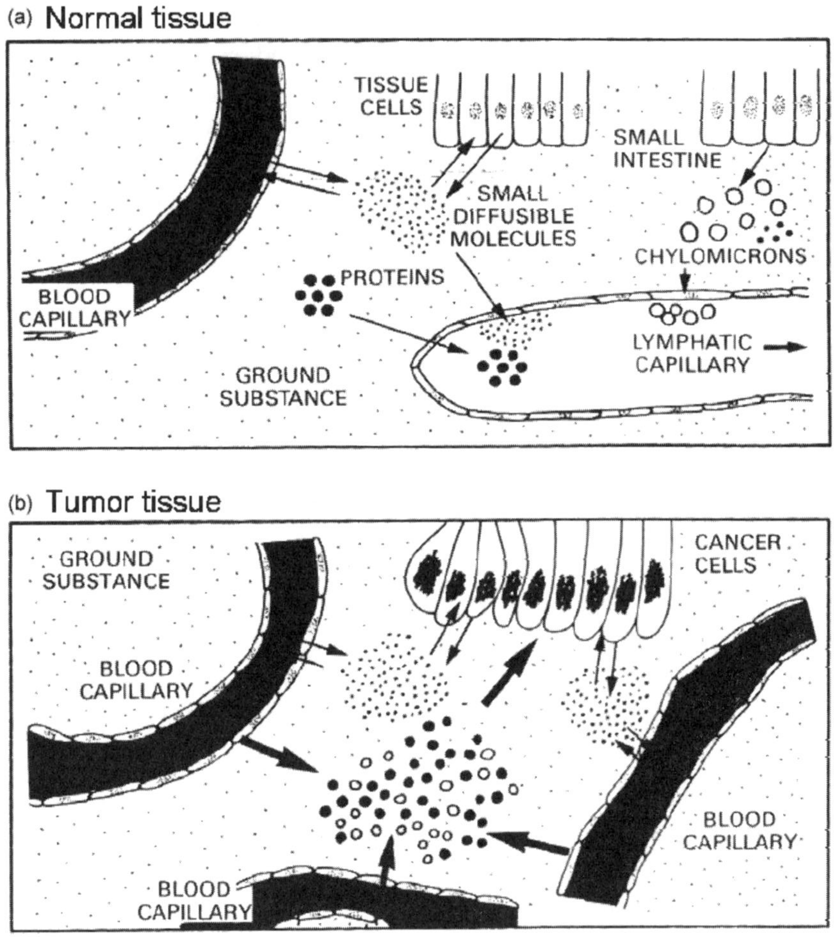


**Fig. 3** Diagrammatic representation of vessel structures and transport of various substances into and out of blood capillaries in normal (a) and tumor (b) tissue. Leakage of macro molecules and nanoparticles from blood capillaries to tumor tissues is enhanced but no lymphatic recovery occurs. Furthermore, in tumor tissues there is no flow of macromolecules back into the blood capillaries, and lymphatic ducts are absent (from Ref. 40).

injection of SMANCS/Lipiodol into tumor tissue via the tumor-feeding artery resulted in drug retention for more than several weeks, and the drug level was 2000-fold higher in tumor tissue than in blood plasma.[4,43,44] In this case, the EPR effect and the first-pass effect work additively. Eventually, the drug disappeared from the tumor together with degraded tumor tissues.

Thus, the advent of SMANCS therapy and the discovery of the EPR effect, which applies to macromolecular anticancer drugs in general, permitted EPR effect-based anticancer drug targeting to solid tumor in becoming a reality.[2,39,40,43–50] For this type of therapy, the plasma level of macromolecules, nanoparticles, or lipidic particles (drugs), as expressed by the area under the concentration curve (AUC), must remain high, preferably for more than 6 hours. Extravasation of these agents into the tumor tissue increases progressively with time (in several hours to days), whereas clearance from tumor does not proceed because of a lack of lymphatic function. It should be noted that one can maintain a high plasma concentration or AUC for many hours, even with low-molecular-weight agents, by slow continuous intravenous infusion. However, low-molecular-weight agents do not exhibit EPR-effect because they can move freely into and out of blood vessels (capillaries), and the plasma level quickly becomes quite low because of excretion. Thus, targeting of drug to tumor does not take place.

Macromolecular conjugates or particles that contain active drugs of low molecular weight are usually required, the active principle being released to react with the molecular target in tumor cells. Therefore, the release rate, enhanced not only tumor delivery, but become a key issue for the development of truly effective anticancer agents. For example, we found that when SMANCS/Lipiodol is infused via the tumor-feeding artery, it is delivered to and deposited throughout the tumor but is not cleared quickly (clearance takes several weeks to months). The concentration of SMANCS administered via the artery was 20–30 μg/g of tumor tissues even 2–3 months after injection of 1 mg/ml (SMANCS/Lipiodol), and this remaining activity was more than 1000 times higher than the minimal inhibitory concentration against tumor cells (i.e. SMANCS had a minimal effective concentration below 0.05 μg/ml).[43] Also, SMANCS is itself biologically active without release from the polymeric matrix. For these reasons, SMANCS therapy is highly effective.

It should be mentioned that not all tumors are highly vascular; some appear to be hypovascular or avascular.[30] Tumors of the pancreas and the prostate, for example, are hypovascular. Many avascular tumor tissues such as the central portion of metastatic tumors demonstrate central necrosis and circulatory insufficiency. Avascular areas are not growing and need little, if any, attention.

The EPR effect functions not only in subcutaneous implanted tumors in mouse tumor models, but also in many other tumors such as VX-II carcinoma implanted in the liver (rabbit); Walker 256 carcinoma which metastasizes to the omentum in rats; and other metastatic liver and colon cancers chemically induced in a mouse model, as well as patients given SMANCS.[31,32,43,44] In the clinical setting, visualization of tumor is possible by use of radioscintigraphy with $^{67}$Ga citrate, a radioactive isotope compound, which binds to transferrin (90 kDa) in blood plasma and selectively accumulates in the tumor by means of the EPR mechanism.[2]

The advantage of polymeric drugs with regard to multidrug resistance has also been discussed.[51–53] The lymphotropic nature of polymeric drugs discussed previously, is of prime importance for controlling lymphatic metastasis, which is the cause of many therapeutic failures.[50,54–57]

## 7. Conclusion

The EPR effect can be considered as a hallmark concept that exploits the anatomical and pathophysiological defects in the tumor vasculature. It plays a critical role in more selective delivery of macromolecular anticancer agents to cancer tissues. Understanding and manipulating the factors contributing to the EPR effect can further improve the selective targeting of high-molecular-weight biocompatible or nanoparticle anticancer drugs to tumor. The EPR effect can be enhanced in many ways, such as by using vascular mediators and by arterial injection to attain concentrations in tumor tissues that can never be possible with conventional intravenous chemotherapeutics. With achievement of the targeting of macromolecular drugs to tumors, the rate of release of free drugs from the composite becomes the key for successful anticancer therapeutics. In summary, EPR-based strategies portend a bright future for cancer chemotherapy.

## References

1. Press Release WHO/52, 28 June 2002. Available online from: http://www.who.int/inf/en/pr-2002-52.html [Accessed 2003 Aug 23].
2. Matsumura Y and Maeda H (1986) A new concept for macromolecular therapeutics in cancer chemotherapy: Mechanism of tumor tropic accumulation of proteins and antitumor agent SMANCS. *Cancer Res* **46**:6387–6392.

3. Noguchi Y, Wu J, Duncan R, Strohalm J, Ulbrich K, Akaike T and Maeda H (1998) Early phase of tumor accumulation of macromolecules: A great difference in clearance rate between tumor and normal tissues. *Jpn J Cancer Res* **89**:307–314.
4. Iwai K, Maeda H and Konno T (1984) Use of oily contrast medium for selective drug targeting to tumor: Enhanced therapeutic effect and X-ray image. *Cancer Res* **44**:2114–2121.
5. Folkman J (1971) Tumor angiogenesis. Therapeutic implications. *N Engl J Med* **285**:1182–1186.
6. Folkman J (1995) Angiogenesis in cancer, vascular, rheumatoid and other disease. *Nat Med* **1**:27–31.
7. Wu J, Akaike T and Maeda H (1998) Modulation of enhanced vascular permeability in tumors by a bradykinin antagonist, a cyclooxygenase inhibitor, and a nitric oxide scavenger. *Cancer Res* **58**:159–165.
8. Maeda H, Akaike T, Wu J, Noguchi Y and Sakata Y (1996) Bradykinin and nitric oxide in infectious disease and cancer. *Immunopharmacology* **33**:222–230.
9. Matsumura Y, Kimura M, Yamamoto T and Maeda H (1988) Involvement of the kinin-generating cascade and enhanced vascular permeability in tumor tissue. *Jpn J Cancer Res* **79**:1327–1334.
10. Maeda H, Matsumura Y and Kato H (1988) Purification and identification of [hydroxyprolyl 3]bradykinin in ascetic fluid from a patient with gastric cancer. *J Biol Chem* **263**:16051–16054.
11. Doi K, Akaike T, Horie H, Noguchi Y, Fujii S, Beppu T, Ogawa M and Maeda H (1996) Excessive production of nitric oxide in rat solid tumor and its implication in rapid tumor growth. *Cancer* **77**:1598–1604.
12. Doi K, Akaike T, Fujii S, Tanaka S, Ikebe N, Beppu T, Shibahara S, Ogawa M and Maeda H (1999) Induction of haem oxygenase-1 by nitric oxide and ischaemia in experimental solid tumors and implications for tumor growth. *Br J Cancer* **80**:1945–1954.
13. Senger DR, Galli SJ, Dvorak AM, Perruzzi CA, Harvey VS and Dvorak HF (1983) Tumor cells secrete a vascular permeability factor that promotes accumulation of ascites fluid. *Science* **219**:983–985.
14. Wu J, Akaike T, Hayashida K, Okamoto T, Okuyama A and Maeda H (2001) Enhanced vascular permeability in solid tumor involving peroxynitite and matrix metalloproteinases. *Jpn J Cancer Res* **92**:439–451.
15. Maeda H, Noguchi Y, Sao K and Akaike T (1994) Enhanced vascular permeability in solid tumor is mediated by nitric oxide and inhibited by both new nitric oxide scavenger and nitric oxide synthase inhibitor. *Jpn J Cancer Res* **85**:331–334.
16. Giordano FJ and Johnson RS (2001) Angiogenesis: The role of the microenvironment in flipping the switch. *Curr Opin Genet Dev* **11**:35–40.

17. Zhang X, Kon T, Wang H, Li F, Huang Q, Rabbani ZN, Kirkpatrick JP, Vujaskovic Z, Dewhirst MW and Li CY (2004) Enhancement of hypoxia-induced tumor cell death *in vitro* and radiation therapy *in vivo* by use of small interfering RNA targeted to hypoxia-inducible factor-1$\alpha$. *Cancer Res* **64**:8139–8142.
18. Skinner S, Tutton P and O'Brine P (1990) Microvascular architecture of experimental colon tumors in the rat. *Cancer Res* **50**:2411–2417.
19. Suzuki M, Takahashi T and Sato T (1987) Medial regression and its functional significance in tumor-supplying host arteries. *Cancer* **59**:444–450.
20. Seymour LW, Miyamoto Y, Maeda H, Brereton M, Strohalm J, Ulbrich K and Duncan R (1995) Influence of molecular weight on passive tumor accumulation of a soluble macromolecular drug carrier. *Eur J Cancer* **31A**:766–770.
21. Duncan R (1999) Polymer conjugates for tumour targeting and intracytoplasmic delivery. The EPR effect as a common gateway? *Pharm Sci Technol Today* **2**:441–449.
22. Davies Cde L, Berk DA, Pluen A and Jain RK (2002) Comparison of IgG diffusion and extracellular matrix composition in rhabdomyosarcomas grown in mice versus *in vitro* as spheroids reveals the role of host stromal cells. *Br J Cancer* **86**:1639–1644.
23. Guyton AC and Hall JE (2000) The body fluids and kidneys. In: *Textbook of Medical Physiology*, 10th ed. Philadelphia: WB Saunders, pp. 358–382.
24. Suzuki M, Hori K, Abe I, Saito S and Sato H (1981) A new approach to cancer chemotherapy: Selective enhancement of tumor blood flow with angiotensin II. *J Natl Cancer Inst* **67**:663–669.
25. Maeda H (2001) Enhanced permeability and retention (EPR) effect in tumor vasculature: The key role of tumor-selective macromolecular drug targeting. *Adv Enzyme Regul* **41**:189–207.
26. Courtice FC (1963) The origin of lipoproteins in lymph. In: *Lymph and Lymphatic system*. Springfield, IL: Charles C Thomas, pp. 89–126.
27. Matsumura Y, Maruo K, Kimura M, Yamamoto T, Konno T and Maeda H (1991) Kinin-generating cascade in advanced cancer patients and *in vitro* study. *Jpn J Cancer Res* **82**:732–741.
28. Okamoto T, Akaike T, Sawa T, Miyamoto Y, van der Vliet A and Maeda H (2001) Activation of matrix metalloproteinases by peroxynitrite-induced protein *S*-glutathiolation via disulfide S-oxide formation. *J Biol Chem* **276**:29596–29602.
29. Maragoudakis ME, Tsopanoglou NE and Andriopoulou P (2002) Mechanism of thrombin-induced angiogenesis. *Biochem Soc Trans* **30**:173–177.
30. Hori K, Saito S, Nihei Y, Suzuki M and Sato Y (1999) Antitumor effects due to irreversible stoppage of tumor tissue blood flow: Evaluation of a novel combretastatin A-4 derivative, AC7700. *Jpn J Cancer Res* **90**:1026–1038.

31. Nagamitsu A, Inuzuka T, Kawasuji M and Maeda H (2003) Recent advances in SMANCS/Lipiodol therapy enhanced targeting and delivery efficacy using vascular modulators. *Drug Deliv Syst* **18**:438–447.
32. Greish K, Fang J, Inutsuka T, Nagamitsu A and Maeda H (2003) Macromolecular therapeutics: Advantages and prospects with special emphasis on solid tumour targeting. *Clin Pharmacokinet* **42**:1089–1105.
33. Jackson RJ, Seed PM, Kircher HC, Willoughby AD and Winker JD (1997) The codependence of angiogenesis and chronic inflammation. *FASEB J* **11**:457–465.
34. Tanaka S, Akaike T, Wu J, Fang J, Sawa T, Ogawa M, Beppu T and Maeda H (2003) Modulation of tumor-selective vascular blood flow and extravasation by the stable prostaglandin $I_2$ analogue beraprost sodium. *J Drug Target* **11**:45–52.
35. Hori K, Saito S, Takahashi H, Sato H, Maeda H and Sato Y (2000) Tumor-selective blood flow decrease induced by an angiotensin converting enzyme inhibitor, temocapril hydrochloride. *Jpn J Cancer Res* **91**:261–269.
36. Li CJ, Miyamoto Y, Kojima Y and Maeda H (1993) Augmentation of tumour delivery of macromolecular drugs with reduced bone marrow delivery by elevating blood pressure. *Br J Cancer* **67**:975–980.
37. Jain R (1994) Barriers to drug delivery in solid tumors. *Sci Am* **271**:58–65.
38. Yuan F, Dellian M, Fujumura D, Leunig M, Berk DA, Torchilin VP and Jain RK (1995) Vascular permeability in a human tumor xenograft: Molecular size dependence and cutoff size. *Cancer Res* **55**:3752–3756.
39. Maeda H, Matsumoto T, Konno T, Iwai K and Ueda M (1984) Tailor-making of protein drugs by polymer conjugation for tumor targeting: A brief review on smancs. *J Protein Chem* **3**:181–193.
40. Maeda H (1991) SMANCS and polymer-conjugated macromolecular drugs: Advantages in cancer chemotherapy. *Adv Drug Deliv Rev* **6**:181–202.
41. Hashizume H, Baluk P, Morikawa S, McLean JW, Thurston G, Roberge S, Jain RK and McDonald DM (2000) Openings between defective endothelial cells explain tumor leakiness. *Am J Pathol* **156**:1363–1380.
42. Yuan F, Salehi HA, Boucher Y, Vasthare US, Tuma RF and Jain RK (1994) Vascular permeability and microcirculation of gliomas and mammary carcinomas transplanted in rat and mouse cranial windows. *Cancer Res* **54**:463–468.
43. Konno T, Maeda H, Iwai K, Maki S, Tashiro S, Uchida M and Miyauchi Y (1984) Selective targeting of anti-cancer drug and simultaneous image enhancement in solid tumors by arterially administered lipid contrast medium. *Cancer* **54**:2367–2374.
44. Maki S, Konno T and Maeda H (1985) Image enhancement in computerized tomography for sensitive diagnosis of liver cancer and semiquantitation of tumor selective drug targeting with oily contrast medium. *Cancer* **56**:751–757.

45. Yokoyama M, Okano T, Sakurai Y, Fukushima S, Okamoto K and Kataoka K (1999) Selective delivery of adriamycin to a solid tumor using a polymeric micelle carrier system. *J Drug Target* **7**:171–186.
46. Kopecek J, Kopeckova P, Minko T and Lu Z (2000) HPMA copolymer-anticancer drug conjugates: Design, activity, and mechanism of action. *Eur J Pharm Biopharm* **50**:61–81.
47. Kopecek J, Kopeckova P, Minko T, Lu ZR and Peterson CM (2001) Water-soluble polymers in tumor targeted delivery. *J Control Rel* **74**:147–158.
48. Gabizon AA (2001) Pegylated liposomal doxorubicin: Metamorphosis of an old drug into a new form of chemotherapy. *Cancer Invest* **19**:424–436.
49. Singer JW, De Vries P, Bhatt R, Tulinsky J, Klein P, Li C, Milas L, Lewis RA and Wallace S (2000) Conjugation of camptothecins to poly(L-glutamic acid). *Ann NY Acad Sci* **922**:136–150.
50. Duncan R (2003) The dawning era of polymer therapeutics. *Nat Rev Drug Discov* **2**:347–360.
51. Miyamoto Y, Oda T and Maeda H (1990) Comparison of the cytotoxic effects of the high- and low-molecular-weight anticancer agents on multidrug-resistant Chinese hamster ovary cells *in vitro*. *Cancer Res* **50**:1571–1575.
52. Minko T, Kopeckova P, Pozharov V and Kopecek J (1998) HPMA copolymer bound Adriamycin overcomes MDR1 gene encoded resistance in a human ovarian carcinoma cell line. *J Control Rel* **54**:223–233.
53. Maeda H, Greish K and Fang J (2005) EPR effect and polymeric drugs: Challenge for a paradigm shift for cancer chemotherapy in the 21st century. *Adv Polym Sci* (in press).
54. Maeda H, Takeshita J and Kanamaru R (1979) A lipophilic derivative of neocarzinostatin. A polymer conjugation of an antitumor protein antibiotic. *Int J Pept Protein Res* **14**:81–87.
55. Maeda H, Takeshita J and Yamashita A (1980) Lymphotropic accumulation of an antitumor antibiotic protein, neocarzinostatin. *Eur J Cancer* **16**:723–731.
56. Takeshita J, Maeda H and Kanamaru R (1982) *In vitro* mode of action, pharmacokinetics, and organ specificity of poly(maleic acid-styrene)-conjugated neocarzinostatin. *SMANCS. Gann* **73**:278–284.
57. Maeda H and Matsumura Y (1989) Tumoritropic and lymphotropic principles of macromolecular drugs. *Crit Rev Ther Drug Carrier Syst* **6**:193–210.

# 4

# Basic Strategies for PEGylation of Peptide and Protein Drugs

*Gianfranco Pasut, Margherita Morpurgo and Francesco M. Veronese*

## Abstract

The term PEGylation defines the modification of a protein, peptide or non-peptide molecule by the chemical linking of one or more poly(ethylene glycol) (PEG) chains. PEG is the polymer of choice for drug conjugation, being non-toxic, non-immunogenic, non-antigenic and highly soluble in water; all properties that are conferred on the conjugated drugs. The PEG derivatives have the following advantages: (1) a prolonged residence in the body due to a reduced kidney clearance, as a consequence of the increased molecular weight; (2) a decreased degradation by proteolytic enzymes, thanks to the shielding effect of the polymer chains; (3) reduction or elimination of protein immunogenicity.

Thanks to these favorable properties, PEGylation now plays an important role in enhancing the use and potential of peptides and proteins as therapeutic agents that, in the native form, often encounter severe limitations in their use. So far, several chemical strategies have been developed to perform the linking of the polymer to a biologically active molecule, and a wide range of methods of which a selection is reported here. A clear demonstration of the potential of PEGylation is the numbers of products that have already reached the market or that will soon be available. In addition to a review of PEGylation methods, this chapter also describes recent achievements in PEGylation of protein and peptide drugs.

## 1. Introduction

Research in polymer-protein conjugation had its beginnings in the sixties through the seventies when dextran was studied as a coupling polymer. Since then, many studies have been carried out using both natural and synthetic polymers and suitable chemical strategies for conjugation have been developed along with the identification of conditions for selective modification and easy purification as well as tailor-made analytical procedures. Among all of the polymers studied so far, PEG emerged as the best candidate for modification, thanks to its uncommon properties that are conferred on the final conjugate. The most interesting characteristics of PEG are: (a) the lack of immunogenicity, antigenicity and toxicity; (b) high solubility in water and in many organic solvents; (c) high flexibility of the chain and; (d) FDA approval for human use. Many pharmaceutical companies are now looking to this technology to improve their products, most of them being proteins or peptides (an increasing portion among all drugs after the human genome sequence release) since PEGylation appears to be powerful solution to many of their shortcomings. Common limits for therapeutic application of peptides and proteins are their tendency to promote an immunological response (in particular when their sequence is not identical to the human protein); their chemical instability (both *in vivo* and in the formulation); and their short body residence time, the last mainly evident for the low molecular weight peptides. Most of these problems are overcomed by PEGylation, thanks to the combination of increased molecular weight, coverage or blockage of critical protein's sites (epitopes or sequences degraded by enzymes) and the enhanced solubility of the conjugates in water. As reported in several articles and reviews (Refs. 1–4), PEGylation increases the hydrodynamic volume of the molecule so that the first common positive effect of conjugation is an extended blood residence time due to reduced kidney clearance. Consequently, the derivatives need a reduced frequency of administration with respect to the unmodified drug. Moreover, PEGylated molecules possess increased solubility. The shielding effect of a PEG chain on specific sites on the protein surface

**Table 1** PEG conjugates that are already on the market.

| Conjugates | Year to market | Disease |
|---|---|---|
| PEG-adenosine deaminase (*Adagen*®)[20] | 1990 | Severe combined immunodeficiency disease (SCID) |
| PEG-asparaginase (*Oncaspar*®)[21] | 1994 | Acute lymphoblastic leukemia |
| Linear PEG-interferon $\alpha$2b (*PEG-Intron*®)[33,77] | 2000 | Hepatitis C and clinical trials for cancer, multiple sclerosis, HIV/AIDS. |
| Branched PEG-interferon $\alpha$2a (*Pegasys*®)[34,35] | 2002 | Hepatitis C |
| PEG-growth hormone receptor antagonist (*Pegvisomant, Somavert*®)[36] | 2002 | Acromegaly |
| PEG-G-CSF (*pegfilgrastim, Neulasta*®)[37] | 2002 | Treating of neutropenia during chemotherapy |
| Branched PEG-anti-VEGF aptamer (*Pegaptanib, Macugen*™)[38] | 2004 | Macular degeneration (age-related) |

prevents or reduces the expression of immunogenicity and antigenicity, and enhances stability towards both chemical and enzymatic degradation.[5–7] For small drugs, it may yield more convenient biodistribution, selected cellular uptake[8–10] and, thanks to a specific linkers or molecules, a controlled drug release or drug targeting into specific organs or cells.[11]

The first application of PEG as a bioconjugation polymer has been proposed in the late 1970s by Professor Frank Davis at Rutgers University.[12] Following this pioneering study, large number of drugs with different structure (proteins, peptides, small molecular weight drugs, polynucleotides) have been PEGylated, thus creating a new class of drugs,[13] some of which have become blockbuster products. In Table 1, PEGylated products that have already reached the market are reported.

## 2. Features of PEG as Bioconjugation Polymer

Raw poly(ethylene glycol) is synthesized by ring opening polymerization of ethylene oxide. The reaction is initiated by methanol or water,

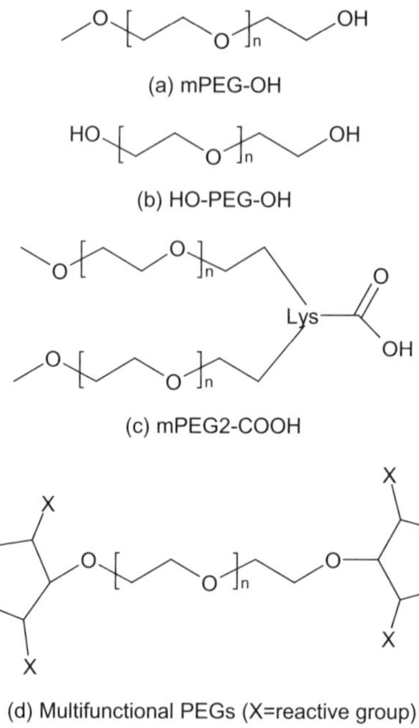


**Fig. 1** PEG structures: (a) linear monomethoxy PEG, (b) linear diol PEG, (c) branched PEG, (d) multifunctional PEGs.

forming polymers with one or two end-chain hydroxyl groups, respectively (mPEG-OH or HO-PEG-OH; Fig. 1). Starting from these simple molecules, a wide series of activated PEGs were developed to address selectively different chemical groups in proteins. Moreover, PEGs with various shapes are now available: branched, multifunctional or heterobifunctional PEGs (Fig. 1). Monofunctional polymers (mPEG-OH), linear or branched, are suitable for protein modification, while those with multiple reactive groups are used to increase the drug/polymer ratio, a strategy that is useful in the case of therapeutics agents with low biological activity that otherwise would require the administration of a large amount of conjugates.

The optimization of both the polymerization procedure and purification process allowed developing PEGs with low polydispersivity, $M_w/M_n$, spanning from 1.01, for PEG below 5 kDa molecular weigh, to 1.1 for PEG with molecular weight as high as 50 kDa.

PEG has unique solvation properties that are due to the coordination of 2–3 water molecules per ethylene oxide unit[6] that, together with the great flexibility of the polymer backbone, are responsible of the PEG biocompatibility and rejecting properties towards protein, which are at the basis of the low immunogenicity and antigenicity of the conjugates.[14] Furthermore, these characteristics give to PEG molecules an apparent molecular weight 5–10 times higher than that of a globular protein of comparable mass, as it can be verified by gel permeation chromatography.[15] Due to this large hydrodynamic volume, a single PEG molecule covers an extended surface of the conjugated protein, preventing degradation by mammalian cells and enzymes.[16]

*In vivo*, PEG undergoes limited chemical degradation and the clearance depends upon its molecular weight: below 20 KDa, it is easily secreted into the urine; while at higher molecular weights, they are eliminated more slowly and the clearance through the liver becomes predominant. The threshold for kidney filtration is about 40–60 KDa (a hydrodynamic radius of approximately 45 Å[17]), that represents the albumin excretion limit. Over this limit, the polymer remains in circulation and it is mainly accumulated in the liver. Alcohol dehydrogenase can degrade low molecular weight PEGs, and chain cleavage can be catalyzed by P450 microsomial enzymes.[18] Some branched PEGs may undergo a molecular weight reduction when the hydrolysis and loss of one polymer chain is catalyzed by anchimeric assistance.[19] Finally, several years of PEG use as an excipient in foods, cosmetics and pharmaceuticals, without toxic effects, are a clear proof of its safety.[16]

The first generation of PEG conjugates was based on low molecular weight polymers ($\leq$12 kDa), most commonly in their linear monomethoxylated form (mPEG). The batches of this polymer often contained a relevant percentage of PEG diol originated from the synthesis, an impurity that, after activation, could act as a potential cross-linking agent. Furthermore, the chemistry employed in mPEG synthesis often presented side reaction products or led to weak and reversible linkages. The drug PEG-adenosine deaminase (Adagen®)[20] and PEG-asparaginase (Oncaspar®)[21] belong to this generation.

The second generation of PEGs is characterized by improved polymer purity (reduction of both polydispersivity and diol content also for high molecular weights PEG) and in a wide selection of activated PEG reagents that allow selectivity of reaction towards different protein sites. The availability of several new PEGs is now widening the opportunities of the PEGylation technology. Among all, we would like to report:

- PEG-propionaldehyde, also in the form of the more stable acetal: the reaction with amino group leads to a Shiff base that, once reduced by sodium ciano borohydride, yields a stable secondary amine that maintains the same neat charge of the parent drug.
- Several PEG-succinimidyl derivatives: highly reactive towards amino group. The reaction rate of these derivatives may significantly change, depending on the length and the composition of the alkyl chain between PEG and the succinimidyl moiety.[22]
- "Y" shaped branched PEG[23] (termed PEG2 or U-PEG) [see Fig. 1(c)]: provides a higher surface shielding effect and it is more effective in protecting the conjugated protein from degradative enzymes and antibodies (Fig. 2). Moreover, proteins modified with this PEG retain higher activity than the same enzyme modified by linear PEGs. This effect is probably due to the hindrance of the branched polymer that prevents the entrance

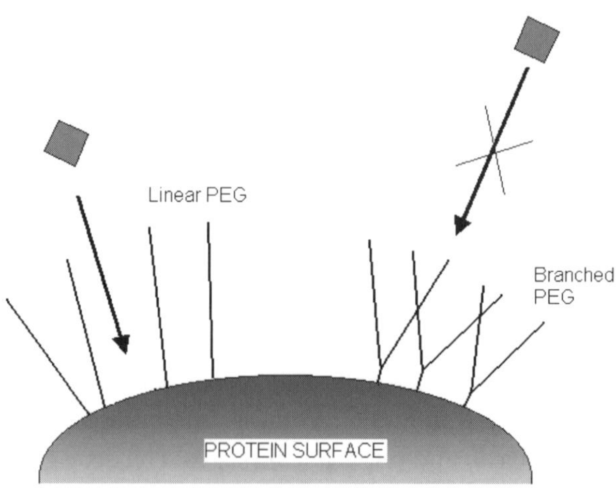


**Fig. 2** Structure of linear and branched PEG on protein surface. The "umbrella like" structure of branched PEG explains the higher capacity in rejecting approaching molecules or cells as compared to linear PEG.

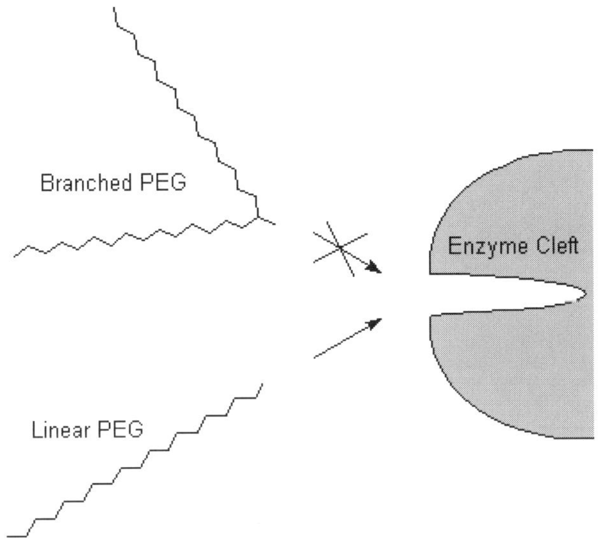


**Fig. 3** Effect of PEG hindrance on the enzyme active site access. The high steric hindrance of branched PEG, that makes difficult the access to active site cleft, may be advocated to explain the lower inactivation of enzymes as compared to linear PEG of the same size.

of PEG inside the enzyme active site cleft or other sites involved in biological activity (Fig. 3).

- PEGs reactive toward thiol groups: PEG-maleimide (MAL-PEG), PEG-vinylsulfone (VS-PEG), PEG-iodoacetamide (IA-PEG) and PEG-orthopyridyl-disulfide (OPSS-PEG) have been specifically developed for this conjugation, but only the last strictly reacts with the thiol groups, avoiding any degree of amino modification that may occur instead in small amount using the other three (Fig. 4).
- Heterobifunctional PEGs[24,25]: these derivatives have different reactive groups at the two polymer ends that allow to link separately two different molecules to the same PEG chain. It is therefore possible to obtain conjugates that carry both a drug and a targeting molecule. Among the proposed and commercially available heterobifunctional PEGs, the ones mainly used are $H_2N$-PEG-COOH, HO-PEG-COOH, $H_2N$-PEG-OH and their reversibly protected and activated forms.

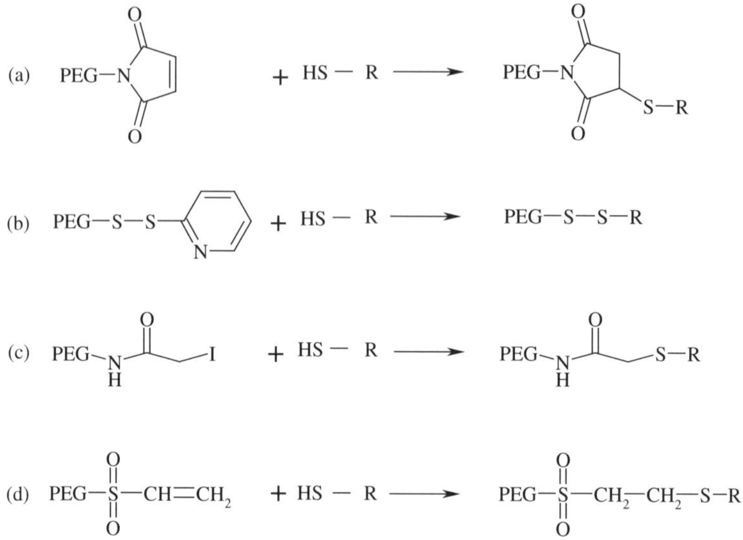


**Fig. 4** Examples of activated PEG molecules reactive towards thiol groups: (a) PEG maleimide, (b) PEG orthopyridyl-disulfide, (c) PEG iodoacetamide, (d) PEG vinylsulfone.

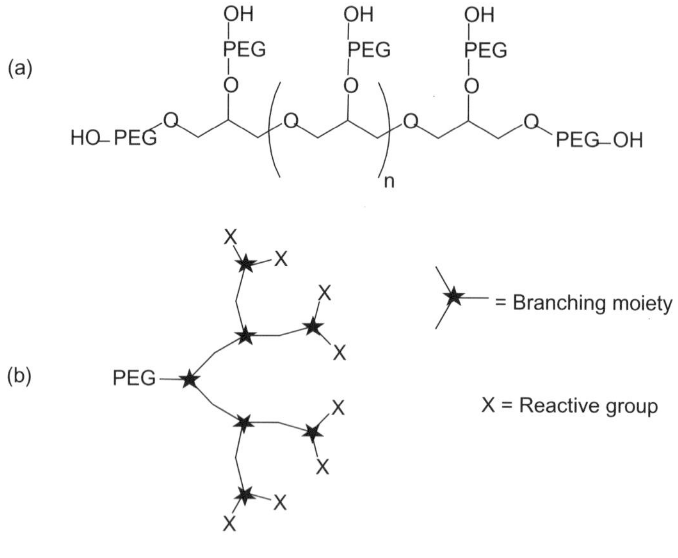


**Fig. 5** Different strategies to achieve multifunctional high loading PEGs: (a) multiarm PEGs, (b) dendronized PEGs, the branching moiety may be a bicarboxylic amino acid, lysine or other bifunctional molecules.

- Multiarm or "dendronized" PEGs (Fig. 5): the former are prepared by linking a linear PEG to a multimeric compound, whereas the latter are linear PEGs with a dentritic structure at one or both chain ends.[26,27] The aim of both derivatives is to increase the drug/polymer molar ratio, overcoming problems of high viscosity that may occur with solutions of mono-functional drug-polymer conjugates. This is particularly true for drugs that require high amount of product for the therapeutic treatments.
- Further advances have been achieved developing PEGs with designed linkers for a controlled release of the conjugated drugs; some of the most exploited linkers are peptides recognized and cleaved by lysosomal enzymes once the conjugates reach the intracellular compartment. Examples of such peptide linkers may be H-Gly-Phe-Leu-Gly-OH or H-Gly-Leu-Phe-Gly-OH.[28,29]
- Alternatively, linkers may respond to pH changes such as the cis-aconityl spacer.[30]
- The linker and the polymer together may be designed to form a double prodrug system, where the drug release is obtained after polymer hydrolysis (first prodrug) that triggers the linker (second prodrug), as reported for the drug delivery system based on 1,6 elimination reaction or trimethyl lock lactonization[31,32] (Scheme 1).

Most of these PEGs, also in the activated form, are now present in the market. Several protein bioconjugates derived from this second generation reached the market, like linear PEG-interferon $\alpha$2b (PEG-Intron®),[33] branched PEG-interferon $\alpha$2a (Pegasys®),[34,35] PEG-growth hormone receptor antagonist (Pegvisomant, Somavert®)[36] and PEG-G-CSF (pegfilgrastim, Neulasta®).[37] A PEG conjugate of a 24-mer oligonucleotide, a branched PEG-anti-VEGF aptamer (Pegaptanib sodium injection, Macugen™)[38] also reached the market, while many other products are under clinical trials and hopefully will be available in the near future.

# 3. PEGylation Chemistry

## 3.1. *Amino group PEGylation*

PEGylation at the level of protein amino groups may be carried out with PEGs having different reactive groups, but although the coupling reaction is based on the same chemistry (for instance acylation), the obtained

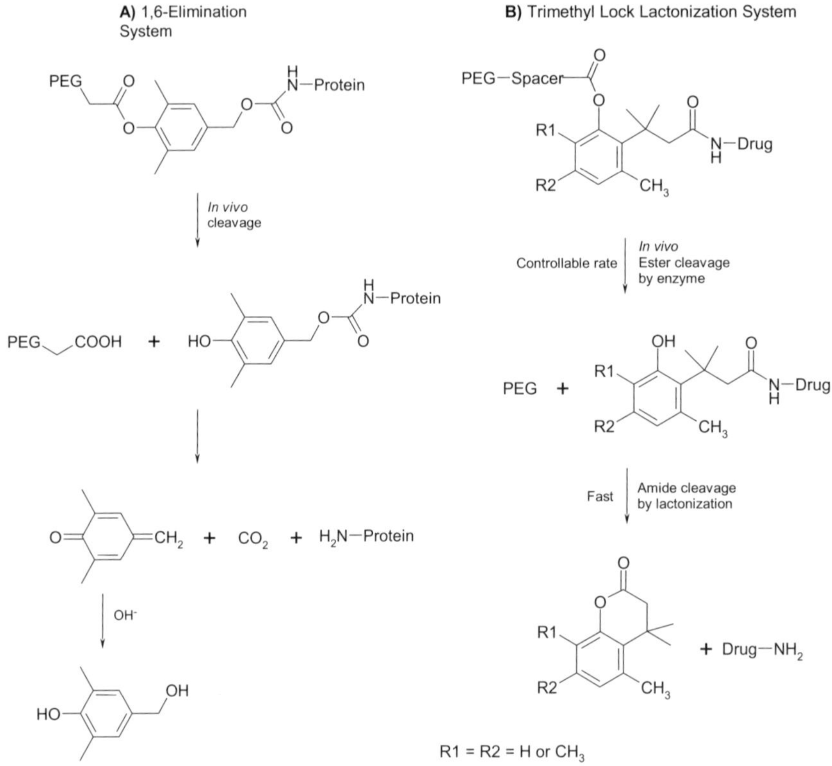


**Scheme 1** Controlled release of active molecules from PEG based on (A) 1,6-elimination system and (B) trimethyl lock lactonization system.

products may be different. The difference can reside in the number of PEG chains linked per protein molecule, in the amino acids involved and in the chemical bond between PEG and protein. This paragraph will report the most common methods for random PEGylation, while procedures for site-direct modification are discussed later.

So far the products present on the market are mainly coming from random PEGylation (see the above mentioned Adagen®, Oncaspar®, PEG-Intron® and Pegasys®) since the FDA Authorities still approve these conjugate mixtures upon demonstration of their reproducibility.

The choice of an activated PEG for amino linking can rely on the wide range of polymers commercially available, the most known are: (a) PEG succinimidyl succinate (SS-PEG), (b) PEG succinimidyl carbonate (SC-PEG), (c) PEG *p*-nitrophenyl carbonate (pNPC-PEG), (d) PEG benzotriazolyl carbonate (BTC-PEG), (e) PEG trichlorophenyl carbonate

(TCP-PEG), (f) PEG carbonylimidazole (CDI-PEG), (g) PEG tresylate, (h) PEG dichlorotriazine and (i) PEG aldehyde (AL-PEG) and a branched form of PEG ($PEG_2$-COOH) (see Fig. 6).

The difference among the above reported PEGs lies in the kinetic rate of amino coupling and in the resulting link between polymer and drug. The derivatives with slower reactivity, such as the carbonate PEGs (pNPC-PEG, CDI-PEG and TCP-PEG) or the aldehyde ones, allow a certain degree of selective conjugation within the amino groups present in a protein, according to their nucleophilicity or accessibility.[7] A difference in reactivity is usually observed between the $\varepsilon$ and the $\alpha$ amino group in proteins due to their p*K*a, of 9.3–9.5 for the $\varepsilon$-amino residue of lysine and 7.6–8 for the $\alpha$-amino group. This difference in reactivity is therefore exploited in the $\alpha$-amino modification performing the conjugation reaction at pH 5.5–6.0, as reported for G-CSF PEGylation with PEG-aldehyde.[37] The $\varepsilon$-amino groups of lysine, possessing high nucleophilicity at high pH, are preferred site of conjugation at pH 8.5–9.

It is noteworthy that the conjugation performed using PEG dichlorotriazine, PEG tresylate and PEG aldehyde (the latter after sodium ciano borohydride reduction) maintains the same total charge on the native protein surface, since these derivatives react through an alkylation reaction, yielding a secondary amine. In contrast, PEGylation conducted with acylating PEGs (i.e. SS-PEG, SC-PEG, pNPC-PEG, CDI-PEG, TCP-PEG and $PEG_2$-COOH) gives weak acidic amide or carbamate linkages with partial loss of positive charge.

The above reported PEG derivatives may sometime give side reactions involving the hydroxyl groups of serine, threonine and tyrosine and the secondary amino group of histidine. These linkages however are generally hydrolytically unstable. The reaction conditions or particular conformational disposition may enhance the percentage of these unconventional PEGylation reactions, for example $\alpha$-interferon was conjugated by SC-PEG or BTC-PEG to His-34 under slightly acidic conditions[39] (the p*K*a value of histidine is between those of the $\alpha$ and $\varepsilon$ amino groups). PEG was found linked to the hydroxyl groups of serine in the decapeptide antide or of tyrosine in epidermal growth factor (EGF) also.[40,41]

As examples of random amino PEGylation two case studies are here reported. *Interferon $\alpha$-2a* (INF) was modified with linear succinimidyl carbonate PEG (SC-PEG; 5 KDa), in pH 10 buffer, through the formation of a urea linkage between protein and polymer. The coupling, performed at

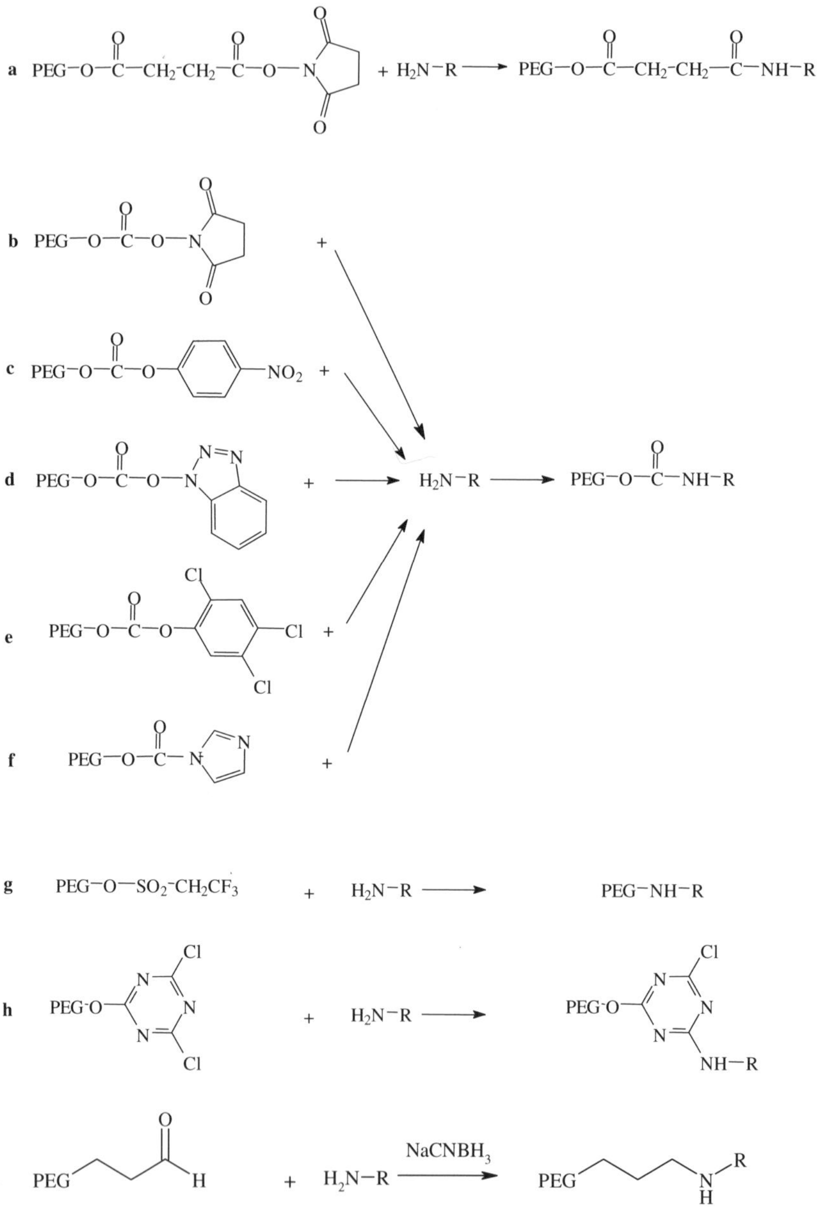


**Fig. 6** Examples of activated PEG molecules reactive towards amino groups: (a) PEG succinimidyl succinate, (b) PEG succinimidyl carbonate, (c) PEG *p*-nitrophenyl carbonate, (d) PEG benzotriazol carbonate, (e) PEG trichlorophenyl carbonate, (f) PEG carbonylimidazole, (g) PEG tresylate and (h) PEG dichlorotriazine and (i) PEG aldehyde (AL-PEG).

equimolar protein/polymer ratio, led mainly to mono-PEGylated isomers and, in small amounts, to di-PEGylated conjugates and free interferon. Characterization of the conjugates indicated that lysine residues were the only site of PEGylation,[42] in agreement with the fact that lysine's amino groups possess higher reactivity than the N-terminal ones at alkaline pHs. The same PEGylation reaction, carried out in phosphate buffer at pH 6.5, demonstrated that derivatives with improved pharmacokinetic profiles and higher activity could be obtained by the conjugation of interferon $\alpha$-2b with SC-PEG (12 KDa). This reaction gave a conjugate at histidine-34, representing approximately 47% of the total PEGylated species (Scheme 2).[43,44] The activity of this interferon preparation was related to the ability to release free and fully active interferon by slow hydrolysis of the labile His-PEG bond.[45] Although the *in vitro* potency of this PEG-interferon is only 1/4 of the free form, its serum residence time is about 6 times longer, allowing less frequent administrations to maintain an efficacy comparable to one of the unmodified interferon.[33,46] These studies brought to the market PEG-Intron® in the year 2000.

A different approach to interferon PEGylation exploited the special properties of branched PEGs. A high molecular weight branched PEG ($PEG_2$, 40 kDa) was chosen on the basis of several preliminary studies, disclosing that: (a) the protein surface protection with a single, long and hindered chain PEG is higher than the one obtained with several small PEG chains linked at different sites[1]; (b) branched PEGs have lower distribution volumes than linear PEG of identical molecular weight, and the delivery to organs such as liver and spleen is faster[47]; (c) proteins modified with branched PEG possess greater stability towards enzymes and pH

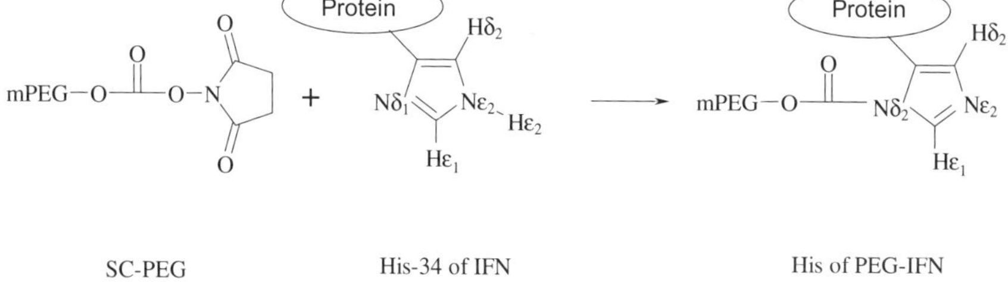


**Scheme 2** Adduct formation at the level of His-34 in interferon $\alpha$-2b using SC-PEG as PEGylating agent.

degradation.[23] The 40 KDa branched succinimidyl PEG ($PEG_2$-NHS) was linked to interferon $\alpha$-2a using a 3 : 1 PEG/protein molar ratio in 50 mM sodium borate buffer pH 9 (Scheme 3).[34] PEGylation under these conditions led to a mixture containing 45–50% of mono-substituted protein, of 5–10% poly-substituted (essentially dimer) and 40–50% of unmodified interferon (Fig. 7). Identification of the major positional isomer within the mono-PEGylated fraction was carried out by a combination of high performance cation exchange chromatography, peptide mapping, amino acid sequencing and mass spectroscopy analysis. It was demonstrated that PEG was attached mainly to one of the following lysines: Lys-31, Lys-121, Lys-131

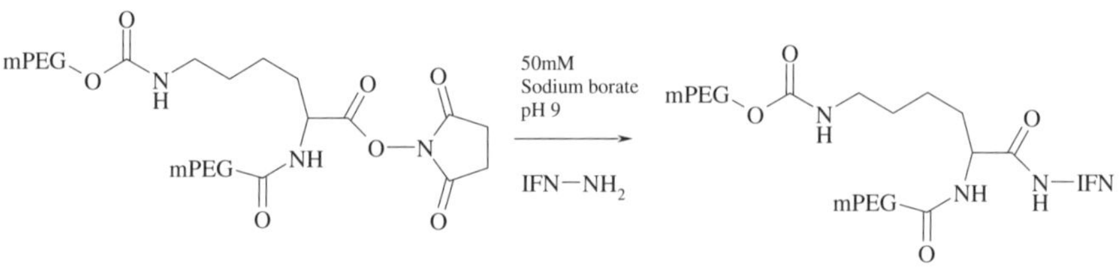


**Scheme 3** PEGylation of interferon $\alpha$-2a by branched $mPEG_2$-COOH (40 kDa).

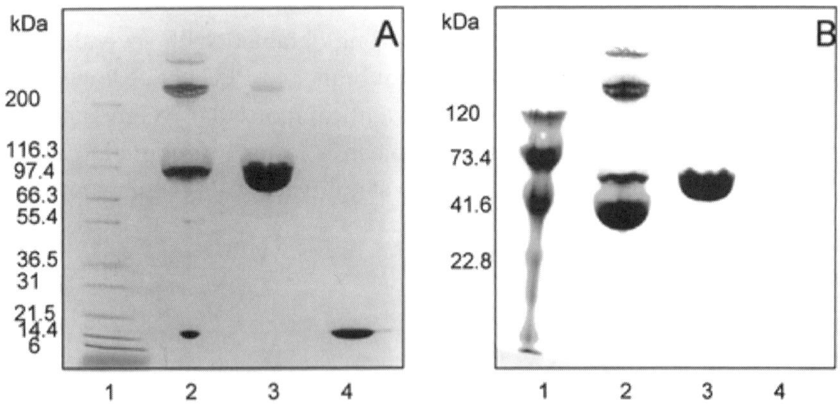


**Fig. 7** SDS-PAGE analysis of the PEGylated interferon $\alpha-2a$ mixture. The conjugates were: (A) specifically stained for protein with Coomassie blue. Lanes: 1, molecular weight marker proteins; 2, PEGylation reaction mixture; 3, purified $PEG_2$-IFN; and 4, interferon −2a. (B) Specifically stained for PEG with iodine. Lanes: 1, molecular weight marker PEGs; 2-4, same as in Fig. 2A. Note that lane 4, containing interferon $\alpha-2a$ in gel B, is not stained by iodine.

**Table 2** Pharmacokinetic properties of Interferon α-2a and its PEGylated form in rats.[35]

| Protein | Half-life (h) | Plasma Residence Time (h) |
|---|---|---|
| Interferon α-2a | 2.1 | 1.0 |
| PEG2 (40 kDa)-interferon α-2a | 15.0 | 20.0 |

or Lys-134.[34] Even though the *in vitro* antiviral activity for PEG2-IFN was greatly reduced (only 7% was maintained with respect the native protein) the *in vivo* activity, measured as the ability to reduce the size of various human tumors, was higher than that of free IFN. The positive result could be related to the extended blood residence time of the conjugated form as shown in Table 2. These studies brought into the market a blood long lasting interferon conjugate, Pegasys®, which is effective in eradicating hepatic and extrahepatic hepatitis C virus (HCV) infection.[35]

## 3.2. *Thiol PEGylation*

The presence of a free cysteine residue represents an optimal opportunity to achieve site direct modification, since its rare occurrence in proteins. PEG derivatives having specific reactivity towards the thiol group, namely MAL-PEG, OPSS-PEG, IA-PEG and VS-PEG (Fig. 4), are commercially available and allow thiol coupling in good yield, everyone presenting some differences in terms of protein-polymer linkage and reaction conditions. Even if the thiol reaction rate of IA-, MAL- or VS-PEGs is very rapid, some degree of amino coupling may also take place, especially if the reaction is carried out at pH values higher than 8. On the other hand, the reaction with OPSS-PEG is very specific for thiol groups, but the obtained conjugates may be cleaved in the presence of reducing agents as such simple thiols or glutathione (also *in vivo*). PEGylation at the level of cysteine allows easier conjugate purification and characterization, since the presence of only one derivatizable site (free cysteine) avoids the formation of positional isomers or products with different degree of substitution, problems common to amino PEGylation. The potential of thiol PEGylation may be further exploited by genetic engineering that allows the introduction of cysteine residue wherever in a protein sequence by insertion or switching with a non essential amino acid.

For example, *hGH* was extensively studied with this strategy; in fact, to overcome the common poly-substitution of amino modification, cysteine muteins were synthesized by recombinant DNA technology. Among all of the possible mutations described in literature and patents, the cysteine addition at the C-terminus of hGH leads to a fully active mutein that allows a site-specific PEGylation using the thiol-reactive PEG-maleimide (PEG-MAL, 8 kDa). It is necessary to treat the rhGH mutein with 1,4-dithio-DL-threitol (DTT) before the coupling step, to convert the carboxyl terminal cysteine in the reduced form and to prevent the formation of scrambled disulphide bridges. After the removal of DTT excess by gel-filtration, the conjugation leads to a mono-PEGylated derivative with over 80% yield.[48]

Another important example of thiol conjugation is the PEGylation of *hemoglobin* (Hb) that allows the prevention of its extravasation. After several unsuccessful attempts through random PEGylation, a site specific modification was performed at Cys-93($\beta$) with maleimidophenyl PEG (MAL-Phe-PEG) (5, 10 and 20 kDa), leading to PEGylated Hb carrying two polymer chains per Hb tetramer.[49] The limit of native Hb is its vasoactivity as a consequence of its extravasation, which can be avoided by PEG coupling that modifies the colligative properties of the derivatives.[15] This product was found to be more efficient than polymerized Hb, the Hb-octamer or -dodecamer.

## 3.3. *Carboxy PEGylation*

PEGylation at the level of protein carboxylic groups needs a previous activation for reacting with an amino PEG. This procedure is not devoid of limitation since undesired intra or intermolecular cross-links may occur because the activated carboxylic groups, besides reacting with the PEG-$NH_2$, react also with the amino groups of the protein itself.

To circumvent this problem, it is possible to use PEG-hydrazide (PEG-CO-NH-$NH_2$) instead of the usual amino PEG. In this case, the protein's COOH groups are activated by water soluble carbodiimide at low pH and following reacted with PEG-hydrazide, where the protein amino groups are protonated and not more suitable for coupling, while hydrazide group, having a low p*Ka*, can still react.[50]

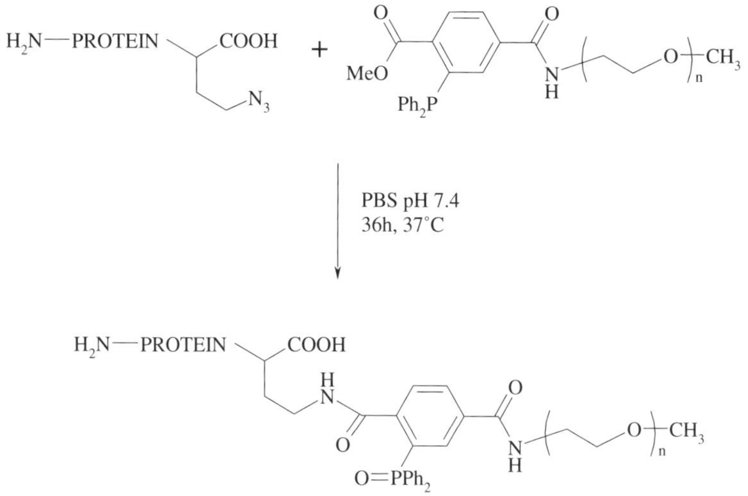


**Scheme 4** Staudinger ligation leading to a C-terminal mono-PEGylated protein by reaction of a mutated protein, containing a C-terminal azido-methionine, with an engineered PEG derivative, methyl-PEG-triarylphosphine.

An alternative method for specific C-terminal PEGylation is based on the Staudinger ligation.[51] The protocol, settled for a truncated thrombomodulin mutant,[52] starts from the *E. coli* expression of a mutated protein containing a C-terminal azido-methionine. This reacts specifically with an engineered PEG derivative, methyl-PEG triarylphosphine, leading to a C-terminal mono-PEGylated protein (Scheme 4). This method, however, implies the preparation of a gene encoding for a protein with a C-terminal linker ending with methionine. Expression in *E. coli* is induced when the transformed bacteria are suspended in a medium where methionine is replaced by the azido-functionalized analogue. Unfortunately, this method is applicable only in the rare case of proteins devoid of methionine in the sequence; otherwise, they will hinder the protein transduction because the azido-analogue does not permit the linking of the following amino acid.

## 4. Strategies in Protein PEGylation

To better exploit the potential of PEGylation several strategies have been developed with the purposes of: (a) obtaining more homogenous products, (b) forming PEGylated conjugates with higher retention of activity,

(c) performing PEGylation in gentle chemical conditions more compatible with easily degradable proteins.

Site selective conjugation is preferable as it allows easier purification and characterization and, most important, a better maintenance of the protein biological activity. In fact, one of the major drawbacks in PEGylation is represented by in loss of protein biological activity, due to factors such as random PEGylation, steric hindrance by PEG chains (interfering with both protein-receptor recognition or enzyme active site accessibility) or changes in protein's secondary-tertiary structure (due to both PEG linkage or to partial denaturation during the PEGylation reaction).

Site direct PEGylation may be achieved by cysteine PEGylation, as described above. However, this is possible only when the amino acid is present in the reduced form in the native protein, a rare case that can be overcome by introducing a cysteine by genetic engineering. In other cases, site directed PEGylation may be obtained by taking advantage of the low p*K*a value of the $\alpha$ amino group, with respect to the lysine's one. As reported, the reaction conducted at neutral or mild acid condition, prevents the PEGylation at the level of lysine, but leaves the N-terminal amino group still able to react.[53] The most successful example of this strategy is the alkylation of r-metHuG-CSF with PEG-aldehyde, proposed by Kinstler. The reaction was carried at pH 5.5 in the presence of sodium cyanoborohydride to reduce the Shiff base initially formed (Scheme 5; Fig. 8).[37,54] The conjugate obtained with a molecule of PEG 20 kDa showed an improved pharmacokinetic profile mainly due to reduced kidney excretion. The PEG-G-CSF conjugate, Pegfilgastrim®, is on the market since 2002.

Preferential site PEGylation can also be achieved by exploiting the different accessibility of the protein amino groups, as reported for a truncated form of growth hormone-releasing hormone ($hGRF_{1\text{-}29}$). It was

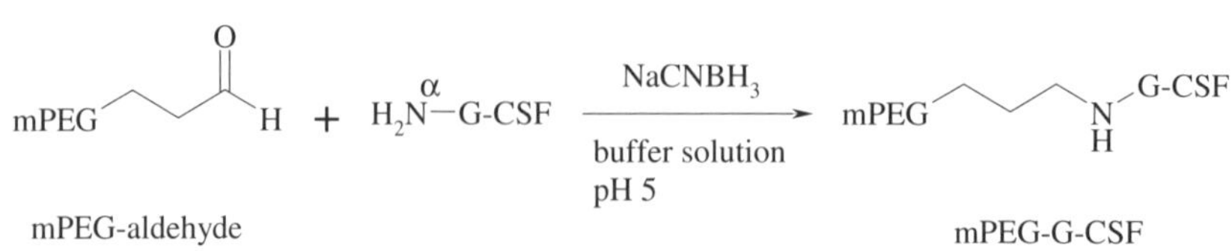


**Scheme 5** Mono-mPEG-G-CSF conjugates were prepared by reductive alkylation of the $\alpha$-amino group of the N-terminal methionine residue of r-metHuG-CSF with mPEG-aldehyde.

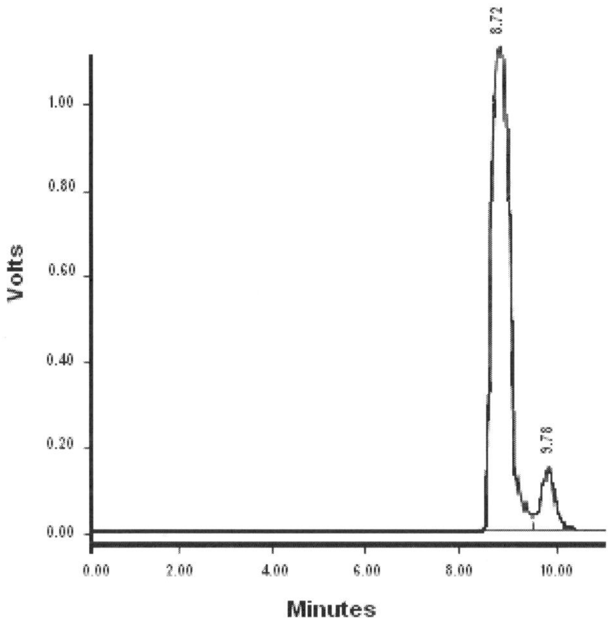


**Fig. 8** Size-exclusion HPLC analysis (UV detector at 280 nm) of the reduced mixture of r-metHuG-CSF reacted with mPEG aldehyde ($M_w = 6$ kDa): *N*-Terminal mono-mPEG–G-CSF conjugate eluted at 8.72 min (92% of total area); Unreacted r-metHuG-CSF eluted at 9.78 min (8% of total area), as from Ref. 37.

demonstrated that, by using an appropriate solvent, it is possible to alter the accessibility and reactivity of the three available amino groups. In fact, NMR and circular dichroism analysis indicated that the percentage of $\alpha$-helix in $hGRF_{1\text{-}29}$, only 20% in water, raises to 90% in structure-promoting solvents such as methanol/water or 2,2,2-trifluoroethanol (TFE), thus facilitating a region-selective modification. When PEGylation was performed in TFE, the monoPEGylated conjugate at the level of Lys-12 reached the 80% of the all PEGylated isomers.[55] Meanwhile, the same reaction conducted in DMSO yielded an almost equimolar mixture of mono-PEGylated Lys-12 and Lys-21 isomers.[56]

Alternatively, specific PEGylation may be performed by blocking some of the reactive groups with a reversible protecting group as reported for insulin. This protein is formed by two polypeptide chains, A and B, and its three amino groups (Gly-A1, Phe-B1 and Lys-B29) are all candidates for PEGylation. Hinds proposed a site-directed PEGylation procedure involving the preliminary preparation of N-BOC (tert-butyl chloroformate) protected insulin.[57] For example, in order to synthesize $N^{\alpha B1}$-PEG-insulin the intermediate $N^{\alpha A1}$, $N^{\varepsilon B29}$-BOC-protected insulin was prepared prior to conjugation with PEG-SPA at the level of free $N^{\alpha B1}$. The final conjugate was obtained upon BOC removal with TFA treatment, forming the $N^{\alpha B1}$-$PEG_{2000}$-insulin conjugate with 83% of the native insulin activity.

In general, in the case of enzyme modification, a requirement for the preservation of activity is that the PEG chains do not have access to the active site and therefore the residues involved in catalysis are not modified. Many strategies have been developed for this goal: (a) the use of branched PEGs that, thanks to their higher hindrance with respect to linear polymers, have reduced accessibility to the active site (Fig. 3); (b) to perform PEGylation in the presence of a substrate or an inhibitor that blocks polymer access to the active site; (c) to conduct the modification after the enzyme is captured on an insoluble affinity resin by substrates or inhibitors linked on it. In the last case, the obtained conjugate is eluted from the resin by changing the pH or adding denaturants. The derivative will possess the active sites and its closer surroundings free of PEG chains (Fig. 9).[58]

A general strategy that may be applied to enzymes or signaling proteins, consists in linking the wanted mass of PEG by using few high molecular weight polymer chains, instead of a high number of low molecular weight ones. In fact, multipoint attachment of PEG on protein surface usually reduces or prevents protein recognition by shielding effect, more than a single PEG chain linked to one location in the protein. The effects of both number and mass of linked PEG chains on recognition and pharmacokinetic parameters are well documented in literature.[59]

A problem that may occur during a protein PEGylation is the low yield, especially when the modification is directed towards a buried or less accessible amino acid. This inconvenience is enhanced when the reaction is performed with high molecular weight PEG due to the high steric hindrance. In the case of *interferon beta* (IFN-$\beta$), conjugation at cysteine 17 could be achieved only with a low MW OPSS-PEG oligomer, but not with a high MW polymer.[60] On the basis of these results, modification with high molecular weight PEGs was successfully attempted via two-steps procedure: in the first step, the protein was modified with a short chain heterobifunctional PEG oligomer; and in the second step, the obtained conjugate was linked

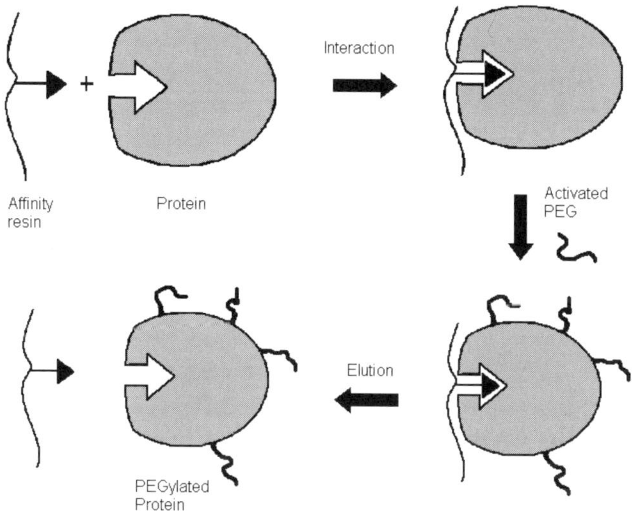


**Fig. 9** Two phase PEGylation strategy for the protection of an enzyme active site from polymer conjugation. The enzyme is first loaded into an affinity resin functionalizated with appropriate ligand. The enzyme's active site binds the ligand, thus protecting the active site itself and the area close to it from PEG modification. After reaction in heterogeneous condition, the modified enzyme is eluted from the column.

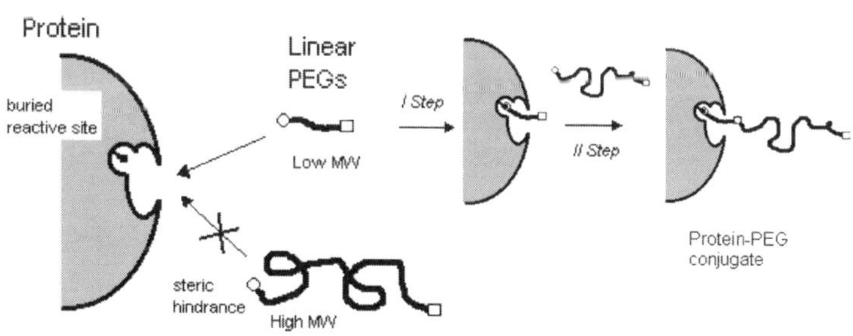


**Fig. 10** Two step tagging PEGylation strategy for a buried SH group in protein. Smaller PEG molecules are more reactive than high molecular weight PEGs towards the buried cysteine.

to a higher molecular weight PEG, possessing specific reactivity towards the terminal end of the first oligomer (Fig. 10). The heterobifunctional PEG oligomer had a thiol reactive group at one end and a hydrazine group at the other (OPSS-PEG-Hz, 2 kDa). As mentioned above, hydrazine is also reactive at low pHs, where protein's amino groups are usually protonated and not reactive, preventing unwanted amino PEGylation of the protein. The INF-SS-PEG-Hz conjugate could therefore be selectively modified with PEG-aldehyde (30 kDa) by reductive alkylation. The overall yield was higher than 80%.

A recent study demonstrated that specific PEGylation at the lone, but not accessible, thiol groups of G-CSF could be achieved upon its exposure in partially denaturant conditions. After modification with OPSS-PEG, the native conformation of G-CSF was recovered by removal of the denaturant.[61]

## 5. Enzymatic Approach for Protein PEGylation

Recently, as alternative solution to obtain homogeneous PEG-protein conjugates in mild reaction conditions, some researchers are investigating the possibilities to exploit the catalysis and specificity of enzymes in PEGylation. The aim is to simplify the purification steps and better preserve the protein activity than the common chemical methods. Since the first report proposed by H. Sato involving the enzyme transglutaminase (TGase),[62] other researchers developed interesting approaches with different enzymes. From the results reported so far, we can hypothesize that these procedures will lead to results of great interests and applicability.

Sato studied two methods for interleukin-2 (IL-2) PEGylation using transglutaminases (TGase) from guinea-pig liver (G-TGase) or from *Streptoverticillium sp.* strain s-8112 (M-TGase). Both enzymes catalyze the transfer of an amino group from a donor (for example PEG-$NH_2$) to a glutamine residue present in a protein (Scheme 6). The two enzymes differ for the required amino acid sequence neighboring the glutamine in the substrate. Several tailor made linear PEGs, differing in molecular weight and in the type of alkylamine at the polymer end, have been synthesized. Among all, the best reactivity was shown by polymers terminating with -$(CH_2)_6$-$NH_2$ group. IL-2 contains six glutamines but none of them is a suitable substrate for the more specific G-TGase. The problem was overcome by preparing a chimeric IL-2 proteins, having at the *N*-terminus, a peptide sequence known to be a good G-TGase substrate, the Pro-Lys-Pro-Gln-Gln-Phe-Met (called TG1) sequence, derived from Substance P[63] (to give rTG1-IL-2) and the Ala-Gln-Gln-Ile-Val-Met (called TG2) sequence,

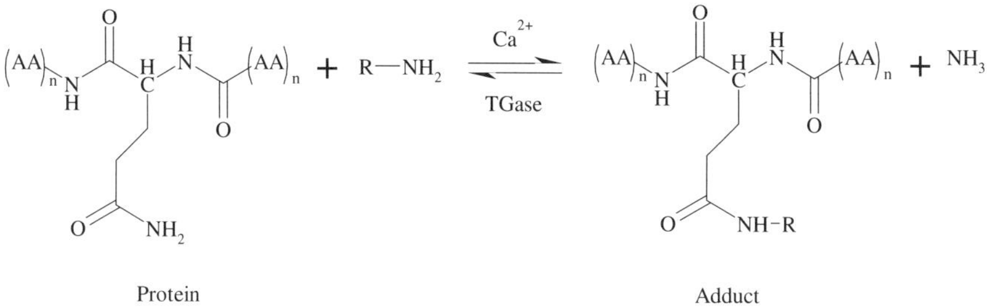


R = lysine of protein, polymer, ect.

**Scheme 6** Reaction catalyzed by TGase between a glutamine residue in a protein and an alkylamine.

derived from fibronectin[64] (to give rTG2-IL-2) have been used. In the former case, a mono-PEGylated conjugate was obtained while a mixture of mono- and di-PEGylated forms resulted from the modification of rTG2-IL-2. The enzymatic coupling was carried out in the very mild conditions of 0.1 M Tris–HCl buffer, pH 7.5 at 25°C for 12 hr in the presence of $CaCl_2$ 10 mM.[65] The derivatives maintained the same activity of the native protein, whereas the classical chemical conjugation with mPEG-NHS yielded products with decreased activity (Table 3). Using the less specific M-TGase, $mPEG_{12000}$-$(CH_2)_6$-$NH_2$ could be directly incorporated into rhIL-2 at the level of Gln-74.[65] Compared to other site-specific chemical PEGylation, such as cysteine coupling or *N*-terminus modification at acidic pH by PEG-aldehyde, the enzyme method produces less undesirable products, namely

**Table 3** Comparison of IL-2 conjugates activities between random PEGylation and site direct PEGylation by TGase.[62]

| Proteins | % activity[a] |
|---|---|
| rhIL-2 | 100 |
| PEG10-rhIL-2 (random PEGylation) | 74 |
| $(PEG10)_2$-rhIL-2 (random PEGylation) | 36 |
| rTG1-IL-2 (chimeric protein for enzymatic PEGylation) | 72 |
| PEG10-rTG1-IL-2 (enzymatic PEGylation) | 69 |
| $(PEG10)_2$-rTG1-IL-2 (enzymatic PEGylation) | 72 |

[a]The amount of % activity was expressed as the percentage of residual bioactivity, compared to the rhIL-2.

protein–protein dimers (due to cysteine oxidation) or $\varepsilon NH_2$ lysine PEGylation (when *N*-terminus PEGylation is performed).

A two step enzymatic PEGylation, called GlycoPEGylation™, was developed. In this case, *E. coli*-expressed proteins were glycosylated at the level of specific serine and threonine amino acids with *N*-acetylgalactosamine (GalNAc) by *in vitro* treatment with the recombinant enzyme O-GalNAc-transferase. The obtained glycosylated proteins were subsequently PEGylated using the O-GalNAc residue as the acceptor site of a sialic acid-PEG, a reaction selectively catalyzed by a sialyltransferase.[66] This enzyme could transfer a PEGylated sialic acid, in the form of cytidine monophosphate derivative (CMP-SA-PEG), to an O-GalNAc residue of glycosylated proteins. The great advance of this technology is the possibility to PEGylate proteins produced in *E. coli* that mimic the mammalian ones, since the PEG chains replace the native sugar moiety at the precise site of glycosylation, forming conjugates that retain the correct structure for receptor recognition plus and the extended plasma half-life.

However, besides the new and interesting aspects of the enzymatic approaches, we have to consider the problems that will arise in an industrial scale production, since the studies so far were limited to a laboratory bench scale and the reactions had a low yield unfortunately.

## 6. PEGylated Protein Purification and Characterization

Theoretically, in a conjugation reaction conducted with an excess of PEG, one could expect that all of the reactive groups of the protein are modified, yielding a single product. However, many factors are involved in PEGylation, such as amino group accessibility and p*Ka*, and usually a mixture of multi-PEG conjugates is obtained. This is true when a lower amount of PEG is employed in order to avoid loss of biological activity. In this case, a mixture of positional isomers is always formed. This last situation requires special attention to maintain reproducibility of the mixture over different batches, since a mixture of products is still accepted by the Authorities, as long as the identification of all adducts is provided.[67,68] In this case, special skill is needed to fractionate the PEG isomers mixture. For this purpose, the possibility to exploit the different isoelectric point of the isomers by ionic exchange chromatography is very useful; while the reverse phase chromatography was found to be less efficient where the gel-filtration separates only the species with different mass, and is unable to distinguish positional isomers.

Once the isomers are separated, it is necessary to identify the localization of the PEGylation site into the primary amino acids sequence. The classical approach involves enzymatic digestion of the polymer-protein derivative, purification of the peptides and their identification by mass spectroscopy or amino acid analysis. A good example is reported in the characterization of PEGylated interferon $\alpha$-2a.[42] The comparison of the peptide fingerprint of the conjugated protein with that of the native protein allows the identification of the region where PEGylation occurred on the basis of the disappeared peptide signal in the peptide fingerprint. Besides the lengthy procedure, the conjugated polymer may interfere by steric hindrance with the proteolytic enzymes, resulting in an incomplete cleavage that complicates the interpretation of the peptide finger printing. To circumvent this inconvenience, a new approach has been recently developed based on the use of tailor-made PEGs, PEG-Met-Nle-COOH or PEG-Met-$\beta$Ala-COOH, that possesses a chemically labile bond in the peptide spacer that can be cleaved at the level of methionine by the treatment with BrCN (Fig. 11), leaving only nor-leucine or $\beta$-alanine tag on the protein. These amino acid tagged peptides are more easily identified by standard sequence investigation methods or by mass spectrometry analysis in the enzymatically-digested mixture.[69]

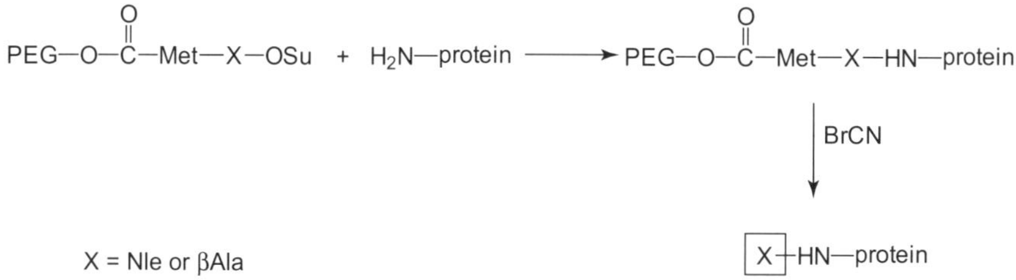


**Fig. 11** Use of PEG-Met-Nle-OSu or PEG-Met-$\beta$Ala-OSu to introduce a reporter amino acid at the PEGylation site: PEG conjugation and the release of PEG by BrCN. Nle or $\beta$Ala that can be identified on the protein by a standard sequence analysis.

## 7. Conclusion

Over the years, PEGylation has become a well developed technology and a wide range of possibilities to perform a protein modification are available these days. One may choose the best condition for the protein of interest and can design a method to obtain a conjugate with the wanted total mass of linked PEG, possibly to a selected amino acid. Furthermore, the combination of PEGylation with genetic engineering exploits new and smart strategies to direct the PEG link towards a preferred site, enabling it to obtain conjugates that better retain the activity of the starting molecule.

Davis and Abuchowsky[14,70] paved the way of this technology with their pioneering works. However, the PEGylation concept, initially proposed and developed for protein modification,[1,2,12] has since then been extended to peptide,[3,71] non-peptide drugs[72,73] and cells.[74]

A remarkable boost in this area came with the second generation of PEGs, when a considerable wide selection of derivatives in high purity and low polydispersity became available. An open problem in PEGylation is still the purification and separation of positional isomers formed during a protein modification. In fact, the similar chemical and physical characteristics of these isomers and their isolation may require the use of more than one chromatographic technique. Therefore, PEGylation methods that prevent the formation of positional isomers and products with a different extent of linked PEG chains are always awaiting new and original solutions. Advancements in this direction may be represented by the recently described PEGylation systems based on enzymatic coupling, in order to allow conjugation to unusual amino acid such as glutamine. Nonetheless, it is important to note that if one can show reproducibility and fine characterization of the conjugation reaction, the product may still be accepted by the Authorities as a well defined mixture of PEGylated isomers. A request that is still expecting satisfactory response is the availability of *in vivo* biodegradable PEG derivatives that release the native protein under controlled conditions. Lastly, the problem of monodisperse or at least very low polydisperse PEG, especially for high molecular weight polymers, still exists, even if the market is now offering only low molecular weight monodisperse PEGs that are below 600 Da.

For a complete summary on PEGylation, one may also be reminded that the technique is now expanding from the protein field to the non-peptide drugs to solve problems beyond the immunogenicity as well as the short residence time in blood, as it was in the early stages of PEGylation. In the area of non-peptide drugs, attention is now dedicated to the potentials of the heterobifunctional PEGs which allow the combination of the advantages of polymer modification with the active targeting capacities of a second molecule linked to the PEG chain. The field of non-peptide drug conjugation is receiving increasing interest,[72] owing to the discovery of numerous new ligands targeted at specific tissues, organs or for the entrapment and the release of drugs into cells.

One can of course ask why is it that only PEG but not other polymers is chosen for protein modification. Actually, very few examples with other macromolecules are on the market, the most successful being the poly(styrene-co-maleic acid) derivative of neocarzinostatin (SMA-NCS), developed by Maeda.[75] The main reason for this situation resides in the mono-functionality of mPEG that avoids cross-linking reaction with the polyfunctionalized proteins, and in the favorable and unique characteristic of PEG. For instance, most of the usual natural or synthetic polymers present multiple points of attachment in the same molecule, as in the case of the natural polysaccharides or albumin, and also for the extensively studied poly(hydroxyethyl acrilamide) (HPMA)[76] that is very promising for small drugs conjugation, but not for proteins.

In conclusion, although PEGylation and conjugation with other polymers are emerging research methodologies, they have already yielded important results and a number of products have already hit the market.

## References


1. Bailon P and Berthold W (1998) Polyethylene glycol-conjugated pharmaceutical proteins. *Pharm Sci Technol Today* **1**:352–356.
2. Zaplisky S and Harris JM (1997) *Chemistry and Biological Applications of Polyethylene Glycol*. American Chemical Society Symposium Series **680**:1–15.
3. Veronese FM (2001) Peptide and protein PEGylation: A review of problems and solutions. *Biomaterials* **22**:405–417.
4. Pasut G, Guiotto A and Veronese FM (2004) Protein, peptide and non-peptide drug PEGylation for therapeutic application: A review. *Exp Op Ther Patents* **14**:859–894.

5. Kopeček J, Kopeckova P, Minko T and Lu Z (2000) HPMA copolymer-anticancer drug conjugates: Design, activity, and mechanism of action. *Eur J Pharm Biopharm* **50**:61–81.
6. Harris JM and Chess RB (2003) Effect of PEGylation on Pharmaceuticals. *Nature Rev Drug Discov* **2**:214–221.
7. Veronese FM and Morpurgo M (1999) Bioconjugation in pharmaceutical chemistry. II *Farmaco* **54**:497–516.
8. Russell-Jones GJ (1996) The potential use of receptor-mediated endocytosis for oral drug delivery. *Adv Drug Delivery Rev* **20**:83–97.
9. Okamoto CT (1998) Endocytosis and transcytosis. *Adv Drug Del Rev* **29**:215–228.
10. Takakura Y, Mahoto RI and Hashida M (1998) Extravasation of macromolecules. *Adv Drug Del Rev* **34**:93–108.
11. Allen TM (2002) Ligand-targeted therapeutics in anticancer therapy. *Nature Rev Cancer* **2**:750–763.
12. Davis FF, Abuchowski A, Van Es T, Palczuk NC, Chen R, Savoca K and Wieder K (1978) Enzyme polyethylene glycol adducts: Modified enzymes with unique properties. *Enzyme Eng* **4**:169–173.
13. Duncan R (2003) The dawning era of polymer therapeutics. *Nature Rev Drug Discov* **2**:347–360.
14. Abuchowski A, Van Es T, Palczuk NC and Davis FF (1977) Alteration of immunological properties of bovine serum albumin by covalent attachment of polyethylene glycol. *J Biol Chem* **252**:3578–3581.
15. Manjula BN, Tsai A, Upadhya R, *et al.* (2003) Site-specific PEGylation of hemoglobin at Cys-93($\beta$): Correlation between the colligative properties of the PEGylated protein and the length of the conjugated PEG chain. *Bioconjug Chem* **14**:464–472.
16. Working PK, Newman SS, Johnson J and Cornacoff JB (1997) Safety of poly(ethylene glycol) derivatives in *Poly(ethylene glycol) Chemistry and Biological Applications*. Harris JM and Zalipsky S (eds.) ACS Books: Washington, DC, pp. 45–54.
17. Petrak K and Goddard P (1989) Transport of macromolecules across the capillary walls. *Adv Drug Del Rev* **3**:191–214.
18. Friman S, Egestad B, Sjovatt J and Svanvik J (1993) Hepatic excretion and metabolism of polyethylene glycols and mannitol in the cat. *J Hepatol* **17**:48–55.
19. Guiotto A, Canevari M, Pozzobon M, Moro S, Orsolini P and Veronese FM (2004) Anchimeric assistance effect on regioselective hydrolysis of branched PEGs: A mechanistic investigation. *Bioorg Med Chem* **12**:5031–5037.

20. Levy Y, Hershfield MS, Fernandez-Mejia C, *et al.* (1988) Adenosine deaminase deficiency with late onset or recurrent infections: Response to treatment with polyethylene glycol modified adenosine deaminase. *J Pediatr* **113**:312–317.
21. Graham LM (2003) PEGASPARAGINASE: A review of clinical studies. *Adv Drug Del Rev* **10**:1293–1302.
22. Harris JM, Guo L, Fang ZH and Morpurgo M (1995) PEG-protein tethering for pharmaceutical applications. *Seventh International Symposium on Recent Advances in Drug Delivery*. Salt Lake City, USA.
23. Monfardini C, Schiavon O, Caliceti P, Morpurgo M, Harris JM and Veronese FM (1995) A branched monomethoxypoly(ethylene glycol) for protein modification. *Bioconjug Chem* **6**:62–69.
24. Akiyama Y, Otsuka H, Nagasaki Y, Kato M and Kataoka K (2000) Selective synthesis of heterobifunctional poly(ethylene glycol) derivatives containing both mercapto and acetal terminals. *Bioconjug Chem* **11**:947–950.
25. Zhang S, Du J, Sun R, *et al.* (2003) Synthesis of heterobifunctional poly(ethylene glycol) with a primary amino group at one end and a carboxylate group at the other end. *React Funct Polym* **56**:17–25.
26. Choe YH, Conover CD, Wu D, *et al.* (2002) Anticancer drug delivery systems: Multi-loaded $N^4$-acyl poly(ethylene glycol) prodrugs of ara-C.: II. Efficacy in ascites and solid tumors. *J Control Rel* **79**:55–70.
27. Schiavon O, Pasut G, Moro S, Orsolini P, Guiotto A and Veronese FM (2004) PEG-Ara-C conjugates for controlled release. *Eur J Med Chem* **39**:123–133.
28. Duncan R, Cable HC, Lloyd JB, Rejmanova P and Kopecek J (1984) Polymers containing enzimatically degradable bonds, 7. Design of oligopeptide side chains in poly N-(2-hydroxypropyl)methacrylamide copolymers to promote degradation by lysosomal enzymes. *Makromol Chem* **184**:1997–2008.
29. Rejmanova P, Pohl J, Baudys M, Kostka V and Kopecek J (1984) Polymers containing enzimatically degradable bonds, 8. Degradation of oligopeptide sequences in N-(2-hydroxypropyl)methacrylamide copolymers by bovine spleen cathepsin B. *Makromol Chem* **184**:2009–2020.
30. Shen WC and Ryser HJP (1981) Cis-aconityl spacer between daunomicin and macromolecula carriers: A model of pH-sensitive linkage releasing drug from a lysosomotropic conjugate. *Biochem Byophysic Res Commun* **102**:1048–1054.
31. Greenwald RB, Yang K, Zhao H, Conover CD, Lee S and Filpula D (2003) Controlled release of proteins from their poly(ethylene glycol) conjugates: Drug delivery system employing 1,6-elimination. *Bioconjug Chem* **14**:395–403.
32. Greenwald RB, Choe YH, Conover CD, Shum K, Wu D and Royzen M (2000) Drug delivery systems based on trimethyl lock lactonization: Poly(ethylene glycol) prodrugs of amino-containing compounds. *J Med Chem* **43**:475–487.

33. Wang YS, Youngster S, Grace M, Bausch J, Bordens R and Wyss DF (2002) Structural and biological characterisation of pegylated recombinant interferon $\alpha$-2b and its therapeutic implications. *Adv Drug Del Rev* **54**:547–570.
34. Bailon P, Palleroni A, Schaffer CA, *et al.* (2001) Rational design of a potent, long lasting form of interferon: A 40KDa branched poly-ethylene glycol-conjugated interferon alpha-2a for the treatment of hepatitis C. *Bioconjug Chem* **12**:195–202.
35. Reddy KR, Modi MW and Pedder S (2002) Use of peginterferon $\alpha$2a (40KD) (Pegasys®) for the treatment of hepatitis C. *Adv Drug Del Rev* **54**:571–586.
36. Trainer PJ, Drake WM, Katznelson L, *et al.* (2000) Treatment of acromegaly with the growth hormone-receptor antagonist pegvisomant. *N Engl J Med* **34**: 1171–1177.
37. Kinstler O, Moulinex G, Treheit M, *et al.* (2002) Mono-N-terminal poly(ethylene glycol)-protein conjugates. *Adv Drug Del Rev* **54**:477–485.
38. The EyeTech Study Group (2002) Preclinical and phase 1A clinical evaluation of an anti-VEGF pegylated aptamer (EYE001) for the treatment of exudative age-related macular degeneration. *Retina* **22**:143–152.
39. ENZON INC.: US patent 5985263 (1999).
40. APPLIED RESEARCH SYSTEM: WO patent 9955376 (1999).
41. Orsatti L and Veronese FM (1999) An unusual coupling of poly(ethylene glycol) to tyrosine residue in epidermal growth factor. *J Bioac Comp Polymer* **14**: 429–436.
42. Monkarsh SP, Ma Y, Aglione A, *et al.* (1997) Positional isomers of mono-pegylated interferon $\alpha$-2a: Isolation, characterization, and biological activity. *Anal Biochem* **247**:434–440.
43. Wylie DC, Voloch M, Lee S, Liu Yh, Cannon-Carlson S, Cutler C and Pramanik B (2001) Carboxyalkylated histidine is a pH-dependent product of pegylation with SC-PEG. *Pharm Res* **18**(9):1354–1360.
44. Wang YS, Youngster S, Bausch J, Zhang R, Mcnemar C and Wyss DF (2000) Identification of the Major Positional Isomer of Pegylated Interferon Alpha-2b. *Biochemistry* **39**:10634–10640.
45. ENZON INC.: US patent 5985263 (1999).
46. Glue P, Fang JWS, Sabo R, *et al.* (1999) Peg-interferon-$\alpha$-2B: pharmacokinetics, pharmacodymanics, safety, and preliminary efficacy data. *Hepatology* **30**(Suppl):A189.
47. Pepinsky RB, Le Page DJ, Gill A, *et al.* (2001) Improved phamacokinetic properties of polyethylene glycol-modified form of interferon-$\beta$-1a with preserved in vitro bioactivity. *J Pharmac Exp Ther* **297**(3):1059–1066.
48. BOLDER BIOTECHNOLOGY INC.: WO patent 9903887 (1999).
49. EISTEIN COLL MED.: US patent 5585484 (1996).

50. Zalipsky S and Menon-Rudolph S (1998) Hydrazine derivatives of poly(ethylene glycol) and their conjugates in: *Poly(ethylene glycol) Chemistry: Biotechnical and Biomedical Applications*, 2nd ed. Harris JM (ed.) Plenum: New York, pp. 319.
51. Saxon E and Bertozzi CR (2000) Cell surface engineering by a modified Staudinger reaction. *Science* **287**:2007–2010.
52. Cazalis CS, Haller CA, Sease-Cargo L and Chaikof EL (2004) C-terminal site-specific PEGylation of truncated thrombomodulin mutant with retention of full bioactivity. *Bioconjug Chem* **15**:1005–1009.
53. Wong SS (1991) Reactive groups of proteins and their modifying agents, in: *Chemistry of Protein Conjugation and Cross-linking*. CRC Press: Boston.
54. Kinstler OB, Brems DN, Lauren SL, Paige AG, Hamburger JB and Treuheit MJ (1995) Characterization and stability of N-terminally pegylated rhG-CSF. *Pharm Res* **13**:996–1002.
55. ARES TRADING SA: WO patent 0228437 (2002).
56. Piquet G, Gatti M, Barbero L, Traversa S, Caccia P and Esposito P (2002) Set-up of a large laboratory scale chromatographic separation of poly(ethylene glycol) derivatives of the growth hormone-releasing factor 1-29 analogue. *J Chrom* **A944**:141–148.
57. Hinds KD and Kim SW (2002) Effects of PEG conjugation on insulin properties. *Adv Drug Del Rev* **54**:505–530.
58. Caliceti P, Schiavon O, Sartore L, Monfardini C and Veronese FM (1993) Active site protection of proteolytic enzymes by poly(ethylene glycol) surface modification. *J Bioact Biocomp Polym* **8**:41–50.
59. Esposito P, Barbero L, Caccia P, *et al.* (2003) PEGylation of growth hormone-releasing hormone (GRF) analogues. *Adv Drug Del Rev* **55**:1279–1291.
60. APPLIED RESEARCH SYSTEM: WO patent 9955377 (1999).
61. Berna M, Spagnolo L and Veronese FM (2005) Sito-specific PEGylation of G-CSF by reversible denaturation. 32nd Meeting Controlled Release Society, 18–22 June. Miami Beach (FL) USA.
62. Sato H (2002) Enzymatic procedure for site-specific pegylation of proteins. *Adv Drug Del Rev* **54**:487–504.
63. Gorman JJ and Folk JE (1980) Transglutaminase amine substrates for photochemical labeling and cleavable cross-linking. *J Biol Chem* **2**:1175–1180.
64. Mcdonagh RP, *et al.* (1981) Amino acid sequence of the factor XIIIa acceptor site in bovine plasma fibronectin. *FEBS Lett* **127**:174–178.
65. Sato H, Hayashi E, Yamada N, Yatagai M and Takahara Y (2001) Further studies on the site-specific protein modification by microbial transglutaminase. *Bioconjug Chem* **12**(5):701–710.
66. NEOSE TECHNOLOGIES INC. (2004) US Patent 2004 132 640.

67. Hooftman G, Herman S and Schacht E (1996) Review: Poly(ethylene glycol)s with reactive endgroups. II. Pratical consideration for the preparation of protein-PEG conjugates. *J Bioact Comp Polymer* **11**:135–159.
68. Delgado C, Francis GE, Malik F, Fisher D and Parkes V (1997) Polymer-derivatized proteins: Analytical and preparative problems. *Pharm Sci* **3**:59–66.
69. Veronese FM, Saccà B, Polverino De Laureto P, *et al.* (2001) New PEGs for peptide and protein modification, suitable for identification of the PEGylation site. *Bioconjug Chem* **1**:62–70.
70. Abuchowski A, Mccoy R, Palczuk NC, Van Es T and Davis FF (1977) Effect of covalent attachment of polyethylene glycol on immunogenicity and circulating life of bovine liver catalase. *J Biol Chem* **252**:3582–3586.
71. Davis FF, Kazo GM, Nucci ML and Abuchowski A (1990) *Peptide and Protein Drug Delivery*. Lee VHL (ed.) Marcel Dekker: New York.
72. Greenwald RB (2001) PEG drugs: An overview. *J Control Rel* **74**:159–171.
73. Greenwald RB, Conover CD and Choe YH (2000) Poly(ethylene glycol) conjugated drugs and prodrugs: A comprehensive review. *Crit Rev Ther Drug Carr Syst* **17**:101–161.
74. Scott MD and Chen AM (2004) Beyonde the red cell: Pegylation of other blood cells and tissues. *Transfusion Clinique et Biologique* **11**:40–46.
75. Maeda H, Ueda M, Morinaga T and Matsumoto T (1985) Conjugation of poly(styrene-co-maleic acid) derivatives to the antitumor protein neocarzinostatin: Pronounced improvements in pharmacological properties. *J Med Chem* **28**(4):455–461.
76. Duncan R, Kopeckova-Rejmanova P, Strohalm J, Hume I, Cable HC, Pohl J, Lloyd JB and Kopecek J (1987) Anticancer agents coupled to N-(2-hydroxypropyl)methacrylamide copolymers. I. Evaluation of daunomycin and puromycin conjugates in vitro. *Br J Cancer* **55**(2):165–174.
77. Bukowski R, *et al.* (2002) PEGylated interferon $\alpha$t2b treatment for patient with solid tumors: A phase I/II study. *J Clin Oncol* **20**:3841–3849.

# 5

# PEGylated Proteins as Cancer Therapeutics

*Margherita Morpurgo, Gianfranco Pasut and Francesco M. Veronese*

## 1. Introduction

In early days, the treatment of cancer could rely only on classic antiproliferative agents which act on the basic principle of killing fast replicating cells. These compounds lack tumor specificity and have several undesired side effects. Radiotherapy is another effective tool which can be used alone or associated with chemical treatment. In this case, localization can be achieved only by physically directing the radiation toward the target organ, thus radiotherapy is generally used to treat solid tumors and in the absence of metastasis.

The lack of tumor specificity by classic treatments has led to the search for more physiological strategies to fight cancer. Modern approaches are based on recent acquisitions on the molecular events associated with cancer initiation and progression. Implementation of natural body defence mechanisms and balancing deregulated processes involved in tumorigenesis, such as regulation of cell cycle progression, angiogenesis, and apoptosis provide rational targets for novel therapies. In most case, interfering with these mechanisms can be achieved by using biomolecules as the active agents, most commonly hormones, proteins or nucleic acids, whose use has become economically advantageous after the advent of the recombinant DNA technology. Despite their potency, these compounds (and mostly proteins and nucleic acids) suffer of intrinsic chemical instability, poor bioavailability and pharmacokinetics that hamper their use as therapeutic agents. PEGylation, namely the chemical modification by covalent attachment of polyethylene glycol, was thus suggested as a strategy to overcome some of these drawbacks.

PEGylation is a well-established technology in the field of biopharmaceutical formulation. It is commonly used to improve either the stability, solubility, bioavailabilty, and immunological properties of bioactive compounds and it is mostly useful in the case of proteins, due to the intrinsic instability, poor bioavailability and antigenic properties of this class of bioactive compounds. Indeed, PEGylation of proteins for pharmaceutical applications has been studied for the past 30 years. As a result, several PEGylated proteins having different therapeutic applications are now available in the market and more are about to come.

Protein PEGylation was initially developed to improve the properties of enzymes. Pioneering studies carried out in the 1970s[1,2] demonstrated that several proteins could be covalently modified with PEG without loosing their bioactivity, while their stability was implemented. Since then, protein PEGylation has been extended to other classes of bioactive proteins, among which also anticancer agents.

As a general rule, PEGylated proteins are less immunogenic and antigenic as compared to their nonmodified parents and display longer body permanence upon administration. PEGylation, however, is often accompanied with diminished protein bioactivity which may depend on the number and location of the PEG chains covalently attached to the protein and the molecular weight and geometry of the polymer (e.g. linear *versus* branched). It turns out that optimization of the coupling strategy and protocols is necessary for each protein under analysis. Covalent attachment of bulky PEG chains can be deleterious for protein activity either because amino acids that are fundamental for substrate recognition may be directly involved in the coupling or because of the polymer steric hindrance in proximity of the recognition surface. The issue of steric hindrance is relevant for those proteins that act upon interaction with large substrates, namely cytokines or antibodies whose activity requires the recognition of membrane receptors or antigens located on the cell surface. Several years of experience have demonstrated that, in this case, site directed or chemically tailored PEGylation is necessary to preserve biological activity. On the other hand, enzymes that act on low molecular weight substrates are less sensitive to inactivation and random modification has often been successfully applied in this case. A more thorough discussion on strategy, potentials and limits of PEGylation is being reported in Chap. 4 of this book and in recent reviews from F.M. Veronese and G. Pasut, published in *Drug Discovery Today* and *Expert Opinion in Therapeutic Patents*.

Presently, two PEGylated proteins — PEG-granulocyte colony stimulating factor (PEG-GCSF or PEG-filgrastim or Neulasta®) and PEG-L-asparaginase (Oncaspar®) — are FDA approved specifically in cancer therapy. Other PEGylated proteins have FDA approval for therapeutic applications, other than cancer, but are undergoing clinical trials to extend their use to certain tumors. Among these are two PEG-Interferons (PEG-IFN), PEG-IFN$\alpha$-2b or (PEG-Intron®) and PEG-IFN$\alpha$-2a (PEGASYS®), presently FDA approved only for the treatment of chronic hepatitis C. Other PEG-proteins are in earlier stages of investigation, among which several PEG-cytokines (namely PEG-IFN1, PEG- r-human megakariocyte growth and development factor -PEG-r-Hu-MGDF-, PEG-interleukines, and PEG-TNF$\alpha$), PEG-enzymes (namely PEG-adenosin deiminase, PEG-methioninase, and PEG-uricase) and PEGylated antibodies against cancer epitopes. This chapter will focus only on PEGylated enzymes and antibodies having potential or established anticancer applications. Some of these products are already marketed, while others are under preclinical or phases I–III clinical trials. Other PEG-proteins that are important in cancer therapy, belonging to the cytokine class, are not described here and are extensively reviewed in Chap. 6.

## 2. Enzymes and PEG-Enzymes in Cancer Therapy

Several enzymes have proven to be useful in the field of cancer therapy by acting through different mechanisms.[3] Metabolite depleting enzymes, such as asparaginase, methioninase, and arginin-deiminase, induce cancer cell death by depleting the environment of essential amino acids for which tumor cells only are auxotrophic. The enzyme chondroitinase AC is able to prevent cell proliferation by removing cell surface chondroitin sulfate proteoglycans that are fundamental for tumor growth, neovascularzation, and metastasis.[4] Other enzyme-related proteins that are being studied in cancer therapy are the hybrid molecules belonging to the antibody directed enzyme prodrug therapy (ADEPT) family, which are composed of an enzymatic and a site-directing moiety linked together. Enzymes are also investigated to reduce or prevent certain chemotherapy side effects. This is the case of uricase, an enzyme that degrades the poorly soluble uric acid whose build-up (that causes gouty arthritis and chronic renal disease) is induced by chemotherapy.[5,6]

Some of these compounds are already available in the clinic in their native form and, in most cases, PEGylation has been investigated to improve their pharmacological properties. PEGylated asparaginase is already FDA approved (Oncaspar®), whereas PEG-argininedeiminase, PEG-methioninase and PEG-uricase are still under preclinical or clinical evaluation. To our knowledge, PEGylation of chondroitinase and ADEPT proteins has not yet been reported in the literature.

## 2.1. *Metabolite depleting enzymes*

Novel strategies for tumor treatment take advantage of different metabolic requirements expressed by tumor cells with respect to healthy tissues. For example, a tumor selective target with high therapeutic potential may be the elevated requirement of a selected metabolite of tumor cells relative to normal cells. Examples are the inability of hematological tumor cells to grow in the absence of asparagine, the dependence of several solid tumors on high level of methionine and the auxothropy for arginine displayed by some melanomas and hepatocellular carcinomas. On the basis of these metabolic differences, enzymes that are able to reduce plasma levels of these metabolites namely asparaginase, methioninase and arginine deiminase, are studied as therapeutic agents in cancer therapy. All of these enzymes have bacterial origin and their clinical use is hampered by poor pharmacokinetics and high risks of immunogenicity. PEGylation has thus become the strategy of choice to solve these problems and allow the therapeutic use of these enzymes.

### 2.1.1. *PEG-asparaginase (PEG-ASNase)*

The enzyme asparaginase catalyzes the hydrolysis of asparagine to aspartic acid and ammonia. The resulting depletion in asparagine is fatal to leukemic lymphoblasts and certain other tumor cells that, by lacking or having very low levels of asparagine synthetase, are unable to synthesize

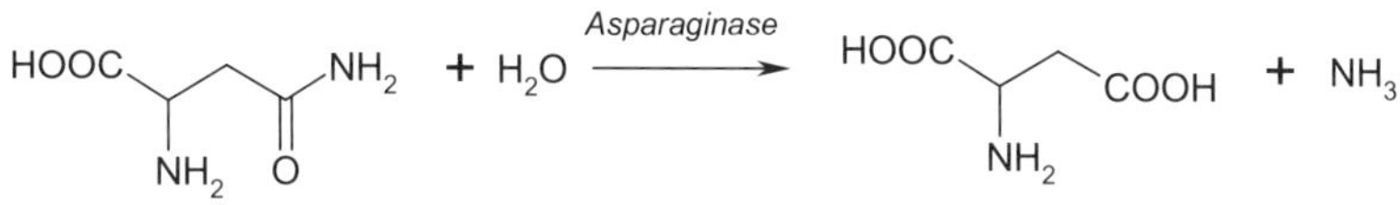


**Scheme 1.** Conversion of asparagine to aspartic acid and ammonia by asparaginase.

asparagine *de novo* and rely on asparagine supplied in the serum for survival.

The potential for asparaginase treatment stemmed from the early observation of Kidd in 1953,[7] who described an activity in guinea pig sera that caused regression of transplanted lymphomas in mice and rats. This observation was later related to presence of asparaginase activity in the guinea pig serum.[8] An asparaginase form was later isolated from *E. coli* which exhibited antitumor activity similar to that found in guinea pig sera[9,10] and provided a practical source of the enzyme for later preclinical and clinical investigations.

Preclinical studies in the late 1950s and early 1960s showed that the enzyme was effective against several tumors. The majority of susceptible tumors were of lymphoid origin, and T-cell lineage was found to be more sensitive than the B-cell one. Several clinical trials demonstrated the efficacy of asparaginase in the treatment of many tumors and FDA approval for the native enzyme was granted in 1978. Since then, asparaginase has become an important chemotherapeutic agent in the treatment of acute lymphoblastic leukemia (ALL) and other lymphoid malignancies. It has demonstrated effectiveness in induction, and in subsequent phases of various multiagent chemotherapeutic regimens. Because this enzyme is generally not myelosuppressive and is not cross-resistant with other antineoplastic agents, it is easily added to combination chemotherapy protocols. New therapeutic indications, as for the treatment of myeloma or solid tumors, are presently being evaluated in clinical trials.[11,12]

Two enzyme forms are presently available in the clinic, one is isolated from *E. coli* (marketed commercially by Merck & Co. as Elspar®), has a molecular weight of 138 000–141 000 Da and is composed of four identical subunits with an active site on each. The second one is isolated from *Erwinia chrysanthemi* (available as *Erwinia* L-asparaginase from Ogden BioServices Pharmaceutical Repository in the United States), it is also a tetramer and has a molecular weight of 138 000 Da. Both enzymes have high activity and stability. They differ in isoelectric point and plasma permanence and lack antigenic cross-reactivity. Both native preparations are approved for use in the therapy of patients in the front line and at relapse.

The major limitation to the use of native asparaginases is clinical hypersensitivity, which develops in 3–78% of patients treated with native forms of the enzyme. Either acute allergic reactions or silent hypersensitivities occur in a significant proportion of patients exposed to multiple courses of asparaginase and the antibodies developed often account for treatment failures. In fact, a clear correlation between anti-L-asparaginase antibody titer and clearance of L-asparaginase (ASNase) was demonstrated.

PEGylation of ASNase was thus investigated as a means to reduce the hypersensitivity problem. The first PEG conjugate of *E. coli*-derived L-asparaginase was already developed in 1979.[13] Asparaginase from *E. coli* was chemically modified with PEG of 5000 MW and the conjugate was shown to cause tumor regression in transplanted mice with less immunogenicity than the native *E. coli*[14–16] form. PEGylation also improves the enzyme chemical stability and its resistance to plasma proteases[17] PEG-asparaginase was entered into clinical trials already in 1984 and has since been administered to thousands of patients with ALL.[18] The PEG-conjugated enzyme is less immunogenic than either of the two native products and can be administered safely to most patients with allergic reactions to *E. coli* or *Erwinia* asparaginases. The longer serum $t_{1/2}$ of PEG-asparaginase allows for a longer interval between doses. PEG-asparaginase was developed by Enzon and was FDA approved in 1994 for use in combination chemotherapy for the treatment of patients with ALL who are hypersensitive to native forms of the enzyme. It is now available commercially from Rhone-Poulenc Rorer as Oncaspar®. The product has similar toxicological profile as the non-modified enzyme. However, PEGylation improves both the immunological and pharmacokinetic properties of the enzyme. The mean serum $t_{1/2}$ of PEG-ASNase is about 15 days as opposed to the 24 hr of the nonmodified *E. coli* enzyme and to the 10 hr of the *Erwinia* form. The longer serum $t_{1/2}$ of PEG-asparaginase allows for a longer interval between administration doses. The rate of total clearance of PEG-asparaginase was found to be 17-fold lower than that of the unmodified enzyme, whereas the volume of distribution was similar for the two preparations. The duration of asparagine serum depletion upon parenteral administration correlated with the serum $t_{1/2}$ of the different enzyme forms. Already early studies showed that ASNase levels were undetectable immediately following the 1-hr infusion of PEG-asparaginase and remained low during the 14-day interval between doses. Interestingly, a recent pharmaco-economic study[19] demonstrated that despite the higher pharmacy cost of PEG-asparaginase *versus* the unmodified enzymes, the overall cost of the treatment with the PEGylated product is similar to the one with the unmodified enzymes.

Since FDA's approval in 1994, drug monitoring has been performed by several phase IV clinical studies and detailed recent reviews are available in the literature.[20,21] Recent studies have been carried out on the rational basis that immunological, pharmacokinetic, and pharmacodynamic factors have a considerable impact on the efficacy of asparaginase therapy. Therefore, investigation is now aimed at defining the optimum dose and dosing schedule of the different asparaginase preparations that are used in the clinic[22] or at correlating antibody levels with pharmacological response.[23] Additional pharmacokinetic studies of old and new asparaginases will improve general understanding of the reasons behind treatment success or failure and allow for the development of more rational dosing schedules in individual patients.

### 2.1.2. *PEG-methioninase (PEG-METase)*

Methionine dependence is a metabolic defect seen only in cancer cells and precludes cells from growing in a methionine-depleted medium. Methioninase (methionine-$\alpha$-deamino-$\gamma$-mercaptomethane-lyase, METase), an enzyme isolated from *Pseudomonas putida* (Hori, 1996), tranforms L-methionine into $\alpha$-ketobutirrate, methanethiol and ammonia and is thus able to induce methionine depletion in the medium. METase is a pyridoxal-L-phosphate-dependent enzyme, it is a homotetramer, each subunit being composed of 398 amino acids. The enzyme was efficiently cloned and purified from a highly expressing *E. coli* system and its potentiality as

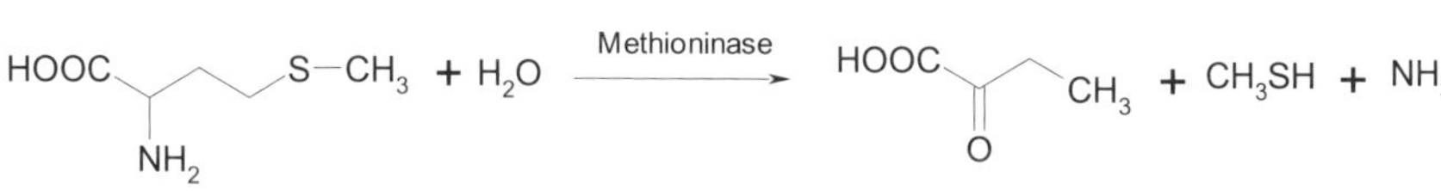


**Scheme 2.** Conversion of methionine to $\alpha$-chetobuttirate, methanthiol and ammonia by methioninase.

an antitumor agent, alone or in combination with other chemotherapeutic agents, was demonstrated in a number of solid tumors, both *in vitro*[24] and in *in vivo* models. Phase I clinical trials have also been carried out and demonstrated the low toxicity of this protein in cancer patients.[25,26] Nevertheless, studies performed in balbC mice demonstrated the immunogenic properties of the recombinant enzyme.[27] This fact, together with its relatively short body permanence upon i.v. administration ($t_{1/2} = 80$–120 min), justified the search for PEGylated derivatives with improved pharmacodynamic properties.

A number of studies carried out by the group of Hoffman demonstrated the superiority of PEGylated forms of this enzyme by a combination of *in vitro* and *in vivo* experiments. The enzyme has been PEGylated with methoxy-PEG of 5000 MW, activated either as succinimidylpropionate[28] or as succinimidylglutarate,[27] at different PEG/methioninase molar ratios. Even if partial loss of enzymatic activity was observed after PEG coupling, the PEGylated enzyme had similar cytotoxic effect on *in vitro* cultured tumor cells[28] and superior properties in *in vivo* experiments, namely a more efficient and longer-lasting serum methionine depletion and significant lower immnunogenicity.[27,29] More precisely, serum half life upon i.v. administration in mice increased from 2 hr for the native protein to 12, 18, and 30 hr for PEGylated forms having, respectively, 4, 6, and 8 PEG molecules attached to each subunit (Fig. 1). Besides, the duration of methionine depletion in serum upon a single i.v. administration of these same compounds was increased from 4 hr (registered for the native enzyme) to 24, 48, and 72 hr respectively. Antibodies (IgG and IgM) against METase were raised by all the enzyme forms, upon a strong immunization protocol carried out in the presence of Freund's adjuvant. Nevertheless, the PEGylated enzymes were significantly less immunogenic as compared to the native enzyme, and the IgG titer raised by the highly PEGylated enzyme was four orders of magnitude less than the one raised by the native enzyme. Less PEGylated samples displayed intermediate immunogenic properties, but in any case, were significantly better than the native enzymes.

A detailed investigation on the pharmacokinetics, antigenicity and toxicity of native and PEGylated-r-METase was recently carried out in mice and primate models.[30,31,29] Native METase and a PEGylated form of the enzyme, modified with PEG-succinimidylglutarate and containing 3–7 molecules of PEG per protein were analyzed. Native r-METase has a biological half life

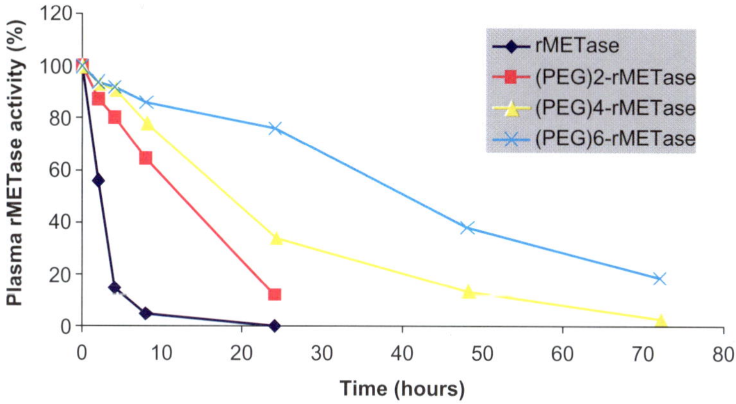


**Fig. 1** Plasma rMETase enzyme activity after i.v. injection of native or PEGylated rMETase in mice. Mice received 80 units of the native rMETase or of the indicated PEGylated form.[27,29]

of 2.5 hr and repeated administrations of the enzyme to primates resulted in severe anaphylactic reactions.[30] Upon PEGylation, the plasma $t_{1/2}$ of the enzyme in its "apo" form increased 36-fold while the active holo form displayed a $t_{1/2}$ of 143 hr only. It was found that the limited circulating half-life of the active holo-PEG-rMETase is due to rapid *in vivo* dissociation of its cofactor pyridoxal-5′-phosphate (PLP).[30] A combination of PEG-rMETase treatment with PLP infusion using an osmotic pump was thus suggested for an effective therapy with a reduced number of protein injections. Some antibodies against anti-PEG-rMETase were produced upon repeated challenges in the primate model. However, the level of such antibodies was 100–1000 fold less than the one elicited by the native enzyme and the antibodies were unable to neutralize the activity of the enzyme so that each challenging dose was effective in depleting serum methionine levels.

### 2.1.3. *PEG-arginine deiminase (PEG-ADI)*

Arginine is one of the nonessential amino acids for humans. Normal cells synthesize it from citrulline using the enzymes argininosuccinatesyntase (ASS) and argininosuccinate lyase. It was found that some cancer cells do not express ASS[32] and are thus unable to synthesize arginine from its precursor. Therefore, it was suggested that an arginine degrading enzyme could be effective against tumors that are auxotrophic for arginine.

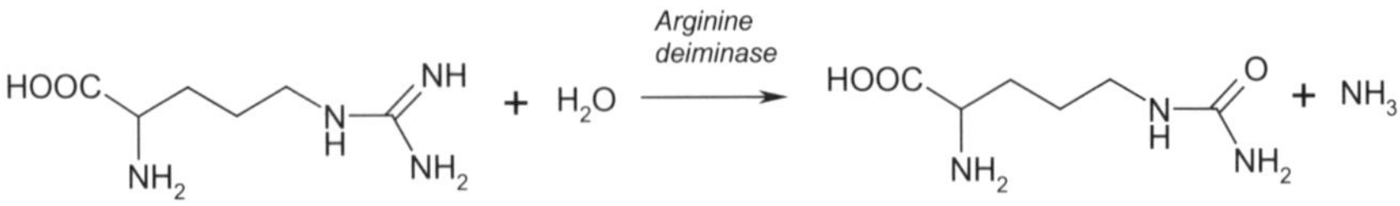


**Scheme 3.** Conversion of arginine to citrulline and ammonia by arginine and deiminase.

Two arginine degrading enzymes have been suggested for this purpose: the first one was arginase,[33] an enzyme that degrades arginine into ornithine and urea. The use of this enzyme was hampered by its weak affinity for arginine ($Km = 45$ mM) and by its nonphysiological optimum pH (>9). Another arginine degrading enzyme is arginine deiminase (ADI), which degrades arginine into citrulline and ammonia. An ADI form isolated from *Mycoplasma homini* having stronger activity as compared to ADIs from other sources[34] has been cloned into an *E. Coli* expression system[35] and studied as a potential antitumor agent. This enzyme was shown to be even more powerful than asparaginase in killing human leukemia cells[36] *in vitro*. ADI acts selectively on arginine and no other amino acid seem to be affected,[37] as opposed to asparaginase that acts on two amino acids, asparagine and glutamine, the last activity been related to undesired side effects.[38,39]

Recently several human melanomas and human hepatocellular carcinomas (HCC) have been tested for their sensitivity to ADI in *in vitro* experiments. All tested cells were sensitive to the enzyme with an $IC_{50}$ ranging between <0.01 and 0.3 μg/ml for melanomas and of <0.01 μg/ml for hepatocarcinomas. Cell sensitivity was found to be related to their inability to express ASS mRNA[37] and apoptosis is thus caused by the arginine starvation induced by ADI. It was found that arginine starvation has to last for a while before cell death occurs. Therefore, low levels of serum arginine must be maintained for a long period of time before cancer arginine-auxotrophic cells begin to die. As a consequence, ADI activity in the serum must be maintained for a long period of time to achieve a therapeutic effect *in vivo*. Native ADI suffers of a short body permanence upon parenteral administration ($t_{1/2} = 5$ hr) together with a high immunogenicity, two classical drawbacks of native proteins from microbial source. PEGylation was thus investigated as a means to solve these issues.[37,40,41] PEGylation was performed with PEGs of different molecular weight, shape, and having different linkers for protein attachment.[40] It was found that the enzyme specific activity decreases as the number of PEG residues attached to the protein increases, independently of PEG MW, shape and of the linker used. Modification of 10–12 lysines/enzyme molecule leads to products having about 50% specific activity as the nonmodified enzyme. Two formulations obtained with PEG of either 5000 or 20 000 Da MW, both with 10–12 PEG chains/enzyme and displaying similar specific acivity *in vitro*, were compared in *in vivo* experiments. Pharmacokinetic studies (Table 1) revealed a superiority of the 20 000 Da MW adduct, independently of the route of adiministration and animal sex.[40] The 20 000 Da MW (ADI-PEG20 000) adduct was compared to native ADI in terms of pharmacokinetics and antitumor efficacy in cancer animal models. A single i.m. administration of ADI-PEG20 000 (5 IU/mouse) was able to lower arginine serum levels below the detection limit for 1 week, as opposed to unmodified ADI that induced a 50% reduction of arginine serum levels which lasted for 1 day only. The PEG adduct showed a half life of about 7 days as opposed to the 5 hr of the native enzyme. Weekly i.m. administrations of PEG-ADI (5 IU/dose, for total 2 weeks) greatly increased the survival rate of animal models growing ADI sensitive tumors, as opposed to the administration of native ADI on a twice-a-week schedule which was found to be ineffective. These promising results opened the way to phases I and II clinical trials for the PEG20 000 adduct in humans with metastatic melanoma and HCC.[41] Nineteen patients with advanced or inoperable HCC were treated. An optimal biological dose (OBD) of 160 IU/m$^2$ to be administered on a weekly base, which was able to reduce the serum arginine levels below detection limits for more than 7 days, was assessed. Out of the 19 patients treated, 16 showed partial to complete response to ADI-PEG20 000 treatment; two patients showed a complete response, seven patients had partial response, seven had stable disease and only three patients showed progressive disease. No significant side effects were observed and the drug was well tolerated. Mild immunogenicity was registered, serum antibodies against ADI-PEG20 000 increased with the duration of treatment. However, none of the plasma samples obtained from any of the patients inhibited the *in vitro* enzymatic activity of ADI-PEG2000 and no patient developed evidence of systemic or local allergic response to ADI-PEG20 000 injections. These results are very promising and further clinical investigations are presently underway.

### 2.1.4. *PEG-uricase (PEG-UK)*

Another enzyme that finds application in cancer treatment is uricase (urate oxidase, EC 1.7.3.3.). This enzyme catalyzes the oxidation of uric acid to

**Table 1** Citrulline AUC as a function of dose and site of injection of ADI-SS-$PEG_{5000}$ and ADI-SS-$PEG_{20000}$.

| Conjugate | Dose ($IU/m^2$) | AUC (mM days)* | |
|---|---|---|---|
| | | i.m. injection | i.v. injection |
| ADI-SS-$PEG_{5000}$ | 80 | $2212 \pm 177$ | $3038 \pm 225$ |
| | 240 | $4250 \pm 319$ | $4423 \pm 258$ |
| ADI-SS-$PEG_{20000}$ | 80 | $4154 \pm 301$ | $5173 \pm 285$ |
| | 240 | $7731 \pm 603$ | $8288 \pm 566$ |

*AUC was calculated on the basis of citrulline levels determinated prior to treatment and on days 1, 3, 5, 7, 9, 13, and 15 post-injection.[40]

yield the more soluble allantoin, which is readily excreted by the kidney. For nearly 40 years, investigators have administered uricase from a variety of animals and micro-organisms to patients to treat hyperuricemia and gout. Hyperuricemia may occur in the course of chemotherapy as a consequence of tumor lysis.

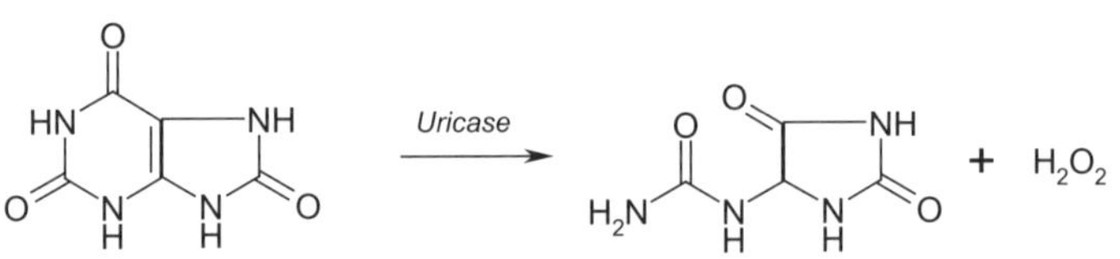


**Scheme 4.** Conversion of uric acid to allantoin and hydrogen peroxide by uricase.

Uricases from different sources have been investigated and a native form of the enzyme purified from *Aspergillus flavus* has been available in France and Italy for over 20 years. Recently this product has been replaced by a recombinant version of the same *A. flavus uricase* produced in the yeast *Saccharomyces cerevisiae* (rasburicase), having improved properties.[42] In 2002, rasburicase became available in the US to treat hyperuricemia caused by tumor lysis syndrome. Rasburicase has been approved only for the initial management of uric acid levels in pediatric patients with leukemia, lymphoma, and solid tumor malignancies who are receiving anticancer therapy expected to result in tumor lysis and subsequent elevation of plasma uric acid. In Europe, rasburicase was approved in 2001 for the treatment and prophylaxis of acute hyperuricemia to prevent acute renal failure in patients with hematologic malignancy with a high tumor burden and patients at risk or a rapid tumor lysis or shrinkage at initiation of chemotherapy.

This enzyme is a tetramer with identical subunits of 34 kDa, where two dimers are superimposed face-to-face to form a tunnel-shaped protein. It has four active sites which are located in a pocket open toward the exterior of the protein. Uricase is present in most mammals but it is lacking in humans because of a nonsense mutation in exon 2. Since humans do not make uricase, all forms of uricases are highly antigenic, and multiple administrations of native uricase have resulted in allergic reactions, anaphylaxis, and even death. As for other proteins, PEGylation has been suggested to overcome these problems.

PEGylation of uricase from different sources has been described in the literature. More precisely, PEGylation has been performed on uricases from *Candida Utilis*,[43–49] from *Arthrobacter protoformiae*,[50] from *Bacillus fastidiosus*[51] and from porcine source.[44,52] It was recently shown that uricase from *C. utilis* has preferable biochemical features compared with uricase from other organisms, including *A. flavus* (rasburicase), hog liver, and *A. protoformiae*. These preferable features include the highest affinity for uric acid and greatest catalytic rate at physiologic pH.

The first PEGylation of uricase from *C. utilis* was described more than 20 years ago by Nishimura.[43,46] The enzyme was modified with a PEG of 5000 Da MW activated as 4,6-dichloro-s-triazine. 43 or 36% of the total 120 amino groups in the enzyme were coupled to the polymer and the modified uricases retained 15 and 45% of the original enzymatic activity, respectively. In both cases, complete loss of the binding ability toward antiuricase serum from rabbit was achieved. The conjugates displayed less immunogenicity and improved plasma residence but, unfortunately, antibodies to the conjugate were raised that recognized to the linker used.[43,46] The same enzyme was later modified with PEG 5000 activated as succinimidyl ester. Modification of 50–70% of the primary amines successfully masked the protein immunogenicity in both mice and humans.[44,45]

Urate oxidase from *C. utilis* is now expressed in *E. coli*. PEGylation of this recombinant enzyme was carried out with succinimidyl succimide activated PEGs having 5000 or 20 000 Da MW[49] and with linear 5000 and branched 10 000 Da PEGs containing a terminal amino acid (nor-leucine or lysine) activated as succinimidyl ester.[47,48] In the work of Bomalaski and colleagues,[49] the 20 000 Da MW conjugate displayed best performance in terms of specific activity and improved plasma residence. PEG was coupled to approximately 66% of the available primary amines and the product retained about 86% of the original activity as compared to 50% residual activity of the 5000 Da adduct. The 20 000 Da MW conjugate was less immunogenic than the 5000 MW and less than other PEG-uricases previously administered to humans. The 20 kDa adduct was administered i.m. in an experimental mouse system and 87% of the drug was recovered in the plasma (Fig. 2). Plasma $t_{1/2}$ was about 3 days. A phase-1 clinical trial was performed for the 20 kDa adduct in patients with hyperuricemia and gout and demonstrated safety and efficacy.[53] A single dose cohort escalation conjugate was performed: plasma residence was maintained for 2–8 days, nontoxicity and no immunogenicity were recorded.[49]

The work carried out by Padua University research group[47,48] compared uricase conjugates obtained with different polymers, namely linear (5000 Da MW) and branched (10 000 Da MW, $PEG_2$) PEGs, PVP, and PAcM. 40–47% of the available primary amines were modified. Active site protection was achieved during the coupling reaction by adding uric acid in the reaction mixture. Residual activity in the conjugates varied between 20 and 80%. Extensive pharmacokinetic[47] and immunogenic evaluation[48]

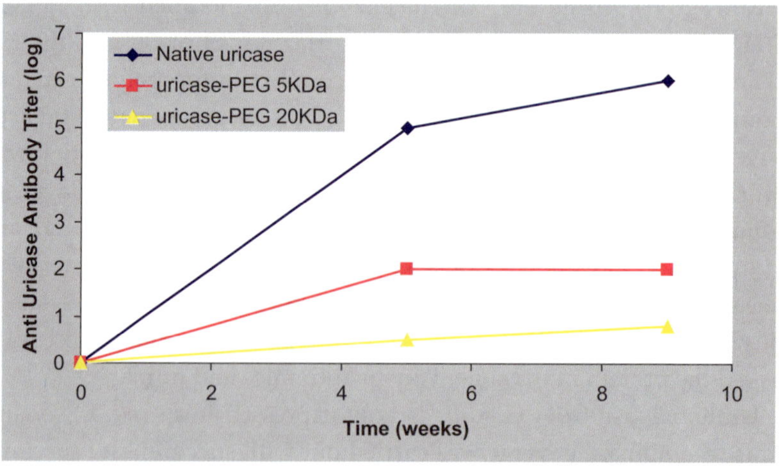


**Fig. 2** Immunogenicity of native urate oxidase, uricase-PEG 20 kDa and uricase-PEG 5 kDa in mice. CD-1 mice were injected i.m. once a week for 8 weeks with 80 IU/m$^2$ of either native or PEG conjugated uricase.[49]

were carried out. Native uricase displayed an elimination $t_{1/2}$ of 6.5 hr and PEGylation improved significantly this parameter as the $PEG_2$ conjugate showed a $t_{1/2}$ of 43 hr while the linear PEG conjugate displayed a $t_{1/2}$ of 56 hr. Differences in tissue distribution were observed between the two conjugates even if in both cases spleen and liver accumulation were preferential. It is, however, noteworthy that the dose administered in this study is well above the therapeutic one.

PEGylation decreased but did not totally eliminate the protein immunogenicity. The $PEG_2$ conjugate was the less immunogenic. Interestingly, antibodies against the conjugated polymer were observed after the second injection.

Uricase from *A. protoformiae* was also modified with PEG of 5000 Da MW and produced by Enzon.[50] About 60% of the proteins primary amines were modified leading to a product devoid of immunogenicity in humans. The conjugate was tested in a single patient that had undergone chemotherapy and had shown exfoliative reaction to alloperidol. PEG-uricase was administered i.m. and uricase activity reached the plasma peak within 24 hr and persisting for about 5 days. No antibodies to either uricase or PEG-uricase developed over a three-week period, during which four doses of the drug had been administered.

Uricase from porcine source has also been PEGylated.[44,52] It was recently shown that increased immunogenicity and accelerated clearance of uricase from this source were due to the presence of large protein aggregates that are undetectable by UV absorbance under normal analytical conditions, but that can be detected by static or dynamic light scattering. The removal of such very large aggregates prior to conjugation with PEG was found to decrease significantly both the immunogenicity of the native protein and the accelerated clearance of the resultant PEG-uricase conjugates. By using a uricase preparation that is devoid of these large aggregates, a lower degree of modification with PEG was found to be necessary to reduce the protein immunogenicity. Lower degree of modification corresponded to a higher bioactivity in the conjugate. Indeed, modification of 12 primary amines with PEG 10 000 led to a product which was devoid of immunogenicity and maintained 75% of the original activity.[52]

Puricase®, a PEGylated form of a uricase from porcine source is now being developed by Savient Pharmaceuticals and recently passed phase-1 clinical investigation. A phase-2 clinical study, an open-label, randomized, multicenter dose-ranging trial involving 41 patients, has recently been concluded. This study compared three dosage levels and two dosage regimens over a twelve-week period. The results are presently being evaluated and will probably be released by mid 2005. Press release from Savient pharmaceuticals Inc. is optimistic and state that data are acceptable for proceeding to more extensive patient exposure in a phase-3 development program.

## 3. PEG-Antibodies

In cancer therapy, antibodies or antibody fragments are most commonly studied as drug targeting agents. In some cases, antibodies and antibody fragments have been developed to have anticancer properties by themselves. Among these, recent examples are the antibodies directed toward the cytokines (e.g. IGF, EGF, and ErbB) receptors,[54–56] or growth factors (e.g. VEGF)[57] that have a direct effect on tumor cell growth by interacting with the receptor-mediated cell proliferation signalling pathway.

This field of investigation comprehends either monoclonal murine or humanized entire antibody molecules, as well as smaller fragments, namely the products of enzymatic degradation Fab′ and $(Fab')_2$ and, more recently, the engineered scFv fragments that, directly expressed in *E. coli* systems, are economically advantageous (Fig. 3).[58–63] Full-length antibodies are

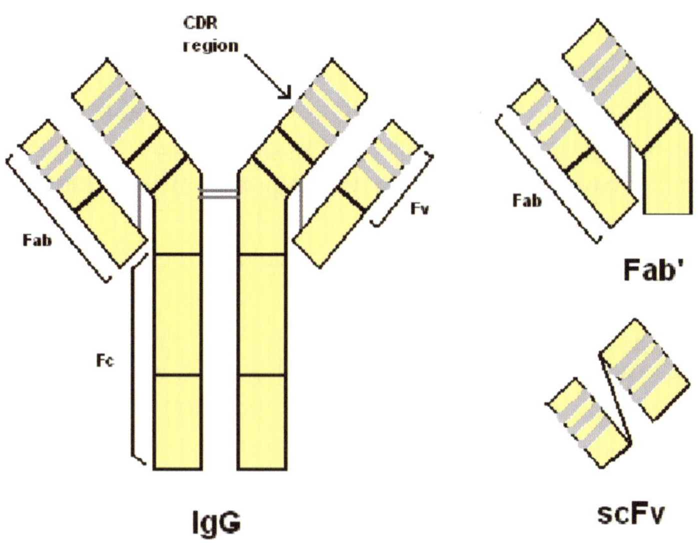


**Fig. 3** Schematic cartoon of an IgG antibody and its fragments

characterized by a long plasma permanence and in the early days, when humanized antibodies were not yet available, rationale for PEGylation was only to diminish the high immunogenic properties of these xenoproteins. Humanized antibodies do not have immunogenic problems and, in principle, PEGylation is not needed. On the other hand, Fab′, $(Fab')_2$ and scFv fragments have very short body permanence upon parenteral administration and PEGylation is necessary to improve this parameter (Table 2). Several studies demonstrated the usefulness of PEGylation to improve plasma residence of Fab′, $(Fab')_2$ and ScFv fragments.[58,64–67] Plasma residence increases with PEG molecular weight and with the number of polymer molecule linked per antibody. However, it appears that it is enough to add one large (40 kDa) MW PEG molecule to restore similar body residence as the one displayed by the entire IgG.[67] Most commonly, when tumor-bearing animal models are injected with PEGylated tumor-epitope-targeting antibodies or antibody fragments, tumor : blood ratio decreases as a consequence of improved plasma residence. However, PEGylated products do show advantages over the nonmodified products, namely an improved tumor uptake,[58,67] probably as a consequence of the polymer induced EPR effect, and minor overall tissue distribution that corresponds also to diminished renal and liver toxicity.[68,69] Despite the positive impact exerted by PEGylation on the pharmacokinetics parameters, several studies performed on both entire antibody molecules and Fab′ fragments also demonstrated that random PEGylation, that is carried out by using amine-directed PEG reagents, has a negative impact on antigen binding properties. Even a small number of PEG molecules/protein may be sufficient to decrease significantly the antigen-binding properties.[67] As expected, loss of binding increases with the number of PEG molecules attached and their

**Table 2** Pharmacokinetic and antigen-binding data for site specific Fab′-PEG conjugates injected in rats.

| Compounds | $t_{1/2}\alpha$ (h) | $t_{1/2}\beta$ (h) | AUC (0−∞) | AUC (% IgG) | Antigen binding (% Fab′) |
|---|---|---|---|---|---|
| IgG | 5.80 | 104 | 8791 | 100 | — |
| Fab′ | 0.33 | 22.7 | 329 | 3.7 | 100 |
| Fab′-PEG (1 × 25 kDa) | 4.97 | 30.5 | 2261 | 26 | 107 |
| Fab′-PEG (1 × 40 kDa) | 5.60 | 45.8 | 4465 | 51 | 102 |
| Fab′-PEG (2 × 25 kDa) | 9.05 | 49.1 | 6999 | 80 | 100 |

molecular weight, consistent with both a steric effect and the direct modification of lysines involved in antigen recognition. Site directed modification has thus become a strategy of choice. A general approach for targeted PEGylation relies on the use of PEG-maleimide for the selective modification of engineered cysteine-bearing antibody fragments. By introducing, by genetic manipulation, the target cysteine on a location far away from the antigen binding site, monoPEGylation of either Fab′[58] or ScFv[61] fragments has been achieved without loss of antigen-binding ability.

These conjugation strategies have been investigated independently of the therapeutic application of the PEGylated antibody. In the specific field of anticancer therapeutics, several studies on PEGylated antibodies have been performed, but no product so far has entered clinical trials. Examples of cancer epitopes that were chosen as targets for PEGylated antibodies are the A33 and CEA antigens and the $ErbB_2$ receptor. The $ErbB_2$ receptor is a molecule that is overexpressed in epithelial cells and in tumors of epithelial origins.[70] Murine monoclonal antibodies and their pepsin digested Fab′ fragments have been modified either with PEG5000 maleimide or with branched PEG20 000NHS ($PEG_2$). PEG-maleimide was conjugated to sulphydryl groups located at the hinge region, which were obtained upon partial reduction of the molecule with dithiotreitol (DTT); modification with $PEG_2$ was randomly targeted towards the amine residues. In the case of the maleimide 5000 Da MW conjugate, modification lead to loss of antigen binding ability, whereas $PEG_2$ adducts maintained reactivity towards their target, as shown by both ELISA and cell culture tests. Besides, experiments carried out in animal tumor models demonstrated that $PEG_2$ adducts were effective in preventing tumor growth and were more efficacious than the unmodified antibodies. On the contrary thiol modified derivatives were less effective.

The A33 antigen[66] is a cell surface glycoprotein expressed on normal gastrointestinal epithelium and on 95% of primary metastatic colon cancers, while being absent in most other normal tissues. Humanized $\alpha$-A33 antibodies, randomly modified with linear 5000 and 20 000 Da PEGs, retained more than 50% of their antigen binding ability and were less immunogenic than the unmodified antibody. Even if the tumor/blood ratio did not improve upon PEGylation, tumor uptake of the modified antibodies was still efficacious.[64]

The CEA antigen is the first oncorelated surface antigen discovered. It is expressed on the surface of many carcinomas and it is extensively investigated as a target for radio-immuno-therapy. The monoclonal $\alpha$-CEA antibody A5B7 and its Fab′ and F(ab′)$_2$ fragments have been PEGylated with PEG of 5000 Da MW and their pharmacokinetics were compared with the unmodified forms in the LS174T colonic xenograft in nude mice[71,72] PEGylation of the intact antibody had little effect on biodistribution, although tumor localization was slightly reduced. In contrast, modification of F(ab′)$_2$ and Fab′ A5B7 significantly prolonged plasma half-life and increased antibody accumulation in the tumor but reduced tissue to blood ratios. Prior to modification, Fab′ A5B7, that has a MW of 50 kDa, was cleared more rapidly from the circulation than the larger F(ab′)$_2$ (MW 10 kDa), but after PEG attachment their biodistributions converged, while the tumor to blood ratios were reduced and resembled that of the intact antibody.

Despite these early investigations, next generation of PEGylated tumor-antigen targeting molecules will probably rely on the recombinant Single chain Fv antibody fragments (ScFv). Not only these product are economically advantageous as compared to entire antibody molecules, but also optimization of their target binding ability is possible by the combination of recent antibody engineering tools and phage display. The generic approach described by Yang for the obtainment of engineered thiol bearing ScFvs suitable for site specific PEGylation[61] has been recently applied to target cancer epitopes.[62,63] Four free-cysteine containing ScFv fragments (Sc-Fv-c) targeting the cancer epitope MUC-1 were obtained by recombinant technology. The proteins were PEGylated using either mono- or bifunctional, maleimide- or *o*-pyridyl-disulfide-, thiol reactive PEGs having a molecular weight ranging from 2 to 40 kDa. The yield of modification varied depending on the polymer molecular weight, functionality (mono- or bivalent) and the reactive group. Monofunctional low molecular weight maleimide derivatives displayed better yields of modification. The most relevant result is that all PEG derivatives maintained the same antigen binding properties as the nonmodified scFv-cs, holding promises for future applications in cancer immunotherapy.

## 4. Conclusions

After almost 40 years of experimentation, PEGylation has become a well-established technology in the field of biopharmaceutical formulation. The remarkable advantages of PEGylated proteins over their nonmodified parents will likely lead to further expansion of this technology to more and more bioactive compounds. Lessons from the past have shown that PEGylation of a bioactive protein has often followed the initial clinical approval of the native compound. Indeed, the number of protein pharmaceutics in clinical trials — among which almost half are indicated for cancer treatment — is increasing dramatically every year[73] and PEGylation will be useful any time poor pharmacokinetics and immunogenic problems will occur. The costs related to the development of PEGylated products is also likely to diminish, mostly thanks to the wide expertise in protein PEGylation acquired in recent years that provided general clues for faster coupling strategy optimization.

## References


1. Abuchowski A, McCoy JR, Palczuk NC, van Es T and Davis FF (1977) Effect of covalent attachment of polyethylene glycol on immunogenicity and circulating life of bovine liver catalase. *J Biol Chem* **252**:3582–3586.
2. Abuchowski A, van Es T, Palczuk NC and Davis FF (1977) Alteration of immunological properties of bovine serum albumin by covalent attachment of polyethylene glycol. *J Biol Chem* **252**:3578–3581.
3. Vellard M (2003) The enzyme as drug: Application of enzymes as pharmaceuticals. *Curr Opin Biotechnol* **14**:444–450.
4. Denholm EM, Lin YQ and Silver PJ (2001) Anti-tumor activities of chondroitinase AC and chondroitinase B: Inhibition of angiogenesis, proliferation and invasion. *Eur J Pharmacol* **416**:213–221.
5. Cairo MS (2002) Prevention and treatment of hyperuricemia in hematological malignancies. *Clin Lymphoma* **3** (Suppl 1):S26–S31.
6. Pui CH (2002) Rasburicase: A potent uricolytic agent. *Exp Opin Pharmacother* **3**:433–442.
7. Kidd JG (1953) Regression of transplanted lymphomas induced *in vivo* by means of normal guinea pig serum. I. Course of transplanted cancers of various kinds in mice and rats given guinea pig serum, horse serum, or rabbit serum. *J Exp Med* **98**:565–582.
8. Broome JD (1963) Evidence that the L-asparaginase of guinea pig serum is responsible for its antilymphoma effects. I. Properties of the L-asparaginase of guinea pig serum in relation to those of the antilymphoma substance. *J Exp Med* **118**:99–120.
9. Campbell HA and Mashburn LT (1969) L-Asparaginase EC-2 from Escherichia coli. Some substrate specificity characteristics. *Biochemistry* **8**:3768–3775.

10. Whelan HA and Wriston JC Jr (1969) Purification and properties of asparaginase from escherichia coli B. *Biochemistry* **8**:2386–2393.
11. Taylor CW, Dorr RT, Fanta P, Hersh EM and Salmon SE (2001) A phase I and pharmacodynamic evaluation of polyethylene glycol-conjugated L-asparaginase in patients with advanced solid tumors. *Cancer Chemother Pharmacol* **47**:83–88.
12. Agrawal NR, Bukowski RM, Rybicki LA, Kurtzberg J, Cohen LJ and Hussein MA (2003) A Phase I-II trial of polyethylene glycol-conjugated L-asparaginase in patients with multiple myeloma. *Cancer* **98**:94–99.
13. Abuchowski A, van Es T, Palczuk NC, McCoy JR and Davis FF (1979) Treatment of L5178Y tumor-bearing BDF1 mice with a nonimmunogenic L-glutaminase-L-asparaginase. *Cancer Treat Rep* **63**:1127–1132.
14. Park YK, Abuchowski A, Davis S and Davis F (1981) Pharmacology of Escherichia coli-L-asparaginase polyethylene glycol adduct. *Anticancer Res* **1**:373–376.
15. Abuchowski A, Kazo GM, Verhoest CR Jr, Van Es T, Kafkewitz D, Nucci ML, Viau AT and Davis FF (1984) Cancer therapy with chemically modified enzymes. I. Antitumor properties of polyethylene glycol-asparaginase conjugates. *Cancer Biochem Biophys* **7**:175–186.
16. Ho DH, Brown NS, Yen A, Holmes R, Keating M, Abuchowski A, Newman RA and Krakoff IH (1986) Clinical pharmacology of polyethylene glycol-L-asparaginase. *Drug Metab Dispos* **14**:349–352.
17. Soares AL, Guimaraes GM, Polakiewicz B, de Moraes Pitombo RN and Abrahao-Neto J (2002) Effects of polyethylene glycol attachment on physicochemical and biological stability of *E. coli* L-asparaginase. *Int J Pharm* **237**:163–170.
18. Kurtzberg J (2000). *Cancer Medicine*, 5th ed. Gansler T (ed.) Decker Inc: Canada.
19. Kurre HA, Ettinger AG, Veenstra DL, Gaynon PS, Franklin J, Sencer SF, Reaman GH, Lange BJ and Holcenberg JS (2002) A pharmacoeconomic analysis of pegaspargase versus native Escherichia coli L-asparaginase for the treatment of children with standard-risk, acute lymphoblastic leukemia: The children's cancer group study (CCG-1962). *J Pediatr Hematol Oncol* **24**:175–181.
20. Davis FF (2003) PEG-adenosine deaminase and PEG-asparaginase. *Adv Exp Med Biol* **519**:51–58.
21. Graham ML (2003) Pegaspargase: A review of clinical studies. *Adv Drug Deliv Rev* **55**:1293–1302.
22. Hawkins DS, Park JR, Thomson BG, Felgenhauer JL, Holcenberg JS, Panosyan EH and Avramis VI (2004) Asparaginase pharmacokinetics after intensive polyethylene glycol-conjugated L-asparaginase therapy for children with relapsed acute lymphoblastic leukemia. *Clin Cancer Res* **10**:5335–5341.

23. Panosyan EH, Seibel NL, Martin-Aragon S, Gaynon PS, Avramis IA, Sather H, Franklin J, Nachman J, Ettinger LJ, La M, Steinherz P, Cohen LJ, Siegel SE and Avramis VI (2004) Asparaginase antibody and asparaginase activity in children with higher-risk acute lymphoblastic leukemia: Children's cancer group study CCG-1961. *J Pediatr Hematol Oncol* **26**:217–226.
24. Tan Y, Xu M, Tan X, Wang X, Saikawa Y, Nagahama T, Sun X, Lenz M and Hoffman RM (1997) Overexpression and large-scale production of recombinant L-methionine-alpha-deamino-gamma-mercaptomethane-lyase for novel anticancer therapy. *Protein Expr Purif* **9**:233–245.
25. Tan Y, Zavala J Sr, Xu M, Zavala J Jr and Hoffman RM (1996) Serum methionine depletion without side effects by methioninase in metastatic breast cancer patients. *Anticancer Res* **16**:3937–3942.
26. Tan Y, Zavala J, Sr, Han Q, Xu M, Sun X, Tan X, Magana R, Geller J and Hoffman RM (1997) Recombinant methioninase infusion reduces the biochemical endpoint of serum methionine with minimal toxicity in high-stage cancer patients. *Anticancer Res* **17**:3857–3860.
27. Sun X, Yang Z, Li S, Tan Y, Zhang N, Wang X, Yagi S, Yoshioka T, Takimoto A, Mitsushima K, Suginaka A, Frenkel EP and Hoffman RM (2003) *In vivo* efficacy of recombinant methioninase is enhanced by the combination of polyethylene glycol conjugation and pyridoxal 5′-phosphate supplementation. *Cancer Res* **63**:8377–8383.
28. Tan Y, Sun X, Xu M, An Z, Tan X, Han Q, Miljkovic DA, Yang M and Hoffman RM (1998) Polyethylene glycol conjugation of recombinant methioninase for cancer therapy. *Protein Expr Purif* **12**:45–52.
29. Yang Z, Wang J, Yoshioka T, Li B, Lu Q, Li S, Sun X, Tan Y, Yagi S, Frenkel EP and Hoffman RM (2004) Pharmacokinetics, methionine depletion, and antigenicity of recombinant methioninase in primates. *Clin Cancer Res* **10**:2131–2138.
30. Yang Z, Sun X, Li S, Tan Y, Wang X, Zhang N, Yagi S, Takakura T, Kobayashi Y, Takimoto A, Yoshioka T, Suginaka A, Frenkel EP and Hoffman RM (2004) Circulating half-life of PEGylated recombinant methioninase holoenzyme is highly dose dependent on cofactor pyridoxal-5′-phosphate. *Cancer Res* **64**:5775–5778.
31. Yang Z, Wang J, Lu Q, Xu J, Kobayashi Y, Takakura T, Takimoto A, Yoshioka T, Lian C, Chen C, Zhang D, Zhang Y, Li S, Sun X, Tan Y, Yagi S, Frenkel EP and Hoffman RM (2004) PEGylation confers greatly extended half-life and attenuated immunogenicity to recombinant methioninase in primates. *Cancer Res* **64**:6673–6678.
32. Sugimura K, Ohno T, Kusuyama T and Azuma I (1992) High sensitivity of human melanoma cell lines to the growth inhibitory activity of mycoplasmal arginine deiminase *in vitro*. *Melanoma Res* **2**:191–196.

33. Savoca KV, Abuchowski A, van Es T, Davis FF and Palczuk NC (1979) Preparation of a non-immunogenic arginase by the covalent attachment of polyethylene glycol. *Biochim Biophys Acta* **578**:47–53.
34. Takaku H, Takase M, Abe S, Hayashi H and Miyazaki K (1992) *In vivo* antitumor activity of arginine deiminase purified from Mycoplasma arginini. *Int J Cancer* **51**:244–249.
35. Misawa S, Aoshima M, Takaku H, Matsumoto M and Hayashi H (1994) High-level expression of Mycoplasma arginine deiminase in Escherichia coli and its efficient renaturation as an anti-tumor enzyme. *J Biotechnol* **36**:145–155.
36. Gong H, Zolzer F, von Recklinghausen G, Havers W and Schweigerer L (2000) Arginine deiminase inhibits proliferation of human leukemia cells more potently than asparaginase by inducing cell cycle arrest and apoptosis. *Leukemia* **14**:826–829.
37. Ensor CM, Holtsberg FW, Bomalaski JS and Clark MA (2002) PEGylated arginine deiminase (ADI-SS PEG20,000 mw) inhibits human melanomas and hepatocellular carcinomas *in vitro* and *in vivo*. *Cancer Res* **62**:5443–5450.
38. Durden DL and Distasio JA (1980) Comparison of the immunosuppressive effects of asparaginases from Escherichia coli and Vibrio succinogenes. *Cancer Res* **40**:1125–1129.
39. Distasio JA, Salazar AM, Nadji M and Durden DL (1982) Glutaminase-free asparaginase from vibrio succinogenes: An antilymphoma enzyme lacking hepatotoxicity. *Int J Cancer* **30**:343–347.
40. Holtsberg FW, Ensor CM, Steiner MR, Bomalaski JS and Clark MA (2002) Poly(ethylene glycol) (PEG) conjugated arginine deiminase: Effects of PEG formulations on its pharmacological properties. *J Control Rel* **80**:259–271.
41. Izzo F, Marra P, Beneduce G, Castello G, Vallone P, De Rosa V, Cremona F, Ensor CM, Holtsberg FW, Bomalaski JS, Clark MA, Ng C and Curley SA (2004) PEGylated arginine deiminase treatment of patients with unresectable hepatocellular carcinoma: Results from phase I/II studies. *J Clin Oncol* **22**:1815–1822.
42. Bayol A, Capdevielle J, Malazzi P, Buzy A, Claude Bonnet M, Colloc'h N, Mornon JP, Loyaux D and Ferrara P (2002) Modification of a reactive cysteine explains differences between rasburicase and Uricozyme, a natural Aspergillus flavus uricase. *Biotechnol Appl Biochem* **36**:21–31.
43. Nishimura H, Ashihara Y, Matsushima A and Inada Y (1979) Modification of yeast uricase with polyethylene glycol: Disappearance of binding ability towards anti-uricase serum. *Enzyme* **24**:261–264.
44. Chen RH, Abuchowski A, Van Es T, Palczuk NC and Davis FF (1981) Properties of two urate oxidases modified by the covalent attachment of poly(ethylene glycol). *Biochim Biophys Acta* **660**:293–298.

45. Davis S, Park YK, Abuchowski A and Davis FF (1981) Hypouricaemic effect of polyethyleneglycol modified urate oxidase. *Lancet* **2**:281–283.
46. Nishimura H, Matsushima A and Inada Y (1981) Improved modification of yeast uricase with polyethylene glycol, accompanied with nonimmunoreactivity towards anti-uricase serum and high enzymic activity. *Enzyme* **26**:49–53.
47. Caliceti P, Schiavon O and Veronese FM (1999) Biopharmaceutical properties of uricase conjugated to neutral and amphiphilic polymers. *Bioconjug Chem* **10**:638–646.
48. Caliceti P, Schiavon O and Veronese FM (2001) Immunological properties of uricase conjugated to neutral soluble polymers. *Bioconjug Chem* **12**:515–522.
49. Bomalaski JS, Holtsberg FW, Ensor CM and Clark MA (2002) Uricase formulated with polyethylene glycol (uricase-PEG 20): Biochemical rationale and preclinical studies. *J Rheumatol* **29**:1942–1949.
50. Chua CC, Greenberg ML, Viau AT, Nucci M, Brenckman WD Jr and Hershfield MS (1988) Use of polyethylene glycol-modified uricase (PEG-uricase) to treat hyperuricemia in a patient with non-Hodgkin lymphoma. *Ann Int Med* **109**:114–117.
51. Schiavon O, Caliceti P, Ferruti P and Veronese FM (2000) Therapeutic proteins: A comparison of chemical and biological properties of uricase conjugated to linear or branched poly(ethylene glycol) and poly(N-acryloylmorpholine). *Farmaco* **55**:264–269.
52. Sherman M, Saifer M and Williams L (2004) Aggregate-free urate oxidase for preparation of non-immunogenic polymer-conjugates. *United States Patent #* **6**:783,965.
53. Bomalaski JS and Clark MA (2004) Serum uric acid-lowering therapies: Where are we heading in management of hyperuricemia and the potential role of uricase. *Curr Rheumatol Rep* **6**:240–247.
54. Klapper LN, Vaisman N, Hurwitz E, Pinkas-Kramarski R, Yarden Y and Sela M (1997) A subclass of tumor-inhibitory monoclonal antibodies to ErbB-2/HER2 blocks crosstalk with growth factor receptors. *Oncogene* **14**:2099–2109.
55. Souriau C, Rothacker J, Hoogenboom HR and Nice E (2004) Human antibody fragments specific for the epidermal growth factor receptor selected from large non-immunised phage display libraries. *Growth Factors* **22**:185–194.
56. Goetsch L, Gonzalez A, Leger O, Beck A, Pauwels PJ, Haeuw JF and Corvaia N (2005) A recombinant humanized anti-insulin-like growth factor receptor type I antibody (h7C10) enhances the antitumor activity of vinorelbine and anti-epidermal growth factor receptor therapy against human cancer xenografts. *Int J Cancer* **113**:316–328.
57. Bergsland EK (2004) Update on clinical trials targeting vascular endothelial growth factor in cancer. *Am J Health Syst Pharm* **61**:S12–S20.

58. Chapman AP, Antoniw P, Spitali M, West S, Stephens S and King DJ (1999) Therapeutic antibody fragments with prolonged *in vivo* half-lives. *Nat Biotechnol* 17:780–783.
59. Peters R and Sikorsky R (1999) PEG antibodies. *Science* **286**:434.
60. Casey JL, Napier MP, King DJ, Pedley RB, Chaplin LC, Weir N, Skelton L, Green AJ, Hope-Stone LD, Yarranton GT and Begent RH (2002) Tumor targeting of humanised cross-linked divalent-Fab′ antibody fragments: A clinical phase I/II study. *Br J Cancer* **86**:1401–1410.
61. Yang K, Basu A, Wang M, Chintala R, Hsieh MC, Liu S, Hua J, Zhang Z, Zhou J, Li M, Phyu H, Petti G, Mendez M, Janjua H, Peng P, Longley C, Borowski V, Mehlig M and Filpula D (2003) Tailoring structure-function and pharmacokinetic properties of single-chain Fv proteins by site-specific PEGylation. *Protein Eng* **16**:761–770.
62. Albrecht H, Burke PA, Natarajan A, Xiong CY, Kalicinsky M, DeNardo GL and DeNardo SJ (2004) Production of soluble ScFvs with C-terminal-free thiol for site-specific conjugation or stable dimeric ScFvs on demand. *Bioconjug Chem* **15**:16–26.
63. Natarajan A, Xiong CY, Albrecht H, Denardo GL and Denardo SJ (2005) Characterization of site-specific ScFv PEGylation for tumor-targeting pharmaceuticals. *Bioconjug Chem* **16**:113–121.
64. Kitamura K, Takahashi T, Yamaguchi T, Noguchi A, Takashina K, Tsurumi H, Inagake M, Toyokuni T and Hakomori S (1991) Chemical engineering of the monoclonal antibody A7 by polyethylene glycol for targeting cancer chemotherapy. *Cancer Res* **51**:4310–4315.
65. Lee LS, Conover C, Shi C, Whitlow M and Filpula D (1999) Prolonged circulating lives of single-chain Fv proteins conjugated with polyethylene glycol: a comparison of conjugation chemistries and compounds. *Bioconjug Chem* **10**:973–981.
66. Deckert PM, Jungbluth A, Montalto N, Clark MA, Finn RD, Williams C, Jr, Richards EC, Panageas KS, Old LJ and Welt S (2000) Pharmacokinetics and microdistribution of polyethylene glycol-modified humanized A33 antibody targeting colon cancer xenografts. *Int J Cancer* **87**:382–390.
67. Chapman AP (2002) PEGylated antibodies and antibody fragments for improved therapy: A review. *Adv Drug Deliv Rev* **54**:531–545.
68. Maeda H, Wu J, Sawa T, Matsumura Y and Hori K (2000) Tumor vascular permeability and the EPR effect in macromolecular therapeutics: A review. *J Control Rel* **65**:271–284.
69. Maeda H (2001) The enhanced permeability and retention (EPR) effect in tumor vasculature: The key role of tumor-selective macromolecular drug targeting. *Adv Enzyme Regul* **41**:189–207.

70. Hurwitz E, Klapper LN, Wilchek M, Yarden Y and Sela M (2000) Inhibition of tumor growth by poly(ethylene glycol) derivatives of anti-ErbB2 antibodies. *Cancer Immunol Immunother* **49**:226–234.
71. Pedley RB, Boden JA, Boden R, Begent RH, Turner A, Haines AM and King DJ (1994) The potential for enhanced tumor localization by poly(ethylene glycol) modification of anti-CEA antibody. *Br J Cancer* **70**:1126–1130.
72. Delgado C, Pedley RB, Herraez A, Boden R, Boden JA, Keep PA, Chester KA, Fisher D, Begent RH and Francis GE (1996) Enhanced tumor specificity of an anti-carcinoembrionic antigen Fab′ fragment by poly(ethylene glycol) (PEG) modification. *Br J Cancer* **73**:175–182.
73. Holmer A (2004) 324 Biotechnology Medicines in Testing Promise to Bolster the Arsenal Against Disease, in *Medicines in Development*, by Pharmaceutical Research and Manufacturers of America (PhRMA), http://www.phrma.org/newmedicines/resources/2004-10-25.145.pdf.

# 6

# PEGylated Proteins in Immunotherapy of Cancer

*James F. Eliason*

## Abstract

The half-lives of proteins *in vivo* can be increased by covalently attaching the polymer polyethylene glycol (PEG) to them. This can significantly decrease their rate of clearance from plasma. The clearance rate is inversely proportional to the total molecular weight of the construct. This approach has been used to design cytokine constructs that can be administered once a week, rather than on a daily or every other day schedule. Two cytokines for which this approach appears to be successful are PEG-interferon-$\alpha_{2a}$ (PEG-IFN-$\alpha_{2a}$; Pegasys™) and PEG-granulocyte colony-stimulating factor (PEG-G-CSF; SD/01). Both use large MW PEG (20–40 kDa) to give sufficiently long duration *in vivo*. With PEG-G-CSF conjugates, the *in vivo* efficacy is directly proportional to molecular weight, whereas the *in vitro* activity is inversely proportional, suggestion that overall duration of contact is more important than affinity of the interaction. Conjugates of a number of other cytokines have been prepared, but until recently, few have used the large molecular weight polymers. As this approach is likely to be taken in the future to make new PEG-cytokine constructs, thorough *in vivo* pharmacokinetic studies will be essential for their development and clinical use.

## 1. Introduction

Historically, the treatment of cancer has involved a variety of cytoreductive therapies, including radiation therapy and chemotherapy. In the initial phases of treatment, these therapies often result in major reductions in tumor burden. However, with the exception of children with certain types of cancers, there has been relatively little impact on survival of cancer patients. Even high-dose therapy requiring hematopoietic stem cell support to counteract lethality to this system can provide long-term survival for 15–20% of patients with metastatic breast cancer, at best. Other approaches are clearly needed to improve the cure rate for advanced adult cancers.

Immunotherapy represents another approach, which has been around for many years but has recently received renewed attention, as a means to eliminate residual disease following cytoreductive therapy. A number of trials in the past provided evidence that this approach can result in dramatic tumor reduction in a small subset of patients with certain types of tumors. However, these studies were usually performed in patients with bulky disease, where this approach is least likely to work. Combinations with conventional therapies and new approaches to stimulating and/or modifying the immune system are now being explored for treatment of cancer.

Cytokines are regulatory proteins, produced by cells, which can stimulate proliferation and/or activation of (1) neighboring cells in a paracrine fashion, (2) the producing cells themselves in an autocrine fashion, or (3) distant cells in a hormonal fashion. They are important components of immune responses, regulating the activities of both the specific immune system, consisting of various lymphocyte subsets, and the nonspecific immune system, composed of myeloid cells such as granulocytes, and macrophages. The latter cell types are not always considered in conjunction with immunotherapy but are important in combating microbial infections in patients receiving cytotoxic therapies that can drastically reduce myeloid cell numbers. A number of cytokines have been tested in the clinical and several are approved for use. The most important of these are the interferon-$\alpha$'s of which three (IFN-$\alpha_{2a}$, IFN-$\alpha_{2b}$, and IFN-consensus) are registered[1] and granulocyte colony-stimulating factor (G-CSF).[2]

One of the practical problems with clinical use of cytokines is that they have to be injected intravenously or subcutaneously. In addition, they have short half-lives in the body, whereas their activities are usually optimal with continuous exposure. Therefore, they have to be administered frequently, usually on a daily basis.

Removal of proteins from circulation is usually through clearance in the kidney. Increasing the molecular weight of proteins by attachment of PEG can significantly reduce their renal clearance, presumably by increasing the size of the conjugates in relation to the pore sizes of the renal vascular bed.[3] A study using radiolabeled PEG molecules (not conjugated to proteins), which had different molecular weights demonstrated that the area under the plasma concentration time curves (AUCs) increased with increasing molecular weight.[4] Another mechanism relevant for removing cytokines from circulation is receptor-mediated endocytosis.[5,6] The effects of PEG on this mechanism are not clear, but if PEGylation significantly decreases the binding affinity of conjugates to their receptors, then clearance will also be decreased. A number of different linkage chemistries have been used for preparing PEG conjugates and these can, in some cases, affect the activity and pharmacokinetics.[7,8]

The pharmacokinetics of PEGylated proteins can also be influenced by the route of their administration. Compared to intravenous or intraperitoneal injection, administration subcutaneously, or intramuscularly may delay the release of proteins into circulation. There is no evidence, however, that PEGylation affects the distribution of circulating cytokines in the body, particularly with respect to retention in tumors.

A further problem is that some of the recombinant proteins can be immunogenic, so resistance can build up due to development of neutralizing antibodies. This has been a problem with IFN-$\alpha$.[1] One approach being applied to overcome these problems is conjugation of the proteins with the polymer polyethylene glycol (PEG), which may hide the antigenic peptide sequences from the immune system. One of the best examples of decreased immunogenicity of PEGylated proteins is the enzyme L-asparaginase, a treatment for acute lymphocytic leukemia. Two versions of this protein have been isolated, one from Escherichia coli and the other from Erwinia Carotovora. Because of their xenobiotic origins, hypersensitivity develops in up to 50% of patients.[9] The PEGylated *E. coli* protein (pegaspargase) has multiple 5 kDa MW chains attached. This conjugate has been registered for use in patients who had reactions to the un-PEGylated protein,[10] demonstrating that PEG can decrease antigenicity of proteins even in patients who are sensitive to the unmodified protein. In addition, this modification increases the plasma half-life from 20 hr for the unmodified protein to 357 hr for the conjugate.[11]

This article will summarize the current status of PEGylated cytokines that may be beneficial for cancer patients.

## 2. PEGylated Interferon-$\alpha$

Interferon-$\alpha$ was the first cytokine produced by recombinant DNA technology to be registered for cancer therapy.[1] A type-I interferon that is produced by fibroblasts was first identified as an antiviral protein. Three species of interferon-$\alpha$ have been registered for treatment of hepatitis B and C virus (HBV and HCV) infections. Two, IFN-$\alpha_{2a}$ and IFN-$\alpha_{2b}$ are naturally occurring forms and the other, consensus IFN (IFN-con1), is a mutein that was designed by using the most frequently observed amino acids found in several of the natural IFN-$\alpha$ subtypes. This usage as an antiviral therapy can also be considered as a cancer prevention therapy because chronic infections with HBV and HCV are important causative factors in hepatocellular carcinoma.

Interferon-$\alpha$ has demonstrated clinical activity against a number of fully developed cancers including hairy cell leukemia, chronic myelogenous leukemia (CML), B- and T-cell lymphomas, myeloma, Kaposi's sarcoma, melanoma, and renal cell carcinoma.[12] In combination with retinoids, IFN-$\alpha$ is also active against squamous cell carcinoma of the skin[13] and cervix.[14]

The first PEGylated IFN to be tested in the clinic was prepared from IFN-$\alpha_{2a}$ by attaching 5 kDa PEG in a 1:1 molar ratio via a urea linkage.[15] This modification increased the half-life of the conjugate only 2-fold compared to the unconjugated cytokine because of its relatively small increase in overall molecular weight and did not provide sufficient improvement to justify further development.[16] Following the experience with high molecular weight conjugates of G-CSF, described in the following section, a new PEG-IFN-$\alpha_{2a}$ conjugate with branched 40 kDa PEG was prepared. This molecule has a 51 hr half-life in rat compared to 0.7 hr for the unmodified protein. This increased half-life resulted in greater antitumor activity against a human renal cell model in athymic nude mice.[17] Several recent reports have demonstrated that this drug, which has the name Pegasys, given once a week is more effective than unmodified IFN-$\alpha_{2a}$ given three times a week in patients with chronic hepatitis C infections.[18,19] Results of a phase I trial in patients with renal cell carcinoma demonstrated that there were no unique toxicities not seen with unmodified IFNs.[20]

A second PEGylated interferon, IFN-$\alpha_{2b}$, has also been developed. It is a conjugate with a 1:1 molar ratio of 12 kDa PEG.[21] The pharmacokinetics and pharmacodynamics of this molecule indicate that it has a shorter half-life than Pegasys as predicted by the lower overall molecular weight.[22] It has activity against CML, but has not been demonstrated to be statistically noninferior to nonPEGylated IFN-$\alpha_{2b}$.[23] This conjugate also has activity to decrease platelet numbers in essential thrombocythemia.[24]

The use of either PEGylated IFN-$\alpha$ in combination of ribavirin for treatment of HCV reduces the relative risk (RR) for remaining infected by 17% with nonPEGylated interferon plus ribavirin.[25] The combination treatment with the PEGylated proteins also results in a more cost effective approach to treating hepatitis C than the combination with unconjugated IFN-$\alpha$.

## 3. PEGylated G-CSF

Granulocyte colony-stimulating factor is a member of the hematopoietic growth factor family that increases the numbers of neutrophilic granulocytes.[2] As such, G-CSF is an important supportive agent for cancer cytotoxic therapy. Although it appears to have no direct or indirect activity on cancer cells, its activity helps alleviate the neutrophilia associated with standard chemotherapy thereby allowing patients to tolerate higher doses of cytotoxic therapy. The so-called dose-dense chemotherapies, which use higher dosages than standard chemotherapy because of hematopoietic growth factor support, have shown promise in treatment of hematological and solid tumors, especially breast cancer.[26] In addition, G-CSF induces migration of hematopoietic stem cells from bone marrow into peripheral blood, making it possible to collect large numbers of stem cells by leukapheresis. This is a much more tolerable source of hematopoietic stem cells from donors, especially in the case of allogeneic transplants, than drawing bone marrow aspirates; although the same is true for obtaining these stem cells from patients who receive them back as autologous stem cell rescue.

Three different molecular forms of recombinant G-CSFs are registered for clinical use in various countries. Filgrastim is produced in *E. coli* and has an N-terminal methionine. Lenograstim is produced in Chinese hamster ovary cells and is therefore glycosylated.[27] Nartograstim is a mutein produced in *E. coli* with five amino acids substitutions in the N-terminal portion that increase its specific activity.[28]

A systematic approach to design a PEG-G-CSF that could be given once a week was undertaken. The goal was to make a molecule that would have pharmacokinetic properties that fit standard cancer chemotherapy practices that are usually designed on weekly schedules.

A number of different conjugates with one to three PEG chains of 5, 10, or 20 kDa per molecule of nartograstim were prepared. These conjugates were tested for their capacity to stimulate proliferation of G-CSF dependent cells *in vitro* and to increase granulocyte numbers in mice.[29] The activity *in vitro* was inversely proportional to the amount of PEG attached. The constructs having the highest the total molecular weight had the lowest the activity on cell proliferation in culture. This result suggests that the larger bulkier constructs may not bind as well to the G-CSF receptors. Interestingly, the *in vivo* activity was directly proportional to the total molecular weight of the constructs. Those molecules with the highest molecular weight had the highest activity in mice.[29] Similar results were evident in studies with cynomolgus monkeys where the higher molecular weight constructs had increased activities.[30] These results indicate that the duration of interaction between PEG-G-CSF and its receptors is more important than the affinity of this interaction.

A once weekly PEG-filgrastim (SD/01) has recently been produced based on the findings with PEG-nartograstim describe above.[31] This molecule has a single PEG chain of 20 kDa molecular weight attached to the N-terminus of filgrastim.[32] In mice, SD/01 causes a similar prolongation of neutrophilia in mice following a single injection and has been shown to reduce the mortality in an experimental murine model of invasive *Candida* albicans infection.[33] In clinical trials, a single administration of SD/01 results in prolonged elevation of neutrophils in patients receiving chemotherapy[34] as well as in normal volunteers.[31] No dose limiting toxicities were seen. In the clinical studies, it was also demonstrated that the peripheral blood stem cell levels were elevated in the patients and volunteers to a comparable extent as with filgrastim injected daily at the same cumulative dose. A trial using a fixed dose of pegfilgrastim demonstrated equivalence to daily filgrastim and significantly reduced incidence of febrile neutropenia than found with filgrastim.[35]

## 4. Other PEGylated Hematopoietic Growth Factors

A number of other hematopoietic growth factors have been PEGylated, but most published studies describe work in preclinical models. One factor that has similar activity to G-CSF is granulocyte macrophage colony-stimulating factor (GM-CSF). It has not received the same wide spread use as G-CSF because it is more toxic. The increased toxicity is due to the fact that it is, as the name implies, less lineage specific and stimulates macrophages as well as granulocytes. However, this additional activity may provide a rationale for its use in immunotherapy as an adjuvant for cancer vaccines. For vaccines to be effective, it is important to have the antigens presented in the proper manner to the immune effector cells. Dendritic cells, which are the most effective antigen presenting cells, can be expanded *in vitro* by GM-CSF.[36] A number of studies in animals have demonstrated that the high GM-CSF levels greatly enhance the antitumor effects of cancer vaccines.[37,38] In a murine study, PEG-GM-CSF has been shown to expand a subpopulation of dendritic cells *in vivo*,[39] so could be an active adjuvant for cancer vaccines.

Another factor with clinical activity is megakaryocyte growth and development factor (MGDF). This is a truncated form of thrombopoietin. MGDF has been shown to enhance platelet recovery in mice following bone marrow transplantation and the PEGylated form has a single 20 kDa PEG chain attached to the amino terminus.[40] This modification helps stabilize the tertiary structure of the truncated protein and gives it approximately 3-fold more active than the unmodified version.[41] The half life *in vivo* is short compared to that of PEG-filgrastim described above, so that that daily dosing is required. Two phase I trials have been reported, one in cancer patients receiving moderately myelosuppressive chemotherapy[42] and the other in breast cancer patients receiving high-dose therapy with autologous peripheral blood stem cell support.[43] Both studies showed that the drug was well tolerated, even at the maximum doses tested of 5 and 10 μg/kg/d, respectively. A recent report,[44] demonstrated PEG-MGDF enhances platelet numbers in cancer patients receiving chemotherapy. Unfortunately, this molecule has not had a significant effect on platelet levels in patients receiving high dose therapy for leukemia or in those receiving stem cell support after chemotherapy.[40]

PEG-MGDF as a single agent mobilizes stem cells, but did not significantly enhance mobilization above that achieved by chemotherapy plus filgrastim in these patients.

Interleukin-6 (IL-6) is another thrombopoietic factor that has been PEGylated.[45] Because its toxicity is greater than that of PEG-MGDF, it seems unlikely that it will play a significant role in clinical practice.

## 5. PEGylated Lymphokines

Another set of cytokines that have been used in cancer treatment trials are called lymphokines because they are produced by lymphocytes and often act on lymphocytes. One, Interleukin-2 (IL-2), formerly known as T-cell growth factor has been used in a number of clinical studies, but it has not become part of standard clinical practice because of its high toxicity. This toxicity has limited its use primarily to major academic medical centers where there is more experience dealing with these toxicities. A phase I clinical trial demonstrated that a PEGylated IL-2 conjugate with 10–20-fold increased half-life had similar toxicities to those of the unmodified protein.[46] Direct injection of PEG-IL-2 into locoregional recurrent head and neck tumors appears to be less toxic than systemic administration, resulting in only temporary local swelling and redness. The response rate was not sufficient to justify additional studies. Interestingly, the one tumor that responded was small.[47] Low dose PEG-IL-2 has been used in patients with HIV with relatively low toxicity and objective stimulation of the immune system.[48,49] A potential therapeutic use for this molecule in cancer may be in minimal residual disease settings, or in combination with anticancer vaccines.

Tumor necrosis factor-$\alpha$ (TNF-$\alpha$), as the name implies, has direct cytotoxic effects on tumor cells. Initial studies in mice suggested that the human factor had little effect on normal cells. However, this was due to lack of species cross-reactivity for one type of TNF receptor, but not another.[50] In fact, TNF-$\alpha$ has considerable toxicity in patients. Preclinical studies with PEG-TNF-$\alpha$ conjugates prepared with different molecular weight PEG chains demonstrated that there was an optimal molecular size of 100–110 kDa for *in vitro* activity in mice.[51]

PEGylation of IFN-$\gamma$, a type II interferon produced by lymphocytes, has been tested in animals and is reported to have a lower affinity for its receptor but higher *in vivo* activity in rats,[52] which is similar to situation with PEG-G-CSF. This IFN has not found wide clinical use because its toxicities are similar to those of IL-2 in relation to potential therapeutic benefit.

## 6. PEGylated Cytokine Inhibitors

In addition to PEGylation of cytokines for therapeutic use, there are also efforts to develop PEGylated protein inhibitors of cytokine action. Because cytokines have powerful effects on the immune system, such as those described above for IL-2 and TNF, inhibition of their activity may be beneficial for indications such as toxic shock and tumor cachexia as well as chronic conditions like arthritis, asthma, Crohn's disease and autoimmune encephalomyelitis. These inhibitors are of two types: (1) soluble cytokine receptors, some of which are produced by cells and appear to be physiological inhibitors; (2) antibodies to the cytokines or to their receptors.

Soluble receptors against TNF, IL-1, IL-4, IL-5, and IFN- $\gamma$ have been studied in preclinical models and in some cases clinical trials.[53] A recombinant TNF receptor hybrid with the Fc portion of human immunoglobulin has been registered for use in rheumatoid arthritis. PEGylated TNF receptor constructs have been reported to be active in a variety of models[54,55] and might be useful for this indication.

Antibody reagents can be directed against the cytokine or the receptor itself. A PEGylated antibody against IL-8 has been prepared with potential application in adult respiratory distress syndrome.[55] An important anticancer antibody, trastuzamab (Herceptin™), is directed against the oncogene ERB-B2 (also called Her-2/Neu), a member of the epidermal growth factor receptor family that is overexpressed in about 30% of breast cancers as well as a number of other tumor types. Treatment of patients having ERB-B2 positive tumors with the humanized antibody trastuzamab leads to regression of the tumors, presumably because the antibody blocks the cell growth stimulus from the receptor and causes the cells to undergo programmed cell death (apoptosis). A PEGylated construct of Herceptin has been tested in animal models and has prolonged activity compared to the unmodified antibody.[56]

## 7. Conclusions and Outlook

After decades of repeated cycles of promises and failures, immunotherapy has finally lead to effective treatments for cancer. At this time, these have primarily come from antibody therapeutics such as trastuzamab and rituximab (Rituxan™), which is directed against the CD20 antigen expressed on some B-cell nonHodgkin's lymphomas.[57] The protein drugs that have the largest sales in the cancer market, however, are for support of standard toxic therapies, namely the hematopoietic growth factors.[58] Studies over the years with hematopoietic cells in culture have shown that the continuous presence of the protein is necessary for proliferation, differentiation, and even survival of these cells.[59–61] These factors all have short half-lives and thus, exogenous injection has limited effectiveness.

PEGylation can significantly extend the plasma half-lives these proteins, not only allowing less frequent administration, but also enhancing their efficacy. Until recently, most of the cytokine constructs had relatively small amounts of PEG attached so the increases in half-life were modest. Modifications of interferon and G-CSF with high molecular weight PEG have shown that such constructs can provide therapeutic benefits. One reason this approach was not taken previously was that constructs were initially selected on the basis of maintaining high activity in various *in vitro* assays. This can lead to selection against the most active forms *in vivo*. Therefore, following the experience with PEG-G-CSFs, it is likely that higher molecular weight constructs with decreased *in vitro* activities will be appearing in the future.

As our knowledge of the immune system increases, our ability to develop effective immunotherapies, including vaccines is enhanced. Hopefully the future will provide a range of agents that will help the body fight cancers as they form.

## References

1. Pfeffer LM, Dinarello CA, Herberman RB, Williams BR, Borden EC, Bordens R, Walter MR, Nagabhushan TL, Trotta PP and Pestka S (1998) Biological properties of recombinant alpha-interferons: 40th anniversary of the discovery of interferons. *Cancer Res* **58**:2489–2499.
2. Welte K, Gabrilove J, Bronchud MH, Platzer E and Morstyn G (1996) Filgrastim (r-metHuG-CSF): The first 10 years. *Blood* **88**:1907–1929.
3. Mehvar R (2000) Modulation of the pharmacokinetics and pharmacodynamics of proteins by polyethylene glycol conjugation. *J Pharm Pharm Sci* **3**:125–136.
4. Yamaoka T, Tabata Y and Ikada Y (1994) Distribution and tissue uptake of poly(ethylene glycol) with different molecular weights after intravenous administration to mice. *J Pharm Sci* **83**:601–606.

5. Kuwabara T, Kobayashi S and Sugiyama Y (1996) Kinetic analysis of receptor-mediated endocytosis of G-CSF derivative, nartograstim, in rat bone marrow cells. *Am J Physiol* **271**:E73–E84.
6. Tushinski RJ, Oliver IT, Guilbert LJ, Tynan PW, Warner JR and Stanley ER (1982) Survival of mononuclear phagocytes depends on a lineage-specific growth factor that the differentiated cells selectively destroy. *Cell* **28**:71–81.
7. Veronese FM (2001) Peptide and protein PEGylation: A review of problems and solutions. *Biomaterials* **22**:405–417.
8. Francis GE, Fisher D, Delgado C, Malik F, Gardiner A and Neale D (1998) PEGylation of cytokines and other therapeutic proteins and peptides: The importance of biological optimization of coupling techniques. *Int J Hematol* **68**:1–18.
9. Keating MJ, Holmes R, Lerner S and Ho DH (1993) L-asparaginase and PEG asparaginase-past, present and future. *Leukemia Lymphoma* **10**:153–157.
10. Ettinger LJ, Kurtzberg J, Voute PA, Jurgens H and Halpern SL (1995) An open-label, multicenter study of polyethylene glycol-L-asparaginase for the treatment of acute lymphoblastic leukemia. *Cancer* **75**:1176–1181.
11. Ho DH, Brown NS, Yen A, Holmes R, Keating M, Abuchowski A, Newman RA and Krakoff IH (1986) Clinical pharmacology of polyethylene glycol-L-asparaginase. *Drug Metab Dispos* **14**:349–352.
12. Gutterman JU (1994) Cytokine therapeutics — lessons from interferon alpha. *Proc Nat Acad Sci USA* **91**:1198–1205.
13. Lippman SM, Parkinson DR, Itri LM, Weber RS, Schantz SP, Ota DM, Schusterman MA, Krakoff IH, Gutterman JU and Hong WK (1992) 13-cis-retinoic acid and interferon alpha-2a — effective combination therapy for advanced squamous cell carcinoma of the skin. *J Nat Cancer Inst* **84**:235–241.
14. Lippman SM, Kavanagh JJ, Paredes-Espinoza M, Delgadillo-Madrueno F, Paredes-Casillas P, Hong WK, Holdener E and Krakoff IH (1992) 13-cis-retinoic acid plus interferon alpha-2a — highly active systemic therapy for squamous cell carcinoma of the cervix. *J Nat Cancer Inst* **84**:241–245.
15. Monkarsh SP, Ma Y, Aglione A, Bailon P, Ciolek D, DeBarbieri B, Graves MC, Hollfelder K, Michel H, Palleroni A, Porter JE, Russoman E, Roy S and Pan YC (1997) Positional isomers of monopegylated interferon alpha-2a: Isolation, characterization and biological activity. *Anal Biochem* **247**:434–440.
16. Nieforth KA, Nadeau R, Patel IH and Mould D (1996) Use of an indirect pharmacodynamic stimulation model of MX protein induction to compare *in vivo* activity of interferon alfa-2a and a polyethylene glycol-modified derivative in healthy subjects. *Clin Pharmacol Ther* **59**:636–646.
17. Bailon P, Palleroni A, Schaffer CA, Spence CL, Fung WJ, Porter JE, Ehrlich GK, Pan W, Xu ZX, Modi MW, Farid A, Berthold W and Graves M (2001) Rational design of a potent, long-lasting form of interferon: A 40 kDa

branched polyethylene glycol-conjugated interferon alpha-2a for the treatment of hepatitis C. *Bioconjug Chem* **12**:195–202.

18. Heathcote EJ, Shiffman ML, Cooksley WG, Dusheiko GM, Lee SS, Balart L, Reindollar R, Reddy RK, Wright TL, Lin A, Hoffman J and De Pamphilis J (2000) Peginterferon alfa-2a in patients with chronic hepatitis C and cirrhosis. *N Engl J Med* **343**:1673–1680.
19. Zeuzem S, Feinman SV, Rasenack J, Heathcote EJ, Lai MY, Gane E, O'Grady J, Reichen J, Diago M, Lin A, Hoffman J and Brunda MJ (2000) Peginterferon alfa-2a in patients with chronic hepatitis C. *N Engl J Med* **343**:1666–1672.
20. Motzer RJ, Rakhit A, Ginsberg M, Rittweger K, Vuky J, Yu R, Fettner S and Hooftman L (2001) Phase I trial of 40-kd branched pegylated interferon alfa-2a for patients with advanced renal cell carcinoma. *J Clin Oncol* **19**:1312–1319.
21. Wang YS, Youngster S, Bausch J, Zhang R, McNemar C and Wyss DF (2000) Identification of the major positional isomer of pegylated interferon alpha-2b. *Biochemistry* **39**:10634–10640.
22. Glue P, Fang JW, Rouzier-Panis R, Raffanel C, Sabo R, Gupta SK, Salfi M and Jacobs S (2000) Pegylated interferon-alpha2b: Pharmacokinetics, pharmacodynamics, safety and preliminary efficacy data. Hepatitis C Intervention Therapy Group. *Clin Pharmacol Ther* **68**:556–567.
23. Michallet M, Maloisel F, Delain M, Hellmann A, Rosas A, Silver RT and Tendler C (2004) Pegylated recombinant interferon alpha-2b vs recombinant interferon alpha-2b for the initial treatment of chronic-phase chronic myelogenous leukemia: A phase III study. *Leukemia* **18**:309–315.
24. Alvarado Y, Cortes J, Verstovsek S, Thomas D, Faderl S, Estrov Z, Kantarjian H and Giles FJ (2003) Pilot study of pegylated interferon-alpha 2b in patients with essential thrombocythemia. *Cancer Chemother Pharmacol* **51**:81–86.
25. Shepherd J, Brodin H, Cave C, Waugh N, Price A and Gabbay J (2004) Pegylated interferon alpha-2a and -2b in combination with ribavirin in the treatment of chronic hepatitis C: A systematic review and economic evaluation. *Health Technol Assess* **8**:iii–iv, 1–125.
26. Citron ML, Berry DA, Cirrincione C, Hudis C, Winer EP, Gradishar WJ, Davidson NE, Martino S, Livingston R, Ingle JN, Perez EA, Carpenter J, Hurd D, Holland JF, Smith BL, Sartor CI, Leung EH, Abrams J, Schilsky RL, Muss HB and Norton L (2003) Randomized trial of dose-dense versus conventionally scheduled and sequential versus concurrent combination chemotherapy as postoperative adjuvant treatment of node-positive primary breast cancer: First report of intergroup trial C9741/cancer and leukemia group B trial 9741. *J Clin Oncol* **21**:1431–1439.
27. Baumann I, Testa NG, Lange C, Dewynter E, Luft T, Dexter TM, Vanhoef MEHM and Howell A (1993) Haemopoietic cells mobilized into the circulation

by lenograstim as alternative to bone marrow for allogeneic transplants. *Lancet* **341**:369.

28. Kuga T, Komatsu Y, Yamasaki M, Sekine S, Miyaji H, Nishi T, Sato M, Yokoo Y, Asano M, Okabe M, Morimoto M and Itoh S (1989) Mutagenesis of human granulocyte colony stimulating factor. *Biochem Biophys Res Commun* **159**:103–111.
29. Bowen S, Tare N, Inoue T, Yamasaki M, Okabe M, Horii I and Eliason JF (1999) Relationship between molecular mass and duration of activity of polyethylene glycol conjugated granulocyte colony-stimulating factor mutein. *Exp Hematol* **27**:425–432.
30. Eliason JF, Greway A, Tare N, Inoue T, Bowen S, Dar M, Yamasaki M, Okabe M and Horii I (2000) Extended activity in cynomolgus monkeys of a granulocyte colony-stimulating factor mutein conjugated with high molecular weight polyethylene glycol. *Stem Cells* **18**:40–45.
31. Molineux G, Kinstler O, Briddell B, Hartley C, McElroy P, Kerzic P, Sutherland W, Stoney G, Kern B, Fletcher FA, Cohen A, Korach E, Ulich T, McNiece I, Lockbaum P, Miller-Messana MA, Gardner S, Hunt T and Schwab G (1999) A new form of Filgrastim with sustained duration *in vivo* and enhanced ability to mobilize PBPC in both mice and humans. *Exp Hematol* **27**:1724–1734.
32. Molineux G (2004) The design and development of pegfilgrastim (PEG-rmetHuG-CSF, Neulasta). *Curr Pharm Des* **10**:1235–1244.
33. van Spriel AB, van den Herik-Oudijk IE and van de Winkel JG (2000) A single injection of polyethylene-glycol granulocyte colony-stimulating factor strongly prolongs survival of mice with systemic candidiasis. *Cytokine* **12**:666–670.
34. Johnston E, Crawford J, Blackwell S, Bjurstrom T, Lockbaum P, Roskos L, Yang BB, Gardner S, Miller-Messana MA, Shoemaker D, Garst J and Schwab G (2000) Randomized, dose-escalation study of SD/01 compared with daily filgrastim in patients receiving chemotherapy. *J Clin Oncol* **18**:2522–2528.
35. Green MD, Koelbl H, Baselga J, Galid A, Guillem V, Gascon P, Siena S, Lalisang RI, Samonigg H, Clemens MR, Zani V, Liang BC, Renwick J and Piccart MJ (2003) A randomized double-blind multicenter phase III study of fixed-dose single-administration pegfilgrastim versus daily filgrastim in patients receiving myelosuppressive chemotherapy. *Ann Oncol* **14**:29–35.
36. Siena S, Di Nicola M, Bregni M, Mortarini R, Anichini A, Lombardi L, Ravagnani F, Parmiani G and Gianni AM (1995) Massive *ex vivo* generation of functional dendritic cells from mobilized CD34+ blood progenitors for anticancer therapy. *Exp Hematol* **23**:1463–1471.
37. Golumbek PT (1993) Controlled release, biodegradable cytokine depots — a new approach in cancer vaccine design. *Cancer Res* **53**:5841–5844.

38. Dranoff G, Jaffee E, Lazenby A, Golumbek P, Levitsky H, Brose K, Jackson V, Hamada H, Pardoll D and Mulligan RC (1993) Vaccination with irradiated tumor cells engineered to secrete murine granulocyte-macrophage colony-stimulating factor stimulates potent, specific, and long-lasting anti-tumor immunity. *Proc Nat Acad Sci USA* **90**:3539–3543.
39. Daro E, Pulendran B, Brasel K, Teepe M, Pettit D, Lynch DH, Vremec D, Robb L, Shortman K, McKenna HJ, Maliszewski CR and Maraskovsky E (2000) Polyethylene glycol-modified GM-CSF expands CD11b(high)CD11c(high) but notCD11b(low)CD11c(high) murine dendritic cells *in vivo*: A comparative analysis with Flt3 ligand. *J Immunol* **165**:49–58.
40. Kuter DJ and Begley CG (2002) Recombinant human thrombopoietin: Basic biology and evaluation of clinical studies. *Blood* **100**:3457–3469.
41. Molineux G, Hartley CA, McElroy P, McCrea C and McNiece IK (1996) Megakaryocyte growth and development factor stimulates enhanced platelet recovery in mice after bone marrow transplantation. *Blood* **88**:1509–1514.
42. Basser RL, Rasko JE, Clarke K, Cebon J, Green MD, Grigg AP, Zalcberg J, Cohen B, O'Byrne J, Menchaca DM, Fox RM and Begley CG (1997) Randomized, blinded, placebo-controlled phase I trial of pegylated recombinant human megakaryocyte growth and development factor with filgrastim after dose-intensive chemotherapy in patients with advanced cancer. *Blood* **89**:3118–3128.
43. Bolwell B, Vredenburgh J, Overmoyer B, Gilbert C, Chap L, Menchaca DM, Cruickshank S and Glaspy J (2000) Phase 1 study of pegylated recombinant human megakaryocyte growth and development factor (PEG-rHuMGDF) in breast cancer patients after autologous peripheral blood progenitor cell (PBPC) transplantation. *Bone Marrow Transplant* **26**:141–145.
44. Basser RL, Underhill C, Davis I, Green MD, Cebon J, Zalcberg J, MacMillan J, Cohen B, Marty J, Fox RM and Begley CG (2000) Enhancement of platelet recovery after myelosuppressive chemotherapy by recombinant human megakaryocyte growth and development factor in patients with advanced cancer. *J Clin Oncol* **18**:2852–2861.
45. Inoue H, Kadoya T, Kabaya K, Tachibana K, Nishi N, Sato M, Ohsawa M, Mikayama T and Mori KJ (1994) A highly enhanced thrombopoietic activity by monomethoxy polyethylene glycol-modified recombinant human interleukin-6. *J Lab Clin Med* **124**:529–536.
46. Meyers FJ, Paradise C, Scudder SA, Goodman G and Konrad M (1991) A phase I study including pharmacokinetics of polyethylene glycol conjugated interleukin-2. *Clin Pharmacol Ther* **49**:307–313.
47. Mattijssen V, De Mulder PH, De Graeff A, Hupperets PS, Joosten F, Ruiter DJ, Bier H, Palmer PA and Van den Broek P (1994) Intratumoral PEG-interleukin-2

therapy in patients with locoregionally recurrent head and neck squamous-cell carcinoma. *Ann Oncol* **5**:957–960.

48. Teppler H, Kaplan G, Smith KA, Montana AL, Meyn P and Cohn ZA (1993) Prolonged immunostimulatory effect of low-dose polyethylene glycol interleukin 2 in patients with human immunodeficiency virus type 1 infection. *J Exp Med* **177**:483–492.
49. Ramachandran R, Katzenstein DA, Winters MA, Kundu SK and Merigan TC (1996) Polyethylene glycol-modified interleukin-2 and thymosin alpha 1 in human immunodeficiency virus type 1 infection. *J Infect Dis* **173**:1005–1008.
50. Lütscher H, Steinmetz M and Lesslaucr W (1991) Tumor necrosis factor: Receptors and inhibitors. *Cancer Cells* **3**:221–226.
51. Tsutsumi Y, Kihira T, Tsunoda S, Kamada H, Nakagawa S, Kaneda Y, Kanamori T and Mayumi T (1996) Molecular design of hybrid tumor necrosis factor-alpha III: Polyethylene glycol-modified tumor necrosis factor-alpha has markedly enhanced antitumor potency due to longer plasma half-life and higher tumor accumulation. *J Pharmacol Exp Ther* **278**:1006–1011.
52. Kita Y, Rohde MF, Arakawa T, Fagin KD, Fish EN and Banerjee K (1990) Characterization of a polyethylene glycol conjugate of recombinant human interferon-gamma. *Drug Des Deliv* **6**:157–167.
53. Fernandez-Botran R (2000) Soluble cytokine receptors: Novel immunotherapeutic agents. *Expert Opin Invest Drugs* **9**:497–514.
54. Edwards CK (1999) PEGylated recombinant human soluble tumour necrosis factor receptor type I (r-Hu-sTNF-RI): novel high affinity TNF receptor designed for chronic inflammatory diseases. *Ann Rheum Dis* **58**(Suppl 1): 173–181.
55. Koumenis IL, Shahrokh Z, Leong S, Hsei V, Deforge L and Zapata G (2000) Modulating pharmacokinetics of an anti-interleukin-8 F(ab')(2) by amine-specific PEGylation with preserved bioactivity. *Int J Pharm* **198**:83–95.
56. Hurwitz E, Klapper LN, Wilchek M, Yarden Y and Sela M (2000) Inhibition of tumor growth by poly(ethylene glycol) derivatives of anti-ErbB2 antibodies. *Cancer Immunol Immunother* **49**:226–234.
57. Reichert J and Pavolu A (2004) Monoclonal antibodies market. *Nat Rev Drug Discov* **3**:383–384.
58. Pavlou AK and Reichert JM (2004) Recombinant protein therapeutics — success rates, market trends and values to 2010. *Nat Biotechnol* **22**:1513–1519.
59. Pélèraux A and Eliason JF (1989) Proliferation of single hemopoietic progenitor cells in the absence of colony-stimulating factors and serum. *Exp Hematol* **17**:1032–1037.

60. Kan O, Baldwin SA and Whetton AD (1994) Apoptosis is regulated by the rate of glucose transport in an IL-3-dependent haemopoietic cell line. *Biochem Soc Trans* **22**:275S.
61. Metcalf D and Merchav S (1982) Effects of GM-CSF deprivation on precursors of granulocytes and macrophages. *J Cell Physiol* **112**:411–418.

# 7

# Silencing Proteins: Nanotechnological Approaches to Deliver siRNA for Cancer Therapy

*Raymond M. Schiffelers, Daan J.A. Crommelin and Gert Storm*

Proteins play an important role in the development of a clinically relevant tumor. Malignant cells are formed in a multistage process in which genes become mutated. Two classes of genes, i.e. proto-oncogenes and tumor suppressor genes, are important candidates for these mutations. These mutated genes are translated into proteins with altered functionalities. The progeny of cells with such mutated proteins can give rise to a cluster of malignant cells.

Subsequently, production of a variety of proteinaceous growth factors, enzymes, adhesion molecules and receptors is responsible for the outgrowth of this small cluster into a clinically relevant tumor, through the stimulation of angiogenesis and cell adhesion. These processes are also implicated in the metastasis of tumor cells. Resistance of tumor cells against chemotherapeutics can arise through the expression of drug inactivating enzymes, drug binding proteins or pumps such as the multi-drug transporters belonging to the superfamily of the ATP-binding cassette proteins.

Silencing the production of representatives of these classes of proteins may be used in cancer treatment. In this chapter, we discuss the delivery of siRNA, as a new therapeutic entity to silence protein expression in tumor cells for cancer therapy.

## 1. Role of Proteins in Tumor Development

Proteins play an important role in the transformation of normal cells to cells with a malignant phenotype, and ultimately in the outgrowth of malignant cells into a clinically relevant tumor and their resistance against treatment. Initially, tumor cells arise in a multistage process in which genes become mutated. Two classes of genes, i.e. proto-oncogenes and tumor suppressor genes, are at the root of the transformation of normal to malignant cells.

### 1.1. *Proto-oncogenes*

In cancer, mutations in proto-oncogenes turn them into oncogenes, which encode proteins that mediate uncontrolled mitosis.[1–4] Examples of oncogene include growth factors (like platelet derived growth factor),[5] their receptors (like c-erbB2),[6] and cell signaling factors including kinases (like src, ras and raf),[7] and transcription factors (like jun, myc and fos).[8] The mutations that turns them into oncogenes, makes these proteins either constitutively active or promotes excessive production. Oncogenes carry dominant mutations causing a mutation in a single allele to predispose the cell to increased mitosis.

### 1.2. *Tumor suppressor genes*

Tumor suppressor genes are genes whose product can inhibit mitosis. An example is p53, which, by binding to a transcription factor known as E2F, can prevent a cell from completing the cell cycle when the cell has suffered (DNA) damage.[9,10] It is estimated that approximately 50% of all human cancers involve reduction in the activity of p53, either through mutation or inactivation of production through methylation of its promoter.[9] Tumor suppressor genes are recessive and require both alleles to be inactivated and/or mutated for a malignant phenotype to appear. As such, this class of proteins is difficult to target with siRNA for therapeutic purposes.

### 1.3. *Pro-angiogenic proteins/pro-metastatic proteins*

Once a cluster of malignant cells is formed, a variety of growth factors, adhesion molecules, receptors and enzymes is engaged in the outgrowth of this cluster into a clinically important tumor. New blood vessels are needed to support the growing tumor's increasing need for oxygen and nutrients.[11,12] The new vasculature is formed through angiogenesis by a local shift in the balance between anti-angiogenic (like platelet factor-4),[13] and pro-angiogenic factors, such as vascular endothelial growth factor (VEGF)[14] and basic fibroblast growth factor (bFGF).[15] These factors stimulate endothelial cells to proliferate, migrate, and form new capillaries. To enable endothelial cells to migrate, a variety of enzymes is synthesized or liberated to digest the extracellular matrix,[16] while integrins and a number of cell adhesion molecules mediate cell-cell and cell-matrix interactions.[17,18]

The newly developed vasculature may be exploited by tumor cells to metastasize. Metastasis is the process by which cancer cells migrate to other tissues. Mortality in cancer is often associated to be caused by metastasis of tumor cells to sites distant from the initial primary tumor site. In addition, tumor cells secrete a mixture of enzymes that can degrade the extracellular matrix and allow them to gain access to the blood stream or lymphatic vessels.[19,20] By attaching to endothelial cells at distant location, other tissues can be colonized.

### 1.4. *Proteins involved in drug resistance*

After diagnosis of a tumor, treatment is often complicated by the expression of specific proteins by tumor cells. ATP Binding Cassette (ABC) proteins are able to actively transport drugs from the cell interior to the exterior. These pumps thereby prevent the built-up of adequate therapeutic concentrations of chemotherapeutic agents inside the cancer cells.[21] In addition, tumor cells can synthesize a variety of proteins that metabolize or inactivate anticancer drugs.[22] Development of drug resistant cancer cells is an important cause of treatment failure in cancer.

## 2. RNA Interference

Cancer therapy can clearly benefit from approaches designed to interfere with protein production at various levels, from the initial malignant transformation of cells to inhibition of proteins that inactivate or hinder chemotherapeutics. The recent demonstration that RNA interference, a process originally identified in *Caenorhabditis elegans*,[23] can also be activated in mammalian cells may represent a promising strategy in this respect.

## 2.1. *Mechanism of RNA interference*

RNAi is a process that results in the silencing of protein expression. It is mediated by double-stranded (ds) RNA with a sequence that is homologous to an mRNA or DNA sequence.[24–26] In *C. elegans*, plants and fungi, the dsRNA can be several hundred base pairs long.[27] In mammalian cells, however, such long dsRNA activates a protein kinase response that results in a non-specific protein production shutdown.[28] Using smaller dsRNA avoids this response. Tuschl and colleagues demonstrated that 19–23 nucleotide long dsRNA, named small interfering (si)RNA, with 2 nucleotide overhangs on the 3′ positions can induce sequence-specific protein silencing in mammalian cells.[29–30]

In the post-transcriptional mode of action, siRNA associates with the RNA induced silencing complex (RISC) in the cytoplasm. In this complex of proteins, siRNA strands are separated and one strand binds to the homologous sequence in the mRNA. The mRNA is subsequently cleaved and consequently protein production is inhibited.[24,31] In the transcriptional mode of action, siRNA acts in the nucleus. Binding of an siRNA strand to CpG islands in the promoter region of a gene activates DNA methyltransferases, resulting in methylation and reduced transcription of the gene. Alternatively, lysine 9 residues in H3 histones may become methylated, which also reduces transcription.[32–34]

One of the most attractive properties of siRNA is its potency to induce gene silencing, compared with other nucleic acids designed to inhibit gene expression such as antisense oligonucleotides and ribozymes. Head-to-head comparisons with antisense oligonucleotides are difficult, as the sequences on mRNA that are the target for antisense or siRNA to achieve maximum inhibition do not overlap. However, in most evaluations, siRNA appears several orders of magnitude more potent than antisense oligonucleotides on a molar basis.[35–38]

In plants and *C. elegans*, systemic spreading of the silencing signal has been noted.[39] At present it is unclear how siRNA can leave the cell where it was generated or delivered to, pass the cell membrane of other cells and silence protein expression. A receptor called SID-1 has been proposed as a receptor for the internalization of dsRNA.[40] It is unknown whether SID-1 homologues are present on mammalian cells and could act as shuttles to transport siRNA over the cell membrane.

## 2.2. *Structural requirements for siRNA*

Several of the structural requirements for successful siRNA molecules have been identified. siRNAs preferably consist of 19 nucleotide duplexes with 2 nucleotide overhangs at the 3′ position.[29,30] The overhang sequence can include mismatches and even deoxynucleotides (like the popular two thymidine overhang), which are thought to be more resistant to nuclease degradation. Replacement of either strand of the siRNA with a strand composed completely of deoxynucleotides abolishes silencing efficacy. Generally, targeting regions within 50–100 bp of the start codon and the terminating codon are avoided, as well as intron sequences. In addition, due to possible stacking of 4 or more of the same bases in a row, these stretches should be avoided as well. As the siRNA strands need to separate, the G-C content of the strands is highly important, G-C contents below 30% and above 60% are suboptimal, especially at the end where the strands are separated first.[41–45]

Reynolds and colleagues have developed a scoring system by which the sequences that fulfill the requirements as described above can be assigned a preference.[46] The scoring system was developed after analyzing the silencing efficiency of 180 siRNAs targeting two mRNAs. The scoring system outlined in Table 1 was obtained.

The tolerance for mismatches is unclear. Some studies have reported that single base mismatches annihilate siRNA-mediated silencing,[47,48] while others have reported that loss of silencing activity is only reached after the introduction of 7 mismatches.[49–51] It seems that mismatches in the center of the duplex are less well tolerated than mismatches closer to the end of the strands. The sequence of the target mRNA also appears to play a role, which could serve as an explanation for the discrepancy between the experimental results.[52–55]

Chemically synthesized siRNA offers precise control over the sequence and amount of siRNA used in the experiment, the possibility for molecular

**Table 1** Scoring system according to Reynolds *et al.* listing the criteria and scoring of siRNA sequences. Higher cumulative scores indicate an increased likelihood of good gene silencing.

| Criterion | Score |
|---|---|
| Moderate to low (30–52%) GC-content | 1 |
| At least 3 A/Us at positions 15–19 of the sense strand | 1 per A or U |
| Lack of internal repeats | 1 |
| A at position 19 (sense) | 1 |
| A at position 3 (sense) | 1 |
| U at position 10 (sense) | 1 |
| G/C at position 19 (sense) | −1 |
| G at position 13 (sense) | −1 |

characterization of the siRNA and the ability to introduce modifications for siRNA-labeling or siRNA-stability. As an alternative, siRNA may be intracellularly produced from encoding DNA using polymerase III promoters, like U6 or H1.[56–58] For plasmid transfection, puromycin-resistance can be included to achieve stable expression, which has been reported to last for two months.[59] The two strands may be produced separately or may be connected via a nucleotide loop, and this type is known as short hairpin (sh)RNA. The stem is formed by 19 nucleotide inverse repeats, and the loop is formed by a critically unimportant number of nucleotides, although 9 seems to be optimal.[60] The 3′ overhang is generally composed of 4 uridine nucleotides. shRNA is enzymatically intracellularly cleaved to yield siRNA.[61] The fact that the two strands are already linked in shRNA is probably the explanation for its enhanced effectiveness, compared with the production of the two separate strands which still need to anneal. The main advantage of DNA-encoded siRNA is that cellular exposure to siRNA can take place for prolonged periods of time, especially important for proteins with a long intracellular half-life.

The various classes of proteins described in the previous section that could be silenced for cancer therapy are evaluated in Table 2, regarding their usefulness as siRNA targets. Two characteristics are important for the feasibility of an siRNA-based therapy: the number of cells that need to be transfected, and the need for additional treatment. Most of the target proteins are expressed in each individual tumor cell, and

**Table 2** Protein targets for siRNA in cancer therapy.

| Protein Class | Target Cell Type | Additional Treatment Needed | Comments |
|---|---|---|---|
| Oncogenes | Tumor cells | No, dominant mutation | siRNA needs to be delivered to all tumor cells |
| Tumor suppressor genes | Tumor cells | | Probably not suitable for siRNA treatment |
| Pro-angiogenic factors | Tumor cells, tumor associated cells, endothelium | No | SiRNA needs to be delivered to enough cells to shift the balance to anti-angiogenesis |
| Pro-metastatic factors | Tumor cells, tumor associated cells, endothelium | Yes, does not necessarily affect primary tumor | Will reduce the likelihood of tumor cell colonization of other organs |
| ABC-transporters | Tumor cells | Yes, chemotherapy | All tumor cells need to be transfected |
| Drug-inactivating proteins | Tumor cells | Yes, chemotherapy | All tumor cells need to be transfected |

they need to be knocked down in each cell to eradicate the tumor. An exception may be the treatment based on inhibition of pro-angiogenic factors. In this case, knocking down production of pro-angiogenic factors in a sufficient number of cells may result in a shift in the balance between pro- and anti-angiogenic mediators towards inhibition of the process.

In addition, many siRNA-based treatment strategies require additional treatment for a therapeutic effect to be observed. For pro-metastatic proteins, inhibition of the spreading of the tumor cells is the main goal, while still requiring alternative treatment for the primary tumor. Silencing of the transporter and drug inactivating proteins do no have a therapeutic effect in themselves and require chemotherapeutic support for their action to become beneficial. Only repression of oncogenes and pro-angiogenic factors may result in direct antitumor effects.

## 3. Approaches for siRNA Delivery

The previous sections deal with the usefulness of siRNA as an anticancer therapeutic. Nevertheless, an important hurdle for siRNA application in cancer therapy remains that of the delivery of siRNA to the target cell type. The pharmacokinetic profile of intravenously administered siRNA shows a fast disappearance from the blood stream.[62] The major organs accounting for this disappearance appear to be the liver and kidneys, leading to important loss of the amount of siRNA available to the target cells (Fig. 1).

To reduce clearance by non-target tissues, many investigators employ cationic carriers for siRNA delivery.[63–70] The carrier forms an electrostatic complex with siRNA, thereby increasing the apparent molecular weight. This avoids glomerular filtration and transport by the "oligonucleotide" channel to the periglomerular space in the brush border of the renal cortex.[71] These complexes should remain intact in biological milieus. Presence of ions and anionic proteins *in vivo* may cause complex dissociation. In addition, although siRNA seems remarkably stable for an RNA molecule, degradation by nucleases in biological fluids may reduce the amount of active siRNA.[41,72–74]

Many applications require the complex to arrive at a target cell type that lies beyond the endothelial cell layer. The continuous endothelial cell layer may form an important barrier especially after intravenous administration. Fortunately, complexes may extravasate in solid tumors through defects

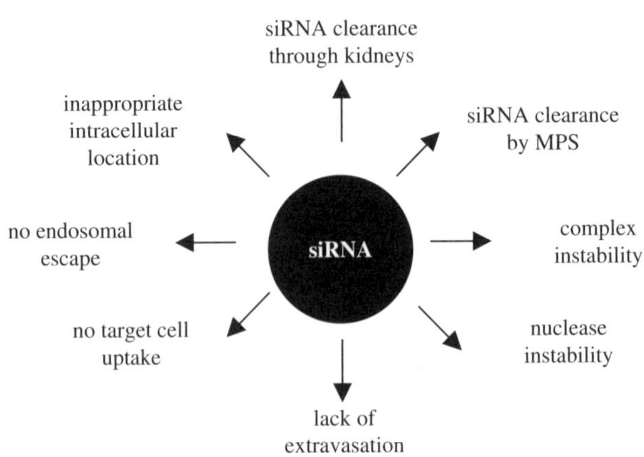


**Fig. 1** Factors that may prevent siRNA from silencing protein production.

present in the endothelium of newly formed blood vessels, but that requires the size of the complex to be approximately 100 nm.[75,76] Finally, the complex needs to be taken up by the target cell type to escape from degradation in the endosome, and to escape over the endosomal membrane, and reach its appropriate intracellular location.

For DNA-based siRNA delivery, many of the requirements and limiting factors are similar. Again, for non-viral DNA delivery, DNA is often complexed but this is primarily done to reduce the hydrodynamic radius of DNA molecules and the negative charge.[77–79] The only additional requirement is that DNA needs to arrive inside the nucleus to be transcribed. This additional passage of the nuclear membrane is an important obstacle for efficient gene transfection.[80–82]

For viral DNA-based siRNA delivery, the efficiency of the process after arriving in the target cell is usually not the limiting factor, nor is carrier stability or nuclease stability an important issue.[83–86] However, viruses usually lack selectivity for the target cell type. Furthermore, inflammatory reactions, immunogenicity, and oncogenic transformations are important safety considerations for viral vectors that need to be addressed.[87–89]

## 4. siRNA Delivery for Cancer Therapy

### 4.1. *siRNA to silence oncogenes*

Most studies on siRNA silencing of oncogenes are performed with *in vitro* cultured cells. These studies circumvent the need for extensive optimization of the delivery step, needed for *in vivo* experiments. *In vivo* studies have been described that employ frequent local injections of siRNA into the tumor or use *ex vivo* transfection of tumor cells followed by implantation. An example of the first approach is the inhibition of the product of the fms-oncogene colony, stimulating factor-receptor and its ligand CSF-1.[90] CSF-1 is thought to be important in the recruitment of macrophages to sites of malignancies, thereby promoting tumor angiogenesis and metastasis. Intratumoral injection of 10 μg siRNA against CSF-1 or the CSF-1 receptor every three days reduced MCF7 mammary tumor growth in nude mice by 40–45%, and silenced target protein expression in tumor lysates. These observations coincided with reduced macrophage infiltration, matrix metalloproteinase and vascular endothelial growth factor A expression and endothelial cell proliferation, resulting in a more than doubled survival time. Nevertheless, frequent local injection or *ex vivo* transfection and implantation are approaches that seem to lack clinical applicability and are primarily useful for proof-of-concept studies.

A study by Kuroda *et al.* describes the use of atelocollagen to deliver siRNA against hWAPL. hWAPL was identified as an oncogen activated by E6/E7 oncoproteins from human papillomavirus. SiHa cervical cancer cells were implanted into nude mice. Silencing of hWAPL by 10 pmol siRNA delivered, mixed with the 2% atelocollagen gel, provided strong inhibition of tumor growth. The siRNA-gel formed by atelocollagen has been reported to increase cellular uptake, nuclease resistance, and prolonged release of siRNA in tumor tissue. The use of the atelocollagen-carrier reduced the number of injections required more than 5-fold, compared with free siRNA.[91]

Another complexing agent was studied by Chien *et al.*[92] They utilized cationic liposomes based on a new cardiolipin analog. The transfection efficiency in mice of reporter plasmids was seven-fold higher than that of the conventional DOTAP-based liposome formulations while showing less toxicity after local administration. In addition, complexation of siRNA against the oncogene c-raf followed by intratumoral injection inhibited tumor growth with 73% in SCID mice bearing human breast xenograft tumors, compared with the group treated with free c-raf siRNA. It is clear that for broader clinical applications, intravenous administration would be preferred. Recently, the same c-raf-siRNA cardiolipin system was injected intravenously in tumor bearing mice. Mice received approximately 150 μg siRNA every 12 hr for 5 days resulting in 50% tumor growth inhibition. "Naked" siRNA failed to inhibit tumor growth.[92,93] It may be anticipated that shielding of the cationic particle may prevent non-target tissue interactions and may increase siRNA accumulation at the target site.

Urban-Klein *et al.* employed one of the longest-standing cationic polymers for nucleic acid delivery: polyethylene imine.[94] In this study, the low molecular weight variant of branched PEI was used. Complexation with PEI promoted siRNA stability and delivery into cells *in vitro*. *In vivo,* intraperitoneal administration was investigated as a route of administration for delivery of siRNA to subcutaneously implanted tumors in mice. Interestingly, only complexed siRNA was shown to be delivered intratumorally, and to result in sequence-specific gene silencing in contrast to "naked" siRNA. These results show that simple cationic complexation may improve delivery to distant sites. The PEI-complexed siRNA at a dose of 60 pmol silenced c-erbB2/neu receptor expression, as evidenced by immunohistochemical staining, was shown to inhibit tumor growth. Again, the cationic complex may be improved by reducing binding to non-target tissues, to improve circulation time and thereby tumor delivery.

The group of Pardridge described a carrier system for intravenous delivery of plasmid-encoded shRNA to silence epidermal growth factor receptor (EGFR) for treatment of brain cancer. Their approach is based on PEGylated immunoliposomes. The lipids are used to encapsulate the nucleic acid. The resulting anionic lipidic particles are shielded by a PEG coating. On the distal end of approximately 1–2% of these PEG-chains, monoclonal antibodies to the transferrin receptor are coupled. As the transferring receptor is expressed on the endothelial cells that form the blood brain barrier and can be expressed on the membrane of brain tumor cells, the monoclonal antibody is thought to promote transport of the carrier across both barriers. Initial studies showed knockdown of luciferase reporter gene expression in a C6 rat glioblastoma model by 90%, persisting for at least 5 days after a single intravenous dose of 10 μg plasmid per rat. Subsequent studies showed that treatment of SCID mice inoculated intracranially with human brain cancer cells was extended by nearly 90% when receiving weekly intravenous injections of the PEGylated immunoliposomes carrying 5 μg plasmids that express shRNA silencing human epidermal growth factor receptor. An additional target demonstrated to be useful for transcytosis of particles across the blood brain barrier is the insulin receptor.[95–97]

Other groups have explored the use of viruses for delivery of DNA encoding shRNA/siRNA. One area of research focuses on inhibition of point mutations and fusion genes. The use of siRNA in this respect is highly interesting, as wild type alleles could be spared as a result of the exquisite specificity of RNAi. The oncogene Tpr-Met is a DNA rearrangement leading to a fusion protein that can be specifically silenced with siRNA, by targeting the junction region leaving the wild type alleles unaffected. When delivered by a lentiviral vector, Tpr-Met oncogene siRNA was effective in specifically suppressing the expression of the mutated form in tumor cells preventing tumorigenesis.[98]

The Met-oncogene was also the target protein of studies by Shinomyia *et al.* They designed an adenovirus for expression of shRNA against Met to disturb the hepatocyte growth factor/scatter factor-Met signaling pathway. Silencing of Met resulted in inhibition of the expression of hepatocyte growth factor/scatter factor inhibiting tumor cell growth and tumor cell viability. Inhibition of the pathway also promoted apoptosis in several tumor cell lines, which was not stimulated in non-malignant cells. *In vivo*, intratumoral infection with $4 \times 10^7$ infectious units of Met-siRNA producing adenovirus significantly reduced tumor growth by approx. 50%.[99]

Survivin, a protein that belongs to the relatively new family of inhibitor of apoptosis proteins (IAPs), plays a regulating role in cell division. DNA encoding siRNA silencing survivin was delivered using an adenoviral vector. Transduction of cancer cells was shown to result in effective gene knockdown effect resulting in apoptosis in a variety of cancer cell lines. Transduced cancer cells were markedly less tumorigenic *in vivo*. In a therapeutic setting, the adenoviral vector was locally injected at a dose of $10^{10}$ virus particles/tumor in a U251 glioma model, and markedly suppressed tumor growth by approximately 70%.[100]

Infection with human herpesvirus-8 may lead to several cancers, like Kaposi sarcoma and primary effusion lymphoma. The virus has several genes that share homology with host genes that help the virus to evade immune responses, prevent cell cycle shutdown and interrupt activation of pro-apoptotic pathways. Silencing of several of these genes at the same time, was shown to promote apoptosis in primary effusion lymphoma cell lines. In an immunodeficient mouse model of effusion lymphoma, lentiviral-mediated RNA interference at a dose of $10^8$ particles per mouse injected i.p. both inhibited development of ascites, and was shown to be an effective treatment for established ascites significantly enhancing mouse survival. Interestingly, the i.p. injected lentiviral vector remained confined to the peritoneal cavity, which is an important safety consideration for viral vectors.[101]

## 4.2. *siRNA to silence mutated tumor suppressor genes*

Loss of the activity of tumor suppressor genes may promote tumor formation. Not surprisingly, silencing of the expression of these proteins is not an option for cancer treatment. However, RNAi may be in this context very useful for dissecting molecular pathways of tumor suppressor genes, like for p53,[102–107] and identifying new tumor suppressor genes like Fbxw7/Cdc4.[108]

## 4.3. *siRNA to silence pro-angiogenic/pro-metastatic genes*

Angiogenesis and metastasis are two intimately associated processes. Therapeutic approaches that are aimed at inhibition of one of these processes often also affect the other.[109,110] Therapeutic intervention with siRNA for these processes has been described. Gondi *et al.* used local injection of plasmids by minipumps in he brain to achieve intracranial production of siRNA.

In a study using DNA-encoding shRNA, Gondi *et al.* silenced metalloproteases uPAR and cathepsin B.[111] In a number of *in vitro* angiogenesis studies silencing of uPAR and cathepsin B were shown to inhibit angiogenic responses of endothelial cells. Moreover, silencing both proteases induced stronger inhibitor effects than silencing each of the enzymes alone. In a therapeutic setting, mice bearing established SNB19 tumor cells intracranially were treated with the plasmid vector. Plasmid was infused in the brain using osmotic mini pumps at a rate of 0.25 μl/h and a total dose of 150 μg. Inhibition of each of the enzymes alone induced regression of tumors to approximately 50% of original tumor volume. More importantly, combination of silencing of both enzymes eliminated the tumor completely.

Two groups made use of intravenous injection of "naked" siRNA. Duxbury *et al.* injected siRNA designed to silence carcinoembryonic antigen-related cell adhesion molecule 6 (CEACAM6), a protein associated with human gastrointestinal cancer.[112] Intravenous injection of 150 μg CEACAM6-specific siRNA per kg body weight twice weekly strongly suppressed tumor growth rate and inhibited angiogenesis. Interestingly, CEACAM6-siRNA completely prevented the occurrence of metastases and thereby enhanced survival. Treatment was not associated with apparent toxicity.

A similar approach was followed by Liang *et al.*[113] They also injected "naked" small interfering RNA duplexes intravenously, however they used a pool of two duplexes designed to silence the same protein, i.e. CXCR4. The pool of siRNAs was used to target multiple sites of the target gene, thereby molecular variants of the same gene with similar functional roles can be inhibited simultaneously to avoid these variants taking over the role of the inhibited variant. The pool of siRNAs was shown to inhibit invasion of breast cancer cells in an *in vitro* Matrigel assay. At a dose of 500 μg/kg administered intravenously twice weekly, CXCR4-siRNA also inhibited metastasis *in vivo* in mice.

These results of the effects of "naked" siRNA are difficult to interpret in view of the physicochemical characteristics of siRNA and its rapid clearance, within minutes, from the circulation. It could be that aspecific immunostimulatory reactions to siRNA administration occur, due to its dsRNA nature.[114] Such reactions have been reported, but that does not account for the sequence-specific knockdown of the target protein reported in these studies. Alternatively, endogenous transport mechanisms may exist to allow siRNA enter cells relatively efficiently, similar to that described in plants, fungi and *C. elegans*.

Other groups have employed carrier systems for siRNA delivery. Sphingosine 1-phosphate receptor 1 (S1P1) is implicated in the stabilization of newly formed vasculature in embryos. Its overexpression has also been demonstrated in tumor vessels. S1P1 was targeted using multiple siRNAs specific for different regions of the mRNA transcript.[115] Cationic liposomes based on 1,2-dioleoyl-3-trimethylammonium-propane (DOTAP) were used to complex siRNA and to augment cell interaction. siRNA silencing S1P1 inhibited endothelial cell migration *in vitro* and infiltration of Matrigel *in vivo*. In mice bearing Lewis lung carcinoma tumors, siRNA complexed to cationic liposomes containing 15 μg siRNA was injected every third day. Tumors exhibited an approximately 50% reduction in tumor growth rate with reduction in vascular density, quantified by CD31 staining.

In our own studies, we have used complexed, shielded and targeted nanoparticles for siRNA delivery to inhibit angiogenesis.[116,117] Self-assembling nanoparticles with siRNA were based on polyethyleneimine that is shielded with a PEG-coating. The PEG-coating was functionalized with an RGD-peptide ligand coupled to the distal end of the polyethylene glycol, to target alpha v-integrins overexpressed on the tumor endothelial cells. The resulting nanoparticles of approximately 90 nm were used on complex siRNAs designed to silence vascular endothelial growth factor receptor-2 expression, a crucial factor in the onset of tumor angiogenesis. Intravenous administration of 40 μg complexed siRNA into tumor-bearing mice every three days resulted in sequence-specific inhibition of VEGFR2-protein expression and inhibition of tumor growth rate. Inhibition was associated with reduced vascularization in the tumor and erratic branching of remaining vasculature. The same system also proved to be successful in a model of angiogenesis caused by herpetic stromal keratitis in the eye.

### 4.4. *siRNA to silence genes that increase tumor drug resistance*

The M2 subunit of ribonucleotide reductase determines the activity of the enzyme and its overexpression is associated with gemcitabine resistance in pancreatic adenocarcinoma.[118,119] Silencing of expression of ribonucleotide reductase M2 subunit with siRNA enhances gemcitabine activity *in vitro* in tumor cells. Intravenously administered siRNA at a dose of 150 μg/kg was able to silence enzyme expression and show synergistic antitumor activity, when co-administered with gemcitabine in an orthotopic model of pancreatic adenocarcinoma in nude mice and prevented metastasis.

## 5. Concluding Remarks

siRNA has made a remarkably rapid advancement from its discovery as the functional mediator of RNA interference to feasibility as *in vivo* therapeutic. Its potency and selectivity have already made it to the agent of choice for gene function validation studies *in vitro*. The availability of DNA-encoded siRNA-vectors that produce siRNA for a prolonged period of time and the development of conditional vectors are important advancements facilitating the study of protein function and tumor cell behavior after implantation in animals.[120–123]

For rational design of *in vivo* therapeutic approaches, several hurdles remain, of which efficient target cell delivery seems amongst the most prominent. Five strategies for delivery of siRNA can be distinguished.

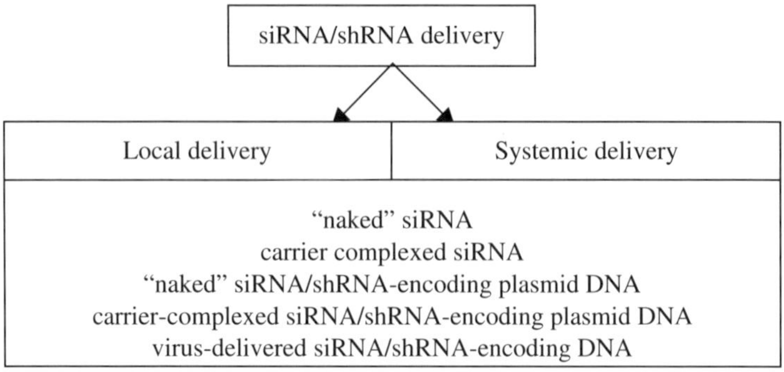


The delivery strategies for (virally or non-virally delivered) DNA-encoding siRNA or shRNA are similar to those described for gene therapy.[124–128] The same issues (target cell binding and uptake, endosomal escape, nuclear transport for non-viral systems vs. safety, oncogenicity, immunogenicity, and target cell delivery for viruses) are encountered for endogenously synthesized siRNA/shRNA as well.

From a delivery viewpoint, the use of (chemically synthesized) siRNA has several advantages compared to siRNA-encoding DNA. These include smaller size and cytosolic site of action. The latter advantage avoids the need for passage of the nuclear membrane. Local delivery of siRNA can have important roles in proof-of-concept studies, but broad clinical applicability is questionable. The attractiveness of the local delivery approach is of course the delivery of the entire dose to the target tissue which improves gene silencing. In contrast to this intuitive advantage, one study reported silencing effects of systemically administered (i.v., i.p. or s.c.) "naked" siRNA in tumor tissue, whereas intratumoral injection failed to silence the luciferase reporter gene.[129] These results are difficult to interpret, especially when considering *in vitro* observations that indicate that "naked" siRNA is only able to induce gene silencing at very high concentrations. These observations may indicate that physical presence at the target site is but one of the factors that need to be addressed to achieve successful RNA interference.

Other studies do report silencing effects of locally injected "naked" siRNA, which would not have been expected based on the *in vitro* observations.[90] It could be that local injection creates a hydrodynamic pressure, which promotes target cell entry.[130–133] Alternatively, siRNA interactions with proteins in the local *in vivo* milieu may facilitate cellular uptake. In accordance with the *in vitro* studies, some studies have shown that siRNA silencing can be improved by complexation to cationic agents, likely through protection against degradation and improvement of cellular uptake.[91,92,94]

After systemic administration of siRNA, kidney and liver take up > 90% of the injected dose, within minutes after injection.[62] Nevertheless, injection of "naked" siRNA has been successfully employed as *in vivo* therapeutic approach.[112,113] This would indicate that only minute quantities of siRNA need to arrive at the target site and, upon arrival, these low amounts are apparently quite efficiently internalized by the target cell type leading to functional gene silencing. It could be that proteinaceous factors exist that enable the siRNA-silencing signal to "travel". Such mechanisms have been described for plants and *C. elegans*[39,134] but suitable candidates for mammalian cells are thus far lacking.

Although injection of "naked" siRNA may seem an attractive, simple solution to the delivery problem, research on antisense has shown that nucleic acids can induce a variety of non-specific effects that can effectuate dramatic therapeutic effects.[135–137] Similar effects may be induced by siRNA which, through its dsRNA character, may have strong immunomodulatory properties.[138,139]

Some, more sophisticated delivery strategies, make use of cationic complexing agents that condense nucleic acids into a net positively charged particle to protect siRNA and to increase cellular uptake.[97,116,117] To prevent promiscuous charge-based interactions, these particles can be shielded with a polymer coating providing steric stabilization and making the particle relatively inert. By coupling a targeting ligand to this polymer layer, site-selective activity is obtained. As siRNA silencing effects should only occur in intended target cells, targeted delivery systems still offer the highest chance of achieving efficient sequence-specific siRNA-mediated gene silencing.

## References

1. Altanerova V (2001) Cancers connected with mutations in RET proto-oncogene. *Neoplasma* **48**:325–331.
2. Dent AL, Vasanwala FH and Toney LM (2002) Regulation of gene expression by the proto-oncogene BCL-6. *Crit Rev Oncol Hematol* **41**:1–9.

3. Dittmer J (2003) The biology of the Ets1 proto-oncogene. *Mol Cancer* **2**:29.
4. Roussel MF and Sherr CJ (2003) Oncogenic potential of the c-FMS proto-oncogene (CSF-1 receptor). *Cell Cycle* **2**:5–6.
5. Billie Au PY, Martin N, Chau H, Moemeni B, Chia M, Liu FF, Minden M and Yeh WC (2005) The oncogene PDGF-B provides a key switch from cell death to survival induced by TNF. *Oncogene* **24**(19):3196–3205.
6. Wu JT (2002) C-erbB2 oncoprotein and its soluble ectodomain: A new potential tumor marker for prognosis early detection and monitoring patients undergoing herceptin treatment. *Clin Chim Acta* **322**:11–19.
7. Porter AC and Vaillancourt RR (1998) Tyrosine kinase receptor-activated signal transduction pathways which lead to oncogenesis. *Oncogene* **17**:1343–1352.
8. Hoffman B, Amanullah A, Shafarenko M and Liebermann DA (2002) The proto-oncogene c-myc in hematopoietic development and leukemogenesis. *Oncogene* **21**:3414–3421.
9. Selivanova G (2004) p53: Fighting cancer. *Curr Cancer Drug Targets* **4**:385–402.
10. Vousden KH and Prives C (2005) P53 and prognosis: New insights and further complexity. *Cell* **120**:7–10.
11. Blagosklonny MV (2004) Antiangiogenic therapy and tumor progression. *Cancer Cell* **5**:13–17.
12. Jain RK (2005) Normalization of tumor vasculature: An emerging concept in antiangiogenic therapy. *Science* **307**:58–62.
13. Bikfalvi A and Gimenez-Gallego G (2004) The control of angiogenesis and tumor invasion by platelet factor-4 and platelet factor-4-derived molecules. *Semin Thromb Hemost* **30**:137–144.
14. Hicklin DJ and Ellis LM (2005) Role of the vascular endothelial growth factor pathway in tumor growth and angiogenesis. *J Clin Oncol* **23**:1011–1027.
15. Dow JK and deVere White RW (2000) Fibroblast growth factor 2: Its structure and property, paracrine function, tumor angiogenesis, and prostate-related mitogenic and oncogenic functions. *Urology* **55**:800–806.
16. Handsley MM and Edwards DR (2005) Metalloproteinases and their inhibitors in tumor angiogenesis. *Int J Cancer*.
17. Hwang R and Varner J (2004) The role of integrins in tumor angiogenesis. *Hematol Oncol Clin North Am* **18**:991–1006, vii.
18. Szekanecz Z and Koch AE (2004) Vascular endothelium and immune responses: Implications for inflammation and angiogenesis. *Rheum Dis Clin North Am* **30**:97–114.
19. Stacker SA, Williams RA and Achen MG (2004) Lymphangiogenic growth factors as markers of tumor metastasis. *Apmis* **112**:539–549.
20. Gassmann P, Enns A and Haier J (2004) Role of tumor cell adhesion and migration in organ-specific metastasis formation. *Onkologie* **27**:577–582.

21. Chang G (2003) Multidrug resistance ABC transporters. *FEBS Lett* **555**: 102–105.
22. Longley DB and Johnston PG (2005) Molecular mechanisms of drug resistance. *J Pathol* **205**:275–292.
23. Fire A, Xu S, Montgomery MK, Kostas SA, Driver SE and Mello CC (1998) Potent and specific genetic interference by double-stranded RNA in Caenorhabditis elegans. *Nature* **391**:806–811.
24. Mello CC and Conte D Jr (2004) Revealing the world of RNA interference. *Nature* **431**:338–342.
25. Agrawal N, Dasaradhi PV, Mohmmed A, Malhotra P, Bhatnagar RK and Mukherjee SK (2003) RNA interference: Biology, mechanism, and applications. *Microbiol Mol Biol Rev* **67**:657–685.
26. Scherr M, Morgan MA and Eder M (2003) Gene silencing mediated by small interfering RNAs in mammalian cells. *Curr Med Chem* **10**:245–256.
27. Cogoni C and Macino G (2000) Post-transcriptional gene silencing across kingdoms. *Curr Opin Genet Dev* **10**:638–643.
28. Clemens MJ and Elia A (1997) The double-stranded RNA-dependent protein kinase PKR: Structure and function. *J Interferon Cytokine Res* **17**:503–524.
29. Elbashir SM, Harborth J, Lendeckel W, Yalcin A, Weber K and Tuschl T (2001) Duplexes of 21-nucleotide RNAs mediate RNA interference in cultured mammalian cells. *Nature* **411**:494–498.
30. Elbashir SM, Lendeckel W and Tuschl T (2001) RNA interference is mediated by 21- and 22-nucleotide RNAs. *Genes Dev* **15**:188–200.
31. Tijsterman M and Plasterk RH (2004) Dicers at RISC; the mechanism of RNAi. *Cell* **117**:1–3.
32. Kawasaki H, Taira K and Morris KV (2005) siRNA induced transcriptional gene silencing in mammalian cells. *Cell Cycle* **4**.
33. Kawasaki H and Taira K (2004) Induction of DNA methylation and gene silencing by short interfering RNAs in human cells. *Nature* **431**:211–217.
34. Morris KV, Chan SW, Jacobsen SE and Looney DJ (2004) Small interfering RNA-induced transcriptional gene silencing in human cells. *Science* **305**: 1289–1292.
35. Miyagishi M, Hayashi M and Taira K (2003) Comparison of the suppressive effects of antisense oligonucleotides and siRNAs directed against the same targets in mammalian cells. *Antisense Nucleic Acid Drug Dev* **13**:1–7.
36. Bertrand JR, Pottier M, Vekris A, Opolon P, Maksimenko A and Malvy C (2002) Comparison of antisense oligonucleotides and siRNAs in cell culture and *in vivo*. *Biochem Biophys Res Commun* **296**:1000–1004.
37. Bilanges B and Stokoe D (2005) Direct comparison of the specificity of gene silencing using antisense oligonucleotides and RNAi. *Biochem J* **388**:573–583.

38. Grunweller A, Wyszko E, Bieber B, Jahnel R, Erdmann VA and Kurreck J (2003) Comparison of different antisense strategies in mammalian cells using locked nucleic acids, 2′-O-methyl RNA, phosphorothioates and small interfering RNA. *Nucleic Acids Res* **31**:3185–3193.
39. May RC and Plasterk RH (2005) RNA interference spreading in C. elegans. *Methods Enzymol* **392**:308–315.
40. Feinberg EH and Hunter CP (2003) Transport of dsRNA into cells by the transmembrane protein SID-1. *Science* **301**:1545–1547.
41. Czauderna F, Fechtner M, Dames S, Aygun H, Klippel A, Pronk GJ, Giese K and Kaufmann J (2003) Structural variations and stabilising modifications of synthetic siRNAs in mammalian cells. *Nucleic Acids Res* **31**:2705–2716.
42. Elbashir SM, Martinez J, Patkaniowska A, Lendeckel W and Tuschl T (2001) Functional anatomy of siRNAs for mediating efficient RNAi in Drosophila melanogaster embryo lysate. *Embo J* **20**:6877–6888.
43. Amarzguioui M, Holen T, Babaie E and Prydz H (2003) Tolerance for mutations and chemical modifications in a siRNA. *Nucleic Acids Res* **31**:589–595.
44. Chiu YL and Rana TM (2003) siRNA function in RNAi: A chemical modification analysis. *Rna* **9**:1034–1048.
45. Harborth J, Elbashir SM, Vandenburgh K, Manninga H, Scaringe SA, Weber K and Tuschl T (2003) Sequence, chemical, and structural variation of small interfering RNAs and short hairpin RNAs and the effect on mammalian gene silencing. *Antisense Nucleic Acid Drug Dev* **13**:83–105.
46. Reynolds A, Leake D, Boese Q, Scaringe S, Marshall WS and Khvorova A (2004) Rational siRNA design for RNA interference. *Nat Biotechnol* **22**:326–330.
47. Yokota T, Miyagishi M, Hino T, Matsumura R, Tasinato A, Urushitani M, Rao RV, Takahashi R, Bredesen DE, Taira K and Mizusawa H (2004) siRNA-based inhibition specific for mutant SOD1 with single nucleotide alternation in familial ALS, compared with ribozyme and DNA enzyme. *Biochem Biophys Res Commun* **314**:283–291.
48. Ding H, Schwarz DS, Keene A, Affar el B, Fenton L, Xia X, Shi Y, Zamore PD and Xu Z (2003) Selective silencing by RNAi of a dominant allele that causes amyotrophic lateral sclerosis. *Aging Cell* **2**:209–217.
49. Mohapatra JK, Sanyal A, Hemadri D, Tosh C, Kumar RM and Bandyopadhyay SK (2005) Evaluation of *in vitro* inhibitory potential of small interfering RNAs directed against various regions of foot-and-mouth disease virus genome. *Biochem Biophys Res Commun* **329**:1133–1138.
50. Holen T, Amarzguioui M, Wiiger MT, Babaie E and Prydz H (2002) Positional effects of short interfering RNAs targeting the human coagulation trigger Tissue Factor. *Nucleic Acids Res* **30**:1757–1766.

51. Saxena S, Jonsson ZO and Dutta A (2003) Small RNAs with imperfect match to endogenous mRNA repress translation. Implications for off-target activity of small inhibitory RNA in mammalian cells. *J Biol Chem* **278**:44312–44319.
52. Amarzguioui M and Prydz H (2004) An algorithm for selection of functional siRNA sequences. *Biochem Biophys Res Commun* **316**:1050–1058.
53. Levenkova N, Gu Q and Rux JJ (2004) Gene specific siRNA selector. *Bioinformatics* **20**:430–432.
54. Pancoska P, Moravek Z and Moll UM (2004) Efficient RNA interference depends on global context of the target sequence: Quantitative analysis of silencing efficiency using Eulerian graph representation of siRNA. *Nucleic Acids Res* **32**:1469–1479.
55. Saetrom P and Snove O Jr (2004) A comparison of siRNA efficacy predictors. *Biochem Biophys Res Commun* **321**:247–253.
56. Xia XG, Zhou H, Ding H, Affar el B, Shi Y and Xu Z (2003) An enhanced U6 promoter for synthesis of short hairpin RNA. *Nucleic Acids Res* **31**:e100.
57. Wadhwa R, Kaul SC, Miyagishi M and Taira K (2004) Vectors for RNA interference. *Curr Opin Mol Ther* **6**:367–372.
58. Miyagishi M and Taira K (2002) U6 promoter-driven siRNAs with four uridine 3′ overhangs efficiently suppress targeted gene expression in mammalian cells. *Nat Biotechnol* **20**:497–500.
59. Brummelkamp TR, Bernards R and Agami R (2002) A system for stable expression of short interfering RNAs in mammalian cells. *Science* **296**:550–553.
60. Miyagishi M, Sumimoto H, Miyoshi H, Kawakami Y and Taira K (2004) Optimization of an siRNA-expression system with an improved hairpin and its significant suppressive effects in mammalian cells. *J Gene Med* **6**:715–723.
61. Tran N, Cairns MJ, Dawes IW and Arndt GM (2003) Expressing functional siRNAs in mammalian cells using convergent transcription. *BMC Biotechnol* **3**:21.
62. Braasch DA, Paroo Z, Constantinescu A, Ren G, Oz OK, Mason RP and Corey DR (2004) Biodistribution of phosphodiester and phosphorothioate siRNA. *Bioorg Med Chem Lett* **14**:1139–1143.
63. Brazas RM and Hagstrom JE (2005) Delivery of small interfering RNA to mammalian cells in culture by using cationic lipid/polymer-based transfection reagents. *Methods Enzymol* **392**:112–124.
64. Hassani Z, Lemkine GF, Erbacher P, Palmier K, Alfama G, Giovannangeli C, Behr JP and Demeneix BA (2005) Lipid-mediated siRNA delivery down-regulates exogenous gene expression in the mouse brain at picomolar levels. *J Gene Med* **7**:198–207.

65. Kakizawa Y, Furukawa S and Kataoka K (2004) Block copolymer-coated calcium phosphate nanoparticles sensing intracellular environment for oligodeoxynucleotide and siRNA delivery. *J Control Release* **97**:345–356.
66. Jiang M, Rubbi CP and Milner J (2004) Gel-based application of siRNA to human epithelial cancer cells induces RNAi-dependent apoptosis. *Oligonucleotides* **14**:239–248.
67. Khan A, Benboubetra M, Sayyed PZ, Ng KW, Fox S, Beck G, Benter IF and Akhtar S (2004) Sustained polymeric delivery of gene silencing antisense ODNs, siRNA, DNAzymes and ribozymes: *In vitro* and *in vivo* studies. *J Drug Target* **12**:393–404.
68. Leng Q, Scaria P, Zhu J, Ambulos N, Campbell P and Mixson AJ (2005) Highly branched HK peptides are effective carriers of siRNA. *J Gene Med* **7**(7):977–986.
69. Lochmann D, Weyermann J, Georgens C, Prassl R and Zimmer A (2005) Albumin-protamine-oligonucleotide nanoparticles as a new antisense delivery system. Part 1: Physicochemical characterization. *Eur J Pharm Biopharm* **59**:419–429.
70. Spagnou S, Miller AD and Keller M (2004) Lipidic carriers of siRNA: Differences in the formulation, cellular uptake, and delivery with plasmid DNA. *Biochemistry* **43**:13348–13356.
71. Rappaport J, Hanss B, Kopp JB, Copeland TD, Bruggeman LA, Coffman TM and Klotman PE (1995) Transport of phosphorothioate oligonucleotides in kidney: Implications for molecular therapy. *Kidney Int* **47**:1462–1469.
72. Braasch DA, Jensen S, Liu Y, Kaur K, Arar K, White MA and Corey DR (2003) RNA interference in mammalian cells by chemically-modified RNA. *Biochemistry* **42**:7967–7975.
73. Elmen J, Thonberg H, Ljungberg K, Frieden M, Westergaard M, Xu Y, Wahren B, Liang Z, Orum H, Koch T and Wahlestedt C (2005) Locked nucleic acid (LNA) mediated improvements in siRNA stability and functionality. *Nucleic Acids Res* **33**:439–447.
74. Kurreck J (2003) *Eur J Biochem* Improvement through novel chemical modifications. **270**:1628–1644.
75. Allen TM and Cullis PR (2004) Drug delivery systems: Entering the mainstream. *Science* **303**:1818–1822.
76. Torchilin VP (2005) Recent advances with liposomes as pharmaceutical carriers. *Nat Rev Drug Discov* **4**:145–160.
77. He S, Arscott PG and Bloomfield VA (2000) Condensation of DNA by multivalent cations: Experimental studies of condensation kinetics. *Biopolymers* **53**:329–341.

78. Kral T, Hof M and Langner M (2002) The effect of spermine on plasmid condensation and dye release observed by fluorescence correlation spectroscopy. *Biol Chem* **383**:331–335.
79. Putnam D, Zelikin AN, Izumrudov VA and Langer R (2003) Polyhistidine-PEG:DNA nanocomposites for gene delivery. *Biomaterials* **24**:4425–4433.
80. Hebert E (2003) Improvement of exogenous DNA nuclear importation by nuclear localization signal-bearing vectors: A promising way for non-viral gene therapy? *Biol Cell* **95**:59–68.
81. Demeneix B, Hassani Z and Behr JP (2004) Towards multifunctional synthetic vectors. *Curr Gene Ther* **4**:445–455.
82. Lechardeur D, Verkman AS and Lukacs GL (2005) Intracellular routing of plasmid DNA during non-viral gene transfer. *Adv Drug Deliv Rev* **57**:755–767.
83. Yi Y, Hahm SH and Lee KH (2005) Retroviral gene therapy: Safety issues and possible solutions. *Curr Gene Ther* **5**:25–35.
84. Palmer DJ and Ng P (2005) Helper-dependent adenoviral vectors for gene therapy. *Hum Gene Ther* **16**:1–16.
85. Wolkowicz R and Nolan GP (2005) Gene therapy progress and prospects: Novel gene therapy approaches for AIDS. *Gene Ther* **12**:467–476.
86. Jia W and Zhou Q (2005) Viral vectors for cancer gene therapy: Viral dissemination and tumor targeting. *Curr Gene Ther* **5**:133–142.
87. Li C, Bowles DE, van Dyke T, Samulski RJ (2005) Adeno-associated virus vectors: Potential applications for cancer gene therapy. *Cancer Gene Ther* **12**(12):913–925.
88. Monahan PE, Jooss K and Sands MS (2002) Safety of adeno-associated virus gene therapy vectors: A current evaluation. *Expert Opin Drug Saf* **1**:79–91.
89. Kappes JC and Wu X (2001) Safety considerations in vector development. *Somat Cell Mol Genet* **26**:147–158.
90. Aharinejad S, Paulus P, Sioud M, Hofmann M, Zins K, Schafer R, Stanley ER and Abraham D (2004) Colony-stimulating factor-1 blockade by antisense oligonucleotides and small interfering RNAs suppresses growth of human mammary tumor xenografts in mice. *Cancer Res* **64**:5378–5384.
91. Kuroda M, Kiyono T, Oikawa K, Yoshida K and Mukai K (2005) The human papillomavirus E6 and E7 inducible oncogene, hWAPL, exhibits potential as a therapeutic target. *Br J Cancer* **92**:290–293.
92. Chien PY, Wang J, Carbonaro D, Lei S, Miller B, Sheikh S, Ali SM, Ahmad MU and Ahmad I (2005) Novel cationic cardiolipin analogue-based liposome for efficient DNA and small interfering RNA delivery *in vitro* and *in vivo*. *Cancer Gene Ther* **12**:321–328.

93. Pal A, Ahmad A, Khan S, Sakabe I, Zhang C, Kasid UN and Ahmad I (2005) Systemic delivery of RafsiRNA using cationic cardiolipin liposomes silences Raf-1 expression and inhibits tumor growth in xenograft model of human prostate cancer. *Int J Oncol* **26**:1087–1091.
94. Urban-Klein B, Werth S, Abuharbeid S, Czubayko F and Aigner A (2005) RNAi-mediated gene-targeting through systemic application of polyethylenimine (PEI)-complexed siRNA *in vivo*. *Gene Ther* **12**:461–466.
95. Zhang Y, Boado RJ and Pardridge WM (2003) *In vivo* knockdown of gene expression in brain cancer with intravenous RNAi in adult rats. *J Gene Med* **5**:1039–1045.
96. Zhang YF, Boado RJ and Pardridge WM (2003) Absence of toxicity of chronic weekly intravenous gene therapy with pegylated immunoliposomes. *Pharm Res* **20**:1779–1785.
97. Zhang Y, Zhang YF, Bryant J, Charles A, Boado RJ and Pardridge WM (2004) Intravenous RNA interference gene therapy targeting the human epidermal growth factor receptor prolongs survival in intracranial brain cancer. *Clin Cancer Res* **10**:3667–3677.
98. Taulli R, Accornero P, Follenzi A, Mangano T, Morotti A, Scuoppo C, Forni PE, Bersani F, Crepaldi T, Chiarle R, Naldini L and Ponzetto C (2005) RNAi technology and lentiviral delivery as a powerful tool to suppress Tpr-Met-mediated tumorigenesis. *Cancer Gene Ther* **12**(5):456–463.
99. Shinomiya N, Gao CF, Xie Q, Gustafson M, Waters DJ, Zhang YW and Vande Woude GF (2004) RNA interference reveals that ligand-independent met activity is required for tumor cell signaling and survival. *Cancer Res* **64**:7962–7970.
100. Uchida H, Tanaka T, Sasaki K, Kato K, Dehari H, Ito Y, Kobune M, Miyagishi M, Taira K, Tahara H and Hamada H (2004) Adenovirus-mediated transfer of siRNA against survivin induced apoptosis and attenuated tumor cell growth *in vitro* and *in vivo*. *Mol Ther* **10**:162–171.
101. Godfrey A, Anderson J, Papanastasiou A, Takeuchi Y and Boshoff C (2005) Inhibiting primary effusion lymphoma by lentiviral vectors encoding short hairpin RNA. *Blood* **105**:2510–2518.
102. Bergamaschi D, Gasco M, Hiller L, Sullivan A, Syed N, Trigiante G, Yulug I, Merlano M, Numico G, Comino A, Attard M, Reelfs O, Gusterson B, Bell AK, Heath V, Tavassoli M, Farrell PJ, Smith P, Lu X and Crook T (2003) p53 polymorphism influences response in cancer chemotherapy via modulation of p73 dependent apoptosis. *Cancer Cell* **3**:387–402.
103. Stoklosa T, Slupianek A, Datta M, Nieborowska-Skorska M, Nowicki MO, Koptyra M and Skorski T (2004) BCR/ABL recruits p53 tumor suppressor protein to induce drug resistance. *Cell Cycle* **3**:1463–1472.

104. Sui G, Affar el B, Shi Y, Brignone C, Wall NR, Yin P, Donohoe M, Luke MP, Calvo D and Grossman SR (2004) Yin Yang 1 is a negative regulator of p53. *Cell* **117**:859–872.
105. Vega FM, Sevilla A and Lazo PA (2004) p53 Stabilization and accumulation induced by human vaccinia-related kinase 1. *Mol Cell Biol* **24**:10366–10380.
106. Porta C, Hadj-Slimane R, Nejmeddine M, Pampin M, Tovey MG, Espert L, Alvarez S and Chelbi-Alix MK (2005) Interferons alpha and gamma induce p53-dependent and p53-independent apoptosis, respectively. *Oncogene* **24**:605–615.
107. Nikolaev AY, Li M, Puskas N, Qin J and Gu W (2003) Parc: A cytoplasmic anchor for p53. *Cell* **112**:29–40.
108. Mao JH, Perez-Losada J, Wu D, Delrosario R, Tsunematsu R, Nakayama KI, Brown K, Bryson S and Balmain A (2004) Fbxw7/Cdc4 is a p53-dependent, haploinsufficient tumour suppressor gene. *Nature* **432**:775–779.
109. Folkman J (2002) Role of angiogenesis in tumor growth and metastasis. *Semin Oncol* **29**:15–18.
110. Kirsch M, Schackert G and Black PM (2004) Metastasis and angiogenesis. *Cancer Treat Res* **117**:285–304.
111. Gondi CS, Lakka SS, Dinh DH, Olivero WC, Gujrati M and Rao JS (2004) RNAi-mediated inhibition of cathepsin B and uPAR leads to decreased cell invasion, angiogenesis and tumor growth in gliomas. *Oncogene* **23**:8486–8496.
112. Duxbury MS, Matros E, Ito H, Zinner MJ, Ashley SW and Whang EE (2004) Systemic siRNA-mediated gene silencing: A new approach to targeted therapy of cancer. *Ann Surg* **240**:667–674; discussion 675–676.
113. Liang Z, Yoon Y, Votaw J, Goodman MM, Williams L and Shim H (2005) Silencing of CXCR4 blocks breast cancer metastasis. *Cancer Res* **65**:967–971.
114. Vorburger SA, Pataer A, Swisher SG and Hunt KK (2004) Genetically targeted cancer therapy: Tumor destruction by PKR activation. *Am J Pharmacogenomics* **4**:189–198.
115. Chae SS, Paik JH, Furneaux H and Hla T (2004) Requirement for sphingosine 1-phosphate receptor-1 in tumor angiogenesis demonstrated by *in vivo* RNA interference. *J Clin Invest* **114**:1082–1089.
116. Schiffelers RM, Ansari A, Xu J, Zhou Q, Tang Q, Storm G, Molema G, Lu PY, Scaria PV and Woodle MC (2004) Cancer siRNA therapy by tumor selective delivery with ligand-targeted sterically stabilized nanoparticle. *Nucleic Acids Res* **32**:e149.
117. Kim B, Tang Q, Biswas PS, Xu J, Schiffelers RM, Xie FY, Ansari AM, Scaria PV, Woodle MC, Lu P and Rouse BT (2004) Inhibition of ocular angiogenesis by siRNA targeting vascular endothelial growth factor pathway

genes: Therapeutic strategy for herpetic stromal keratitis. *Am J Pathol* **165**: 2177–2185.

118. Duxbury MS, Ito H, Benoit E, Zinner MJ, Ashley SW and Whang EE (2004) Retrovirally mediated RNA interference targeting the M2 subunit of ribonucleotide reductase: A novel therapeutic strategy in pancreatic cancer. *Surgery* **136**:261–269.
119. Duxbury MS, Ito H, Zinner MJ, Ashley SW and Whang EE (2004) RNA interference targeting the M2 subunit of ribonucleotide reductase enhances pancreatic adenocarcinoma chemosensitivity to gemcitabine. *Oncogene* **23**:1539–1548.
120. Matsukura S, Jones PA and Takai D (2003) Establishment of conditional vectors for hairpin siRNA knockdowns. *Nucleic Acids Res* **31**:e77.
121. Higuchi M, Tsutsumi R, Higashi H and Hatakeyama M (2004) Conditional gene silencing utilizing the lac repressor reveals a role of SHP-2 in cagA-positive Helicobacter pylori pathogenicity. *Cancer Sci* **95**:442–447.
122. Czauderna F, Santel A, Hinz M, Fechtner M, Durieux B, Fisch G, Leenders F, Arnold W, Giese K, Klippel A and Kaufmann J (2003) Inducible shRNA expression for application in a prostate cancer mouse model. *Nucleic Acids Res* **31**:e127.
123. Chang HS, Lin CH, Chen YC and Yu WC (2004) Using siRNA technique to generate transgenic animals with spatiotemporal and conditional gene knockdown. *Am J Pathol* **165**:1535–1541.
124. Wagner E (2004) Strategies to improve DNA polyplexes for *in vivo* gene transfer: Will "artificial viruses" be the answer? *Pharm Res* **21**:8–14.
125. Merdan T, Kopecek J and Kissel T (2002) Prospects for cationic polymers in gene and oligonucleotide therapy against cancer. *Adv Drug Deliv Rev* **54**: 715–758.
126. De Smedt SC, Demeester J and Hennink WE (2000) Cationic polymer based gene delivery systems. *Pharm Res* **17**:113–126.
127. Buning H, Braun-Falco M and Hallek M (2004) Progress in the use of adeno-associated viral vectors for gene therapy. *Cells Tissues Organs* **177**:139–150.
128. Tomanin R and Scarpa M (2004) Why do we need new gene therapy viral vectors? Characteristics, limitations and future perspectives of viral vector transduction. *Curr Gene Ther* **4**:357–372.
129. Filleur S, Courtin A, Ait-Si-Ali S, Guglielmi J, Merle C, Harel-Bellan A, Clezardin P and Cabon F (2003) SiRNA-mediated inhibition of vascular endothelial growth factor severely limits tumor resistance to antiangiogenic thrombospondin-1 and slows tumor vascularization and growth. *Cancer Res* **63**:3919–3922.
130. Hamar P, Song E, Kokeny G, Chen A, Ouyang N and Lieberman J (2004) Small interfering RNA targeting Fas protects mice against renal ischemia-reperfusion injury. *Proc Natl Acad Sci USA* **101**:14883–14888.

131. Kobayashi N, Matsui Y, Kawase A, Hirata K, Miyagishi M, Taira K, Nishikawa M and Takakura Y (2004) Vector-based *in vivo* RNA interference: Dose- and time-dependent suppression of transgene expression. *J Pharmacol Exp Ther* **308**:688–693.
132. Lewis DL and Wolff JA (2005) Delivery of siRNA and siRNA expression constructs to adult mammals by hydrodynamic intravascular injection. *Methods Enzymol* **392**:336–350.
133. Giladi H, Ketzinel-Gilad M, Rivkin L, Felig Y, Nussbaum O and Galun E (2003) Small interfering RNA inhibits hepatitis B virus replication in mice. *Mol Ther* **8**:769–776.
134. Van Houdt H, Bleys A and Depicker A (2003) RNA target sequences promote spreading of RNA silencing. *Plant Physiol* **131**:245–253.
135. Lebedeva I and Stein CA (2001) Antisense oligonucleotides: Promise and reality. *Annu Rev Pharmacol Toxicol* **41**:403–419.
136. Stein CA and Cheng YC (1993) Antisense oligonucleotides as therapeutic agents — Is the bullet really magical? *Science* **261**:1004–1012.
137. Stein CA (1992) Anti-sense oligodeoxynucleotides — Promises and pitfalls. *Leukemia* **6**:967–974.
138. Sledz CA, Holko M, de Veer MJ, Silverman RH and Williams BR (2003) Activation of the interferon system by short-interfering RNAs. *Nat Cell Biol* **5**:834–839.
139. Snove O Jr and Holen T (2004) Many commonly used siRNAs risk off-target activity. *Biochem Biophys Res Commun* **319**:256–263.

# 8

# Anti-Cancer Proteins and Peptides in Liposomes

*Vladimir Torchilin*

## 1. Proteins and Peptides as Cancer Therapeutics

Numerous proteins and peptides possess biological activity that makes them potent anticancer agents. Among protein drugs, enzymes represent an important and, probably, the best investigated group.[1–3] L-asparaginase, the most frequently used antitumor enzyme[4] hydrolyzes asparagine via desamination of the amino acid with the formation of aspartic acid, and its therapeutic action is based on the high requirement of some tumors, such as acute lymphoblastic leukemia, for asparagine. As a result, L-asparaginase became a standard tool in the treatment of leukemia.[5] Other enzymes of interest[6] include glutaminase, cystein desulfatase; cystein aminotransferase, cysteine oxidase, arginase, arginine deaminase and arginine decarboxylase. Interesting approaches involve also the use of folate-degrading enzymes,[7] ribonucleases and exonucleases.[8]

Antibodies against certain cancer-specific ligands represent another example of protein drugs. Clinically used are Trastuzumab (a humanized antibody against HER-2/neu, extracellular domain of the HER-2 gene expression product characteristic of certain types of breast cancer) and

Rituxan (a humanized antibody against the B-cell specific antigen CD20). Trastuzumab induces immune-mediated responses, down-regulates HER-2 receptor, and promotes the production of cell-cycle inhibitors.[9] This agent is well tolerated by patients, and tumor remission was observed in 60–70% patients after the combined treatment with Transtuzumab and paclitaxel.[10] Rituxan is effective against hematologic malignancies.[11]

Peptides such as somatostatin analogs (octretide, lanreotide, vapreotide) are clinically used for the treatment of pituitary and gastrointestinal tumors.[12] Peptide inhibitors of angiogenesis including endostatin also show a great promise for cancer treatment.[13,14] Research on depsipeptides has also produced several potential anticancer agents.[15]

Those are just few examples. Solid-phase peptide synthesis as well as the recombinant DNA and hybridoma technology allow for the production of large quantities of clinical grade peptides and proteins. The completion of the human genome sequence and advances in proteomics provide a powerful tool for the further progress in the identification and future clinical development of peptide- and protein-based anticancer drugs.

## 2. Stability, Delivery and Distribution Issues for Peptide and Protein Drugs

Although, quite a few biologically active peptides and proteins have been identified as potential anticancer agents with some of them are being successfully used in clinical medicine, the potential of this group of drugs is far from being completely utilized. The use of proteins and peptides as therapeutic agents is hampered by several important factors, such as their rapid elimination from the circulation mostly because of renal filtration; enzymatic degradation, uptake by the reticuloendothelial system (RES), accumulation in nontargeted organs and tissues, and inefficient cell entry in cases when their targets are located inside the cell. Foreign proteins can also provoke an immune response in the host. Specific antibodies against a given protein can rise during the life span or as a result of the repeated administration of a therapeutic protein (peptide). In any case it leads to protein inactivation or even to allergic reactions, which very often make its further application impossible.

In addition, proteins and peptides possess the whole set of "inconvenient" intrinsic properties as complex macromolecules, which are, as a rule, foreign to the recipient organism. These properties include low stability of peptide and especially protein drugs at physiological pH values and temperatures, particularly when these proteins have to be active in other than their normal environment. Different processes leading to the inactivation of various biologically-active proteins and peptides *in vivo* are: conformational intramolecular protein transformation into inactive conformation from the effect of temperature, pH, high salt concentration, or detergents; the dissociation of protein subunits or, in case of enzymes, enzyme-cofactor complexes, and the association of protein or peptide molecules with the formation of inactive associates; noncovalent complexation with ions or low-molecular-weight and high-molecular-weight compounds, affecting the native structure of the protein or peptide; proteolytic degradation under the action of endogenous proteases; chemical modification by different compounds in solution (for example, oxidation of SH-groups in sulfhydryl enzymes and Fe (II) atoms in heme-containing proteins by oxygen; thiol-disulfide exchange, destruction of labile side-groups like tryptophan and metionine).

Various approaches to overcome rapid elimination and nonspecific biodistribution of conventional drugs have been developed and can be adapted for the delivery of anticancer peptides and proteins. The simplest way to compensate the rapid elimination of biologically active substances is to inject the substance of interest at a controllable rate using an external pump. Delivery of a drug to a target organ can also be achieved by administration of the drug directly into the area of interest. Both methods, however, require skilled medical personnel, often restrict the patient to a hospital setting for a long time, and are not feasible when a drug is very toxic or the site of the desired administration is difficult to reach or delocalized. Implantable devices[16–18] ranging from erodible polymeric gels[16] to microfabricated chips[19,20] were proposed to overcome low stability and nonspecific biodistribution of potential therapeutic agents. However, the majority of implantable devices are rather costly. An additional disadvantage is that their insertion and, if needed, removal require surgical intervention. Their use is limited to sites of delivery with a size that can accommodate the object. Moreover, the use of implantable devices does not address the problem of intracellular delivery of pharmaceuticals.

## 3. Possible Solutions. Liposomes as Pharmaceutical Carriers

The variety of methods has been developed ("protein immobilization") to achieve protein and peptide protection from various denaturating effects and impart desired properties to such drugs. The most important of these properties are: (1) increased stability and prolonged activity of the protein/peptide drug in the body; (2) decreased immunogenicity and affinity to specific antibodies, which permits to administer repeated therapy; (3) decreased affinity to natural inhibitors; (4) the possibility to administer the whole therapeutic dose of a protein/peptide drug in a single injection, a therapeutic procedure the most simple and convenient both for the patient and the physician; (5) a decrease in the total quantity of a protein/peptide drug needed for the treatment, which makes the treatment more economical. Parenterally administered immobilized/modified protein/peptide drug should exist in the active form for a long period of time. It should also reach and penetrate an affected tissue and demonstrate improved pharmacokinetic properties and an increase in the bioavailability.

Among many known methods to stabilize protein/peptide drugs and favorably change their bioavailability and pharmacokinetics, the encapsulation of such drugs into artificial microreservoir type carriers (liposomes, micelles, nanocapsules, etc.) is frequently applied. Inside such carriers an enzyme is protected from many aggressive influences of the external medium and, in turn, does not act on different external objects. An additional advantage of these carriers is that a single carrier particle is capable of delivering multiple drug species. All microparticulates have a size, which excludes the possibility of their loss by renal filtration. Certain problems with microreservoir type carriers are associated with their tendency to be taken up by the RES cells primarily in liver and spleen.[21]

One of the most popular and well-elaborated technologies to immobilize/stabilize protein drugs and improve their pharmacological properties is their incorporation by into artificial phospholipid vesicles — liposomes. Liposomes are spherical self-closed structures formed by one or several concentric lipid bilayers with aqueous phase inside and between those bilayers. They possess a whole set of attractive pharmacological properties: (1) Liposomes are biocompatible; (2) they cause no or very little antigenic, pyrogenic, allergic and toxic reactions, and easily undergo biodegradation;

(3) they can entrap water-soluble (hydrophilic) pharmaceutical agents into their internal water compartment and water-insoluble (hydrophobic) pharmaceuticals into the membrane; (4) liposome-incorporated pharmaceuticals are protected from the inactivating effect of external conditions, yet do not cause undesirable side-reactions; (5) liposomes provide a unique opportunity to deliver pharmaceuticals into cells or even inside individual cellular compartments. Size, charge and surface properties of liposomes can be easily changed simply by adding new ingredients to the lipid mixture before liposome preparation and/or by variation of preparation methods. Different methods have been suggested to prepare liposomes of different sizes, structure and size distribution. The most frequently used methods are ultrasonication, reverse phase evaporation, and detergent removal from mixed lipid-detergent micelles by dialysis or gel-filtration. To increase liposome stability towards the action of an aggressive physiological environment, cholesterol is incorporated into the liposomal membrane (sometimes up to 50% mol). The size of liposomes depends on their composition and preparation method and can vary from around 80 nm to greater than 1 μm in diameter. The encapsulation efficacy for different substances is also variable depending on the liposome composition, size, charge, and preparation method. The use of the reverse phase evaporation method[22] permits inclusion of 50 and more percent of the substance to be encapsulated from the water phase into the liposomes. Besides, a variety of methods have been developed to obtain lyophilized liposomal preparations possessing good storage stability.[23] The *in vitro* release rate of different compounds from liposomes, including proteins of moderate molecular weight, such as lysozyme or insulin, is usually under 1% per hr, under the condition that the incubation temperature sufficiently differs from the phase transition temperature of a given phospholipid. Maximal permeability of liposomes is usually observed at temperatures close to the phase transition temperature of the liposomal phospholipid. *In vivo*, this parameter can vary within wide limits from minutes to hours and depends on the liposome membrane composition, cholesterol content, and liposome location in the body. Liposome preparation, their physico-chemical properties and possible biomedical application have already been discussed in many monographs; see for example.[24–26]

To increase liposome accumulation in the desired areas and the efficacy of liposomal pharmaceutical agents, the use of targeted liposomes with surface-attached ligand capable of recognition and binding to tissues or cells of interest has been suggested. Liposomes with a specific affinity for an affected tissue were believed to increase the efficacy of liposomal pharmaceutical agents. Immunoglobulins, primarily of the IgG class, and their fragments are the most widely used targeting moieties for liposomes. Still, despite improvements in the targeting efficacy, the majority of immunoliposomes ended in the liver as a consequence of insufficient time for the interaction between the target and targeted liposome. Different methods have been suggested to achieve long circulation of liposomes *in vivo*, including coating the liposome surface with inert, biocompatible polymers, such as polyethylene glycol (PEG), which form a protective layer over the liposome surface and slows down the liposome recognition by opsonins and subsequent clearance.[27,28] The longevity phenomenon is based on the decreased rate of plasma protein adsorption on the hydrophilic surface of PEGylated liposomes.[29] An important feature of protective polymers is their flexibility, which allows a relatively small number of their surface-grafted molecules to create dense "clouds" over the liposome surface.[30,31] As a result, long-circulating liposomes demonstrate dose-independent, non-saturable, log-linear kinetics, and increased bioavailability due to the inhibition of liposome destruction and their capture by the RES.[32] Although, PEG remains the gold standard in liposome steric protection, attempts continue to identify other polymers that could be used to prepare long-circulating liposomes. Earlier studies with various water-soluble flexible polymers have been summarized in.[31,33] More recent papers describe long-circulating liposomes prepared using poly[N-(2-hydroxypropyl)methacrylamide)],[34] poly-N-vinylpyrrolidones,[35] L-amino acid-based biodegradable polymer-lipid conjugates,[36] and polyvinyl alcohol.[37]

The further development of the concept of long-circulating liposomes involves the attempt to combine the properties of long-circulating liposomes and immunoliposomes in one preparation, i.e. simultaneously attach both the protecting polymer and the targeting moiety on the liposome.[38–40] To achieve better selective targeting of PEG-coated liposomes, it is advantageous to attach the targeting ligand to the particles not directly, but via a PEG spacer arm, so that the ligand is extended outside of the dense PEG brush excluding steric hindrances for its binding to the target receptors. With this in mind, PEG has to be modified with a phospholipid on one side and with the targeting ligand on the other. Currently, various advanced technologies are used, and the targeting moiety is usually attached above the protecting polymer layer, by coupling it with the distal water-exposed terminus of activated liposome-grafted polymer molecule.[39,41]

It has been shown with a broad variety of examples that, similar to macromolecules, liposomes are capable of accumulating in tumors of various origins *via* the enhanced permeability and retention (EPR) effect.[42,43] High-molecular-weight (40 kDa or higher), long-circulated macromolecules, including peptides and proteins conjugated with water-soluble polymers, are capable of spontaneous accumulations in solid tumors via the EPR effect.[44,45] This effect is based on certain specificities of tumor tissues. The first specificity is that the tumor vasculature, unlike the vasculature of healthy tissues, is permeable to macromolecules with a molecular weight of 50 kDa or even higher. This allows macromolecules to enter into the interstitial tumor space. Another specific feature of tumors is that the lymphatic system, responsible for the drainage of macromolecules from normal tissues, is virtually not working in case of many tumors as the result of the disease.[45] Because of this, macromolecules having entered tumor tissues are retained there for a prolonged time. Unlike macromolecules, low-molecular-weight conventional pharmaceutics are not retained in tumors because of their ability to return to the circulation by diffusion.[45] Further details on the mechanism of the EPR effect and the list of preparations for which this effect has been observed can be found in several recent review articles.[44,45]

Clinical application of liposomes already has its own history, with some currently approved liposomal drugs as well as some in clinical trials. The initial success with many of these drugs has fueled further clinical investigations. To mention just few the most recent ones, the research on the use of doxorubicin in PEG-liposomes for the treatment of solid tumors concentrated recently around the selective delivery of the drug in patients with breast carcinoma metastases and survival improvement.[46,48] Same set of indications was targeted by the combination therapy involving liposomal doxorubicin and paclitaxel[49] or Caelyx (doxorubicin in PEG-liposomes) and carboplatin.[50] Caelyx is also used for patients with squamous cell cancer of the head and neck[51] and ovarian cancer.[52] Patient research showed the impressive effect of doxorubicin in PEG-liposomes against unresectable hepatocellular carcinoma,[53] cutaneous T-cell lymphoma,[54] and sarcoma.[55]

Liposomal lurtotecan was found to be effective in patients with topotecan resistant ovarian cancer.[56]

Liposomes served as carriers of proteins and peptides in a huge variety of cases. The main goals of encapsulating proteiuns and peptides into liposomes include: (1) Incroporation of protein and peptide drugs into liposomes in order to improve their therapeutic activity in a very broad sense of this word and to diminish various drawbacks and side effects frequently characteristic for such drugs; (2) Incorporation of various proteins and peptides into liposomes in order to modulate the immune response towards these proteins and peptides or to other antigens (for example, by protein- or peptide-modulated activation of certain components of immune systems or certain steps of the immune response development). Liposomes loaded with protein/peptide drugs (proteoliposome) can incorporate these drugs in a variety of fashions: water-soluble proteins are entrapped into the liposomal inner aqueous space (and, in case of multilammellar liposomes, into the aqueous space between bilayers), while less soluble proteins and peptides may be incorporated into the phospholipid membrane.

## 4. Liposomal Proteins and Peptides for Cancer Treatment

The following parameters are usually considered as a proof of a protein/peptide incorporation into the inner aqueous phase of liposomes or its firm and irreversible association with the liposomal membrane: the possibility of chromatographic separation of liposome-encapsulated and free protein/peptide; the latency of liposome-encapsulated protein/peptide; the correlation between protein/peptide incorporation and a change in net surface charge of the lipid bilayer.

L-asparaginase used for the treatment of asparagines-dependent tumors was a frequent candidate for liposome encapsulation. Thus, the increase in the circulation half-life of the liposomal L-asparaginase (in lecithine-dicetylphosphate liposomes, a negative charge of the liposomes was used to inhibit the liposome interaction with cells and to keep them in the blood longer) and the decrease in its antigenicity and susceptibility towards the proteolytic degradation together with the increase in the efficacy of experimental tumor therapy in mice have been shown.[57] The longevity of the liposomal L-asparaginase (in nonlong-circulating liposomes) depends also on the liposome size: in large liposomes (ca. 1200 nm) the circulation time was decreased, while in small liposomes (ca. 170 nm) it was prolonged 10-fold compared to free enzyme.[58] The use of the liposome-encapsulated asparaginase improves the survival of animals with P1534 tumors compared to free enzyme. It is also important that the encapsulation into liposomes prevents the production of anti-asparaginase antibodies. Palmitoyl-L-asparaginase was also incorporated into liposomes and demonstrated prolongation of blood life by almost 10-fold, decrease in acute toxicity and improved antitumor activity *in vivo*.[59]

Cell lines (including human osteosarcoma) treated with liposome-incorporated nucleoside kinases demonstrated an increased sensitivity towards nucleoside analogues.[60]

The use of liposomes for the transfer of therapeutic enzymes through the blood-brain barrier, which permits to deliver these enzymes into cells of the central nervous system also seems very attractive. It has been already shown that horse raddish peroxidase, encapsulated into liposomes made of phosphatidylcholine, cholesterol and phosphatidic acid (7:2:1 molar ratio), acquires the ability to cross the hemato-encephalic barrier, whereas the native enzyme cannot. The presence of peroxidase in brain cells was proved by histochemical methods.[61] The same authors have shown that after injection of the liposomal glucose oxidase into the rat's tail vein, up to 5% of the enzymatic activity can be discovered in brain tissues.[62]

A very interesting approach to the use of liposomal enzymes is their application for antibody-directed enzyme prodrug therapy (ADEPT). ADEPT approach allows for the specific generation of active cytotoxic molecules from their inactive precursors or prodrugs in the vicinity of tumor cells. For this purpose, a conjugate of a tumor-specific antibody with an enzyme responsible for the conversion of a prodrug into the active drug is targeted towards tumor, accumulates there, and convert inactive prodrug into a cytotoxic molecule right at the site of its action.[63–65] In order to increase the efficiency of the required enzyme in the tumor, it was suggested to use not just "straight" antibody-enzyme conjugates, but rather imuuno-liposomes loaded with the required enzyme (immuno-enzymosomes).[66,67] Experiments have been performed with tumor-specific liposomes bearing $\beta$-glucuronidase capable of activating anthracycline prodrugs.[66,68]

Cytokines are frequent candidates for liposomal dosage forms that have been expected to extend their lifetime in the body. Thus, the incorporation of recombinant interleukin-2 into liposomes increased its blood circulation time by 8-fold.[69] Liposomal muramyl tripeptide was successfully used in patients with relapsed osteosarcoma.[70] Mannosilated liposomes with muramyl dipeptide significantly inhibited liver metastases in tumor-bearing mice.[71] The possibility of using TNF encapsulated into long-circulating PEG-liposomes for the treatment of solid tumors, possibly, in combination with other cytotoxic agents was discussed in the literature.[72] PEG-coated liposomes have also been proposed for the oral delivery of recombinant human epidermal growth factor.[73] Liposomal delivery of the peptide inhibitor of the transcription factor nuclear factor-kappaB was shown to significantly inhibit the proliferation of vascular smooth muscle cells.[74]

The authors of [75,76] described an interesting approach to the use of liposomal drugs in cancer therapy. They suggested the destruction of malignant cells by hyper-activated macrophages of the host himself. For this purpose, macrophages have been stimulated with different lymphokines. In order to increase the quantity of lymphokines delivered to macrophages, the former can be preliminarily encapsulated into liposomes. In experiments on mice with inoculated tumors it was shown that liposomal lymphokines are capable of preventing metastasis formation in 70% of the animals, in comparison with 10% prevention when native lymphokines were used.

Liposomal systems have also been considered as a cytokine supplement in tumor cells vaccines since they, may provide a cytokine reservoir at the antigen presentation site,[77,78] and the benefits of the liposomal interferon-gamma in the generation of systemic immune responses in B16 melanoma model have been clearly demonstrated by these authors. Some peptides (pH-sensitive) might enhance drug release from folate-targeted liposomes at endosomal pHs[79] thus improving drug delivery into the cell cytoplasm.

A special attention has been paid to the immunological properties of liposome-immobilized proteins/peptides. The ability of liposomes to demonstrate in some cases adjuvant properties is already well known. The authors of [80] demonstrated the possibility of enhanced immune response development in mice receiving intravenously liposomes with immobilized bovine $\beta$-glucuronidase, despite the expected protection of the intraliposomal enzyme from contact with immunocompetent cells.

## 5. Tumor Cell Targeting Using Specific Ligand-Bearing Liposomes

The use of specific "vector" molecules can further enhance tumor targeting of peptide/protein carries or make them the EPR-effect independent. The latter is especially important for the cases of tumors with immature vasculature, such as tumors in the early stages of their development, and for delocalized tumors. Vector molecules (those having affinity toward ligands characteristic for target tissues) capable of recognizing tumors were found among antibodies, peptides, lectines, saccharides, hormones, and some low molecular weight compounds.[81] From this list, antibodies and their fragments provide the most universal opportunity to target various for targeting and have the highest potential specificity.

Antibodies capable of recognizing specific antigens have been derived for the majority of known tumors. Among them are antibodies against ovarian cancer, prostate cancer, colorectal cancer, etc.[82] Humanized versions of antibodies and their fragments in which rodent-derived binding sites and human conservative regions are combined using recombinant technology are available.[83,84] Among antibodies with anticancer specificity, recently described monoclonal antinuclear autoantibodies (ANAs) with nucleosome-restricted specificity are of particular interest.[85,86] Unlike the majority of anticancer antibodies that are capable of recognizing only specific tumors, ANAs possess specificity against a broad variety of tumors. Nonpathogenic ANAs represent a subclass of natural anticancer antibodies, and several of them differentiate between tumor and normal cells.[85] The monoclonal nonpathogenic ANAs, 2C5 and 1G3, were shown to recognize the surface of numerous tumor, but not normal, cells.[85,86] Nucleosome-restricted specificity was found for these ANAs, and tumor cell surface-bound intact nucleosomes originating from apoptotically dying neighboring tumor cells have been shown to be their universal molecular target on the surface of the variety of tumor cells.[86,87] In addition to their anticancer effect, these antibodies allow for development of delivery systems with specificity universal across different tumor types.

Although various monoclonal antibodies have been shown to deliver liposomes to many targets, the search for the optimization of properties of immunoliposomes still continues. Internalizing antibodies are required to achieve a really improved therapeutic efficacy of antibody-targeted liposomal drugs as was shown using B-lymphoma cells and internalizable epitopes (CD19) as an example.[88] PEG-modified long-circulating doxorubicin-containing immunoliposomes targeted with anti-HER-2/neu monoclonal antibody fragments represent a recent example of increased efficiency of the targeted delivery systems.[89] In all studied HER2-overexpressing models, immunoliposomes showed potent anticancer activity superior to that of control nontargeted liposomes.[89] In part, this superior activity was attributed to the ability of the immunoliposomes to deliver their load inside the target cells *via* the receptor-mediated endocytosis,[89] which is obviously important if the drug's site of action sites locates inside the cell. Antibody CC52 against rat colon adenocarcinoma CC531 attached to PEGylated liposomes provided specific accumulation of liposomes in rat model of metastatic CC531.[90] Nucleosome-specific antibodies capable of recognition of various tumor cells via tumor cell surface-bound nucleosomes improve Doxil® targeting to tumor cells and increase its cytotoxicity.[91] GD2-targeted immunoliposomes with novel antitumoral drug, fenretinide, inducing apoptosis in neuroblastoma and melanoma cell lines, demonstrated strong anti-neuroblastoma activity both *in vitro* and *in vivo* in mice.[92] Combination of immunoliposome and endosome-disruptive peptide improves cytosolic delivery of liposomal drug, increases cytotoxicity, and opens new approach to constructing targeted liposomal systems as shown with diphteria toxin A chain incorporated together with pH-dependent fusogenic peptide diINF-7 into liposomes specific towards ovarian carcinoma.[93]

Targeting tumors with folate-modified liposomes becomes a popular approach, since folate receptor (FR) expression is frequently overexpressed in many tumor cells. Studies demonstrated the possibility of delivery of macromolecules[94] and then liposomes[95] into living cells utilizing FR endocytosis, which could bypass multidrug resistance. Current interest to folate-targeted drug delivery by liposomes is quite high.[96,97] Liposomal daunorubicin[98] and doxorubicin[99] were delivered into various tumor cells via folate receptor and demonstrated increased cytotoxicity. Folate-targeted liposomes have been suggested as delivery vehicles for targeting tumors with haptens for tumor immunotherapy.[100] Since transferrin (Tf) receptors (TfR) are also overexpressed on the surface of many tumor cells, antibodies against TfR as well as Tf itself are among popular ligands for liposome targeting to tumors and inside tumor cells.[101] The coupling of Tf to PEG on PEGylated liposomes was performed to combine longevity and targetability for drug delivery into solid tumors.[102] Tf-coupled doxorubicine-loaded liposomes demonstrate increased binding and toxicity against C6 glioma.[103]

All other ligands for liposome targeting are also directed towards specific receptors overexpressed on certain cancer cells. Thus, liposome targeting to tumors has been achieved by using vitamin and growth factor receptors.[104] Vasoactive intestinal peptide (VIP) was used to target PEG-liposomes with radionuclides to VIP-receptors of the tumor, which resulted in an enhanced breast cancer inhibition in rats.[105] Doxorubicin-loaded PEG-liposomes were targeted by RGD peptides to integrins of tumor vasculature and demonstrated increased efficiency against C26 colon carcinoma in murine model.[106] Similar angiogenic homing peptide was used for targeted delivery to vascular endothelium of drug-loaded liposomes in experimental treatment of tumors in mice.[107] Epidermal growth factor receptor (EGFR)-targeted immunoliposomes were specifically delivered to variety of tumor cells overexpressing EGFR.[108] Mitomycin C in long-circulating hyaluronan-targeted liposomes increases its activity against tumors overexpress hyaluronan receptors.[109] Galactosylated liposomes were used to target drugs to the liver for therapy of liver tumors or metastases.[110] Since the increased level of chondroitin sulfate expression was found in many tumor cells, cisplatin-loaded cationic nonPEGylated and PEGylated liposomes specifically binding chondroitin sulfate were used for successful suppression of tumor growth and metastases suppression *in vivo*.[111]

All these targeting approaches can certainly be used for the targeted liposomes-mediated delivery of protein and peptide anti-cancer drugs into tumors.

## 6. Transmembrane Delivery of Protein and Peptide Drugs into Tumor Cells

In a search of new anticancer drugs, a shift is gradually occurring from the semi-empirical approach based on evaluation of efficacy of candidate compounds against cultured cancer cells and animal tumor models to molecular mechanism-based drug-discovery.[112] The information regarding cellular metabolic and signaling pathways essential for tumorogenesis allows for the identification of protein/peptide drug targets, and the creation of a working draft of the human genome sequence[113,114] in combination with high-throughput methods of molecular biology promises continued rapid growth in identifying such targets.[115,116] However, not all potential targets are considered suitable for drug development, often because of their intracellular location. Myc/Max dimerization, Src homology-2 domain interaction, and Ras/Raf association have been identified[112] as promising interference targets for the inhibition of tumor development.[117,118] Peptide inhibitors of these interactions, which are relatively easy to identify using the phage display technique[119] or combinatorial peptide libraries,[120] were not considered for drug development apparently because of typical for peptides poor pharmacokinetics and their inability to reach their molecular targets inside the cells. Sometimes, tumor results from malfunctions of tumor suppressors genes and the lack of activity of the proteins they encode.[116,121] In this case, the delivery into tumor cells of working copies of proteins obtained by recombinant methods would provide indispensable tools for validation of gene functions and potential development of protein or gene therapy-based methods of treatment. The use of these proteins for molecular target validation and eventual development of anticancer drugs is again hampered by low permeability of cell membranes.

Thus, the successful delivery of anticancer drugs including proteins and peptides into tumors, for example by means of liposomes and immunoliposomes, does, however, solve only a part of a general efficiency problem. For many protein and peptide drugs, the next task is to achieve their intracellular delivery, since many targets for anticancer drugs are located inside cells.

The very nature of cell membranes prevents the peptide or protein entry unless there is an active transport mechanism, which is usually the case for very short peptides.[122] As mentioned above, vector molecules promote the delivery of associated drug-carriers inside the cells *via* receptor-mediated endocytosis.[89] This process involves attachment of the vector molecule and an associated drug carrier to specific ligands on target cell membranes, followed by endosome formation. An efficient cellular uptake *via* endocytosis is generally observed, but the delivery of intact peptides and proteins is compromised by an insufficient endosomal escape and subsequent lysosomal degradation. Enhanced endosomal escape can be achieved through the use of, for example, lytic peptides,[93,123,124] pH-sensitive polymers[125] or swellable dendritic polymers.[126] These agents have provided encouraging results in overcoming limitations of endocytosis-based cytoplasmic delivery, but there is still a need for further improvement or consideration of alternative delivery strategies.

Different methods of liposomal content delivery into the cytoplasm have been elaborated.[127] According to one of these methods, the liposome is made of pH-sensitive components and, after being endocytosed in the intact form, it fuses with the endovacuolar membrane under the action of lowered pH inside the endosome, releasing its content into the cytoplasm. Attemps have been made to combine the liposomal pH-sensitivity with longevity and ligand-mediated targeting. Thus, long-circulating PEGylated pH-sensitive liposomes, although have a somewhat decreased pH-sensitivity, still effectively deliver their contents into cytoplasm (recent review in Ref. [128]). Antisense oligonucleotides are delivered into cells by anionic pH-sensitive PE-containing liposomes stable in the blood, however, undergoing phase transition at acidic endosomal pH and facilitating oligo release into cell cytoplasm (recent review in Ref. 129 New pH-sensitive liposomal additives were recently described including oleyl alcohol[130] and pH-sensitive mono-stearoyl derivatives of morpholine.[131] Serum stable, long-circulating PEGylated pH-sensitive liposomes were also prepared using, on the same liposome, the combination of PEG and pH-sensitive terminally alkylated copolymer of N-isopropylacrylamide and methacrylic.[132] Combination of liposome pH-sensitivity and specific ligand targeting for cytosolic drug delivery utilizing decreased endosomal pH values was described for both folate and Tf-targeted liposomes.[133–136]

An approach has recently emerged, which allows for a much more straightforward and efficient way for delivery of peptides and proteins to the cytoplasm. This approach is based on the phenomenon called transduction, and uses the ability of certain peptides to ferry conjugated macromolecules, such as proteins[137] and DNA[138] and even nanoparticles,[139–142] across cell membranes directly into cytoplasm. Peptides that cause transduction (protein transduction domains, PTDs, or cell-penetrating peptides, CPPs) were derived from proteins of viruses and *Drosophila Antennapedia* transcription factor and can be as short as 10-to-16-mer.[137,143,144] The details of CPP action on cell membranes are not completely elucidated, however the recent data assume more than one mechanism for CPPs and

CPP-mediated intracellular delivery of various molecules and particles. TAT-mediated intracellular delivery of large molecules and nanoparticles was proved to proceed via the energy-dependent macropinocytosis with subsequent enhanced escape from endosome into the cell cytoplasm,[145–147] while individual CPPs or CPP-conjugated small molecules penetrate cells via electrostatic interactions and hydrogen bonding and do not seem to depend on the energy.[146,148,149] Since traversal through cellular membranes represents a major barrier for efficient delivery of macromolecules into cells, CPPs, whatever their mechanism of action is, may serve to transport various drugs and even drug-loaded pharmaceutical carriers into mammalian cells *in vitro* and *in vivo*.

Several proteins including those involved in oncogenesis, cancer related signal transductuction and cell proliferation pathways have been delivered in active form into various human cells *in vitro* using fused PTD peptides.[150–153] It has also been shown that TAT PTD allows delivery of biologically active proteins into various cells *in vivo*.[154] The recent review of the current status and perspectives of the PTD-mediated delivery of proteins and peptides into cancer cells can be found in.[146] These results open new avenues in the development of peptide- and protein-based anticancer therapeutics with intracellular molecular targets. The use of PTD-modified liposomes should provide a combination of the advantages of microreservoir carriers, enhanced accumulation in tumors *via* the EPR effect and cytoplasmic drug delivery. It was demonstrated that liposomes, could be delivered into various cells by multiple TAT-peptide or other CPP molecules attached to the liposome surface.[141,142,155] Complexes of TAT-peptide-liposomes with a plasmid (plasmid pEGFP-N1 encoding for the Green Fluorescence Protein, GFP) were used for successful *in vitro* transfection of various tumor and normal cells as well as for *in vivo* transfection of tumor cells in mice bearing Lewis lung carcinoma.[156] Whatever mechanisms underlay the TAT-mediated delivery of large cargo, such as liposomes, into cells, the covalent coupling of TAT-peptides to microparticulate drug carriers may provide an efficient tool for cytosolic delivery of various drugs including proteins and peptides *in vitro* and even *in vivo* in certain protocols of the local treatment.

Summarizing, liposomes can successfully serve for nonspecific (*via* the EPR effect) and specific (by the ligand-mediated targeting) delivery of protein and peptide drugs into tumors and even incide individual cancer cells.

The full clinical potential of this approach is expected to be revealed in the near future.

## References


1. Wolf M and Ransberger K (1972) *Enzyme Therapy*, New York: Vantage Press.
2. Holcenberg JS and Roberts J (eds.) (1981) *Enzyme as Drugs*, New York: John Wiley.
3. Torchilin VP (1991) *Immobilized Enzymes in Medicine*, Springer-Verlag: Berlin.
4. Capizzi RL and Cheng Y-C (1981) Therapy of neoplasia with asparaginase in *Enzymes as Drugs*. Holcenberg JS and Roberts J (eds.) John Wiley: New York.
5. Asselin BL (1999) The three asparaginases. Comparative pharmacology and optimal use in childhood leukemia. *Adv Exp Med Biol* **457**:621.
6. Roberts J (1981) Therapy of neoplasia by deprivation of essential amino acids in *Enzymes as Drugs*. Holcenberg JS and Roberts J (eds.) John Wiley: New York.
7. Kalghatgi KK and Bertino JR (1981) Folale-degrading enzymes: A review with special emphasis on carboxypeptidase G in *Enzymes as Drugs*. Holcenberg JS and Roberts J (eds.) John Wiley: New York.
8. Levy CC and Karpetsky TP (1981) Human ribonucleases in *Enzymes as Drugs*. Holcenberg JS and Roberts J (eds.) John Wiley: New York.
9. Baselga J and Albanell J (2001) Mechanism of action of anti-HER2 monoclonal antibodies. *Ann Oncol* **12**(Suppl 1): S35.
10. Harries M and Smith I (2002) The development and clinical use of trastuzumab (Herceptin). *Endocr Relat Cancer* **9**:75.
11. Marshall H (2001) Anti-CD20 antibody therapy is highly effective in the treatment of follicular lymphoma. *Trends Immunol* **22**:183.
12. Froidevaux S and Eberle AN (2002) Somatostatin analogs and radiopeptides in cancer therapy. *Biopolymers* **66**:161.
13. Figg WD, Kruger WDEA, Price DK, Kim S and Dahut WD (2002) Inhibition of angiogenesis: Treatment options for patients with metastatic prostate cancer. *Invest New Drugs* **20**:183.
14. Kerbel R and Folkman J (2002) Clinical translation of angiogenesis inhibitors. *Nat Rev Cancer* **2**:727.
15. Ballard CE, Yu H and Wang B (2002) Recent developments in depsipeptide research. *Curr Med Chem* **9**:471.
16. Langer R (1998) Drug delivery and targeting, *Nature* **392**(Suppl 6679): 5–10.
17. Buchwald H and Rohde TD (1992) Implantable pumps. Recent progress and anticipated future advances. *Asaio J* **38**:772.

18. Kohudic MA (ed.) (1994) *Advances in Controlled Delivery of Drugs*, Technomic Publishing, Co.: Lancaster.
19. Santini JT, Jr, Cima MJ and Langer R (1999) A controlled-release microchip. *Nature* **397**:335.
20. LaVan DA, Lynn DM and Langer R (2002) Moving smaller in drug discovery and delivery. *Nat Rev Drug Discov* **1**:77.
21. J. H. 26. Senior, Fate and behavior of liposomes *in vivo*: A review of controlling factors, *Crit Rev Ther Drug Carrier Syst* **3**:123 (1987).
22. Szoka F and Papahadjopoulos D (1980) Comparative properties and methods of preparation of lipid vesicles (liposomes). *Ann Rev Biophys Bioeng* **9**:467.
23. Madden TD, Bally MB, Hope MJ, Cullis PR, Schieren HP and Janoff AS (1985) Protection of large unilamellar vesicles by trehalose during dehydration: retention of vesicle contents. *Biochim Biophys Acta* **817**:67.
24. Papahadjopoulos D (ed.) (1978) *Liposomes and Their Use in Biology and Medicine*, Ann NY Acad Sci: New York.
25. Lasic DD and Papahadjopoulos D (eds.) (2003) *Medical Applications of Liposomes*, Amsterdam: Elsevier.
26. Torchilin VP and Weissig V (eds.) (2003) *Liposomes: A Practical Approach*, Oxford University Press: UK.
27. Klibanov AL, Maruyama K, Torchilin VP and Huang L (1990) Amphipatic polyethyleneglycols effectively prolong the circulation time of liposomes. *FEBS Lett* **268**:235.
28. Blume G and Cevc G (1993) Molecular mechanism of the lipid vesicle longevity *in vivo*. *Biochim Biophys Acta* **1146**:157.
29. Woodle MC, Newman MS and Cohen JA (1994) Sterically stabilized liposomes: Physical and biological properties. *J Drug Targ* **5**:397.
30. Torchilin VP, Omelyanenko VG, Papisov MI, Bogdanov AA, Jr, Trubetskoy VS, Herron JH and Gentry CA (1994) Poly(ethylene glycol) on the liposome surface: On the mechanism of polymer-coated liposome longevity. *Biochim Biophys Act* **1195**:11.
31. Torchilin VP and Trubetskoy VS (1995) Which polymers can make nanoparticulate drug carriers long-circulating? *Adv Drug Deliv Rev* **16**:141.
32. Allen TM and Hansen C (1991) Pharmacokinetics of stealth versus conventional liposomes: Effect of dose. *Biochim Biophys Acta* **1068**:133.
33. Woodle MC (1998) Controlling liposome blood clearance by surface-grafted polymers. *Adv Drug Deliv Rev* **32**:139.
34. Whiteman KR, Subr V, Ulbrich K and Torchilin VP (2001) Poly(HPMA)-coated liposomes demonstrate prolonged circulation in mice. *Liposome Res* **11**:153.

35. Torchilin VP, Levchenko TS, Whiteman KR, Yaroslavov AA, Tsatsakis AM, Rizos AK, Michailova EV and Shtilman MI (2001) Amphiphilic poly-N-vinylpyrrolidones: Synthesis, properties and liposome surface modification. *Biomaterials* **22**:3035.
36. Metselaar JM, Bruin P, de Boer LW, de Vringer T, Snel C, Oussoren C, Wauben MH, Crommelin DJ, Storm G and Hennink WE (2003) A novel family of L-amino acid-based biodegradable polymer-lipid conjugates for the development of long-circulating liposomes with effective drug-targeting capacity. *Bioconj Chem* **14**:1156.
37. Takeuchi H, Kojima H, Yamamoto H and Kawashima (2001) Evaluation of circulation profiles of liposomes coated with hydrophilic polymers having different molecular weights in rats. *J Control Rel* **75**:83.
38. Torchilin VP, Klibanov AL, Huang L, O'Donnell S, Nossif ND and Khaw BA (1992) Targeted accumulation of polyethylene glycol-coated immunoliposomes in infarcted rabbit myocardium, *FASEB J* **6**:2716.
39. Blume G, Cevc G, Crommelin MDJA, Bakker-Woundenberg IAJM, Kluft C and Storm G (1993) Specific targeting with poly(ethylene glycol)-modified liposomes: Coupling of homing devices to the ends of polymeric chains combines effective target binding with long circulation times. *Biochim Biophys Acta* **1149**:180.
40. Abra RM, Bankert RB, Chen F, Egilmez NK, Huang K, Saville R, Slater JL, Sugano M and Yokota SJ (2002) The next generation of liposome delivery systems: recent experience with tumor-targeted, sterically-stabilized immunoliposomes and active-loading gradients. *J Liposome Res* **12**:1.
41. Torchilin VP, Levchenko TS, Lukyanov AN, Khaw BA, Klibanov AL, Rammohan R, Samokhin G and Whiteman KR (2001) p-Nitrophenylcarbonyl-PEG-PE-liposomes: Fast and simple attachment of specific ligands, including monoclonal antibodies, to distal ends of PEG chains via p-nitrophenylcarbonyl groups. *Biochim Biophys Acta* **1511**:397.
42. Yuan F, Leunig M, Huang SK, Berk DA, Papahadjopoulos D and Jain RK (1994) Microvascular permeability and interstitial penetration of sterically stabilized (stealth) liposomes in a human tumor xenograft. *Cancer Res* **54**:3352.
43. Gabizon AA (2001) Pegylated liposomal doxorubicin: Metamorphosis of an old drug into a new form of chemotherapy, *Cancer Inv* **19**:424.
44. Maeda H (2001) SMANCS and polymer-conjugated macromolecular drugs: Advantages in cancer chemotherapy, *Adv Drug Deliv Rev* **46**:169.
45. Maeda H, Sawa HT and Konno T (2001) Mechanism of tumor-targeted delivery of macromolecular drugs, including the EPR effect in solid tumor and clinical overview of the prototype polymeric drug SMANCS, *J Control Rel* **74**:47.

46. Symon Z, Peyser A, Tzemach D, Lyass O, Sucher E, Shezen E and Gabizon A (1999) Selective delivery of doxorubicin to patients with breast carcinoma metastases by stealth liposomes. *Cancer* **86**:72.
47. Perez AT, Domenech GH, Frankel C and Vogel CL (2002) Pegylated liposomal doxorubicin (Doxil) for metastatic breast cancer: The Cancer Research Network, Inc., experience. *Cancer Invest* **20**(Suppl 2).
48. O'Shaughnessy JA (2003) Pegylated liposomal doxorubicin in the treatment of breast cancer. *Clin Breast Cancer* **4**:318.
49. Schwonzen M, Kurbacher CM and Mallmann P (2000) Liposomal doxorubicin and weekly paclitaxel in the treatment of metastatic breast cancer. *Anticancer Drugs* **11**:681.
50. Goncalves A, Braud AC, Viret F, Genre D, Gravis G, Tarpin C, Giovannini M, Maraninchi D and Viens P (2003) Phase I study of pegylated liposomal doxorubicin (Caelyx) in combination with carboplatin in patients with advanced solid tumors. *Anticancer Res* **23**:3543.
51. Harrington KJ, Lewanski C, Northcote AD, Whittaker J, Peters AM, Vile RG and Stewart JS (2001) Phase II study of pegylated liposomal doxorubicin (Caelyx) as induction chemotherapy for patients with squamous cell cancer of the head and neck. *Eur J Cancer* **37**:2015.
52. Johnston SR and Gore ME (2001) Caelyx: Phase II studies in ovarian cancer. *Eur J Cancer* **37**(Suppl 9):S8.
53. Schmidinger M, Wenzel C, Locker GJ, Muehlbacher F, Steininger R, Gnant M, Crevenna R and Budinsky AC (2001) Pilot study with pegylated liposomal doxorubicin for advanced or unresectable hepatocellular carcinoma. *Br J Cancer* **85**:1850.
54. Wollina U, Dummer R, Brockmeyer NH, Konrad H, Busch JO, Kaatz M, Knopf B, Koch HJ and Hauschild A (2003) Multicenter study of pegylated liposomal doxorubicin in patients with cutaneous T-cell lymphoma. *Cancer* **98**:993.
55. Skubitz KM (2003) Phase II trial of pegylated-liposomal doxorubicin (Doxil) in sarcoma. *Cancer Invest* **21**:167.
56. Seiden MV, Muggia F, Astrow A, Matulonis U, Campos S, Roche M, Sivret J, Rusk J and Barrett E (2004) A phase II study of liposomal lurtotecan (OSI-211) in patients with topotecan resistant ovarian cancer. *Gynecol Oncol* **93**:229.
57. Fishman Y and Citri N (1975) L-asparaginase entrapped in liposomes: Preparation and properties. *FEES Lett* **60**:17.
58. Gaspar MM, Perez-Soler R and Cruz ME (1996) Biological characterization of L-asparaginase liposomal formulations. *Cancer Chemother Pharmacol* **38**:373.

59. Jorge JC, Perez-Soler R, Morais JG and Cruz ME (1994) Liposomal palmitoyl-L-asparaginase: Characterization and biological activity. *Cancer Chemother Pharmacol* **34**:230.
60. Zheng X, Lundberg M, Karlsson A and Johansson M (2003) Lipid-mediated protein delivery of suicide nucleoside kinases. *Cancer Res* **63**:6909.
61. Yagi K, Naoi M, Sakai H, Abe H, Konishi H and Arichi S (1982) Incorporation of enzyme into the brain by means of liposomes of novel composition. *J Appl Biochem* **4**:121.
62. Naoi M and Yagi K (1980) Incorporation of enzyme through blood-brain barrier into the brain by means of liposomes. *Biochem Intl* **1**:591
63. Bagshawe KD, Springer CJ, Searle F, Antoniw P, Sharma SK, Melton RG and Sherwood RF (1988) A cytotoxic agent can be generated selectively at cancer sites. *Br J Cancer* **58**:700.
64. Senter PD, Saulnier MG, Schreiber GJ, Hirschberg DL, Brown JP, Hellstrom and Hellstrom KE (1988) Anti-tumor effects of antibody-alkaline phosphatase conjugates in combination with etoposide phosphate. *Proc Natl Acad Sci USA* **85**:4842.
65. Deonarian MP and Epenetos AA (1994) Targeting enzymes in cancer therapy: Old enzymes in new roles. *Br J Cancer* **70**:786.
66. Storm G, Vingerhoeds MH, Crommelin DJA and Haisma HJ (1997) Immunoliposomes bearing enzymes (immuno-enzymosomes) for site-specific activation of anticancer prodrugs. *Adv Drug Deliv Rev* **24**:225.
67. Fonseca MJ, Jagtenberg JC, Haisma HJ and Storm G (2003) Liposome-mediated targeting of enzymes to cancer cells for site-specific activation of prodrugs: Comparison with the corresponding antibody-enzyme conjugate. *Pharm Res* **20**:423.
68. Vingerhoeds MH, Haisma HJ, Belliot SO, Smit RH, Crommelin DJ and Storm G (1996) Immunoliposomes as enzyme-carriers (immuno-enzymosomes) for antibody-directed enzyme prodrug therapy (ADEPT): Optimization of prodrug activating capacity. *Pharm Res* **13**:604.
69. Kanaoka E, Takahashi K, Yoshikawa T, Jizomoto H, Nishihara Y and Hirano K (2001) A novel and simple type of liposome carrier for recombinant interleukin-2. *J Pharm Pharmacol* **53**:295–302.
70. Kleinerman ES, Gano JB, Johnston DA, Benjamin RS and Jaffe N (1995) Efficacy of liposomal muramyl tripeptide (CGP 19835A) in the treatment of relapsed osteosarcoma. *Am J Clin Oncol* **18**:93.
71. Opanasopit P, Sakai M, Nishikawa M, Kawakami S, Yamashita F and Hashida M (2002) Inhibition of liver metastasis by targeting immunomodulators using mannosylated liposome carriers. *J Contr Rel* **80**:283.

72. ten Hagen TLM and Eggermont AMM (1997) A rat extremity soft tissue sarcoma model for the study of systemic treatment with Stealth® liposome-encapsulated tumor necrosis factor-$\alpha$ and cytotoxic agents. *Adv Drug Deliv Rev* **24**:245.
73. Li H, Song JH, Park JS and Han K (2003) Polyethylene glycol-coated liposomes for oral delivery of recombinant human epidermal growth factor. *Int J Pharm* **258**:11.
74. Selzman CH, Shames BD, Reznikov LL, Miller SA, Meng X, Barton HA, Werman A, Harken AH, Dinarello CA and Banerjee A (1999) Liposomal delivery of purified inhibitory-kappaBalpha inhibits tumor necrosis factor-alpha-induced human vascular smooth muscle proliferation. *Circ Res* **84**:867.
75. Fidler IJ (1980) Therapy of spontaneous metastases by intravenous injection of liposomes containing lymphokines. *Science* **208**:1469.
76. Kleinerman ES, Fogier WE and Fidler IJ (1985) Intracellular activation of human and rodenl macrophages by human lymphokines encapsulated in liposomes. *J Leucocyte Biol* **37**:571.
77. van Slooten ML, Visser AJ, van Hoek A, Storm G, Crommelin DJ and Jiskoot W (2000) Conformational stability of human interferon-gamma on association with and dissociation from liposomes. *J Pharm Sci* **89**:1605.
78. van Slooten ML, Storm G, Zoephel A, Kupcu Z, Boerman O, Crommelin DJ, Wagner E and Kircheis R (2000) Liposomes containing interferon-gamma as adjuvant in tumor cell vaccines. *Pharm Res* **17**:42.
79. Turk MJ, Reddy JA, Chmielewski JA and Low PS (2002) Characterization of a novel pH-sensitive peptide that enhances drug release from folate-targeted liposomes at endosomal pHs. *Biochim Biophys Acta* **1559**:56.
80. Hudson LDS, Fiddler MB and Desnick RJ (1979) Enzyme therapy. X. Immune response induced by enzyme and buffer-loaded liposomes in C3H/HeJ Gush mice. *J Pharmacol Exp Ther* **208**:507.
81. Gregoriadis G (1977) Targeting of drugs. *Nature* **265**:407.
82. Torchilin VP (ed.) (1995) *Handbook of Targeted Delivery of Imaging Agents*, CRS Press: Boca Raton.
83. Jurcic JG, Scheinberg DA and Houghton AN (1997) Monoclonal antibody therapy of cancer. *Cancer Chemother Biol Response Modif* **17**:195.
84. Dillman RO (2001) Monoclonal antibodies in the treatment of malignancy: Basic concepts and recent developments. *Cancer Inv* **19**:833.
85. Iakoubov L, Rokhlin O and Torchilin VP (1995) Anti-nuclear autoantibodies of the aged reactive against the surface of tumor but not normal cells. *Immunol Lett* **47**:147.

86. Iakoubov LZ and Torchilin VP (1997) A novel class of antitumor antibodies: Nucleosome-restricted antinuclear autoantibodies (ANA) from healthy aged nonautoimmune mice. *Oncol Res* **9**:439.
87. Iakoubov LZ and Torchilin VP (1998) Nucleosome-releasing treatment makes surviving tumor cells better targets for nucleosome-specific anticancer antibodies. *Cancer Detect Prev* **22**:470.
88. Sapra P and Allen TM (2002) Internalizing antibodies are necessary for improved therapeutic efficacy of antibody-targeted liposomal drugs. *Cancer Res* **62**:7190.
89. Park JW, Kirpotin DB, Hong K, Shalaby R, Shao Y, Nielsen UB, Marks JD, Papahadjopoulos D and Benz CC (2001) Tumor targeting using anti-her2 immunoliposomes. *J Control Rel* **74**:95.
90. Kamps JA, Koning GA, Velinova MJ, Morselt HW, Wilkens M, Gorter A, Donga J and Scherphof GL (2000) Uptake of long-circulating immunoliposomes, directed against colon adenocarcinoma cells, by liver metastases of colon cancer. *J Drug Targ* **8**:235.
91. Lukyanov AN, Elbayoumi TA, Chakilam AR and Torchilin VP (2004) Tumor-targeted liposomes: Doxorubicin-loaded long-circulating liposomes modified with anti-cancer antibody. *J Control Rel* **100**:135.
92. Raffaghello L, Pagnan G, Pastorino F, Cosimo E, Brignole C, Marimpietri D, Bogenmann E, Ponzoni M and Montaldo PG (2003) Immunoliposomal fenretinide: A novel antitumoral drug for human neuroblastoma. *Cancer Lett* **197**:151.
93. Mastrobattista E, Koning GA, van Bloois L, Filipe AC, Jiskoot W and Storm G (2002) Functional characterization of an endosome-disruptive peptide and its application in cytosolic delivery of immunoliposome-entrapped proteins. *J Biol Chem* **277**:27135.
94. Leamon CP and Low PS (1991) Delivery of macromolecules into living cells: A method that exploits folate receptor endocytosis. *Proc Natl Acad Sci USA* **88**:5572.
95. Lee RJ and Low PS (1994) Delivery of liposomes into cultured KB cells via folate receptor-mediated endocytosis. *J Biol Chem* **269**:3198.
96. Lu Y and Low PS (2002) Folate-mediated delivery of macromolecular anticancer therapeutic agents. *Adv Drug Deliv Rev* **54**:675
97. Gabizon A, Shmeeda H, Horowitz AT and Zalipsky S (2004) Tumor cell targeting of liposome-entrapped drugs with phospholipid-anchored folic acid-PEG conjugates. *Adv Drug Deliv Rev* **56**:1177.
98. Ni S, Stephenson SM and Lee RJ (2002) Folate receptor targeted delivery of liposomal daunorubicin into tumor cells. *Anticancer Res* **22**:2131.

99. Pan XQ, Wang H and Lee (2003) Antitumor activity of folate receptor-targeted liposomal doxorubicin in a KB oral carcinoma murine xenograft model. *Pharm Res* **20**:417.
100. Lu Y and Low PS (2002) Folate targeting of haptens to cancer cell surfaces mediates immunotherapy of syngeneic murine tumors. *Cancer Immunol Immunother* **51**:153.
101. Hatakeyama H, Akita H, Maruyama K, Suhara T and Harashima H (2004) Factors governing the *in vivo* tissue uptake of transferrin-coupled polyethylene glycol liposomes *in vivo*. *Int J Pharm* **281**:25.
102. Ishida O, Maruyama K, Tanahashi H, Iwatsuru M, Sasaki K, Eriguchi M and Yanagie H (2001) Liposomes bearing polyethyleneglycol-coupled transferrin with intracellular targeting property to the solid tumors *in vivo*. *Pharm Res* **18**:1042.
103. Eavarone DA, Yu X and Bellamkonda RV (2000) Targeted drug delivery to C6 glioma by transferrin-coupled liposomes. *J Biomed Mater Res* **51**:10.
104. Drummond DC, Hong K, Park JW, Benz CC and Kirpotin DB (2000) Liposome targeting to tumors using vitamin and growth factor receptors. *Vitam Horm* **60**:285.
105. Dagar S, Krishnadas A, Rubinstein I, Blend MJ and Onyuksel H (2003) VIP grafted sterically stabilized liposomes for targeted imaging of breast cancer: *In vivo* studies. *J Control Rel* **91**:123.
106. Schiffelers RM, Koning GA, ten Hagen TL, Fens MH, Schraa AJ, Janssen AP, Kok KJ, Molema G and Storm G (2003) Anti-tumor efficacy of tumor vasculature-targeted liposomal doxorubicin. *J Control Rel* **91**:115.
107. Asai T, Shimizu K, Kondo M, Kuromi K, Watanabe K, Ogino K, Taki T, Shuto S, Matsuda A and Oku N (2002) Anti-neovascular therapy by liposomal DPP-CNDAC targeted to angiogenic vessels. *FEBS Lett* **520**:167.
108. Mamot C, Drummond DC, Greiser U, Hong K, Kirpotin DB, Marks JD and Park JW (2003) Epidermal growth factor receptor (EGFR)-targeted immunoliposomes mediate specific and efficient drug delivery to EGFR- and EGFRvIII-overexpressing tumor cells. *Cancer Res* **63**:3154.
109. Peer D and Margalit R (2004) Loading mitomycin C inside long circulating hyaluronan targeted nano-liposomes increases its antitumor activity in three mice tumor models. *Int J Cancer* **108**:780.
110. Matsuda I, Konno H, Tanaka T and Nakamura S (2001) Antimetastatic effect of hepatotropic liposomal adriamycin on human metastatic liver tumors. *Surg Today* **31**:414.
111. Lee CM, Tanaka T, Murai T, Kondo M, Kimura J, Su W, Kitagawa T, Ito T, Matsuda H and Miyasaka M (2002) Novel chondroitin sulfate-binding cationic

liposomes loaded with cisplatin efficiently suppress the local growth and liver metastasis of tumor cells *in vivo*. *Cancer Res* **62**:4282.

112. Gibbs JB (2000) Mechanism-based target identification and drug discovery in cancer research. *Science* **287**:1969.
113. Venter JC, *et al.* (2001) The sequence of the human genome. *Science* **291**:1304.
114. Lander ES, *et al.* (2001) Initial sequencing and analysis of the human genome. *Nature* **409**:860.
115. Workman P (2001) New drug targets for genomic cancer therapy: Successes, limitations, opportunities and future challenges. *Curr Cancer Drug Targets* **1**:3.
116. Balmain A (2001) Cancer genetics: From Boveri and Mendel to microarrays. *Nat Rev Cancer* **1**:77.
117. Gibbs JB and Oliff A (1994) Pharmaceutical research in molecular oncology. *Cell* **79**:193.
118. Sawyer TK (1998) Src homology-2 domains: Structure, mechanisms, and drug discovery. *Biopolymers* **47**:243.
119. Katz DA (1997) Structural and mechanistic determinants of affinity and specificity of ligands discovered or engineered by phage display. *Annu Rev Biophys Biomol Struct* **26**:27.
120. Cortese R, (ed.) (1995) *Combinatorial Libraries: Synthesis, Screening and Application Potential*, Walter de Gruyter, Inc: New York.
121. Hussain SP, Hofseth LJ, Harris CC (2001) Tumor suppressor genes: At the crossroads of molecular carcinogenesis, molecular epidemiology and human risk assessment. *Lung Cancer* **34**(Suppl 2):S7.
122. Egleton RD and Davis TP (1997) Bioavailability and transport of peptides and peptide drugs into the brain. *Peptides* **18**:1431.
123. Kamata H, Yagisawa, Takahashi S and Hirata H (1994) Amphiphilic peptides enhance the efficiency of liposome-mediated DNA transfection. *Nucleic Acids Res* **22**:536.
124. Midoux P, Kichler A, Boutin V, Maurizot JC and Monsigny M (1998) Membrane permeabilization and efficient gene transfer by a peptide containing several histidines. *Bioconjug Chem* **9**:260.
125. Lackey CA, Press OW, Hoffman AS and Stayton PS (2002) A biomimetic pH-responsive polymer directs endosomal release and intracellular delivery of an endocytosed antibody complex. *Bioconjug Chem* **13**:996.
126. Padilla De Jesus OL, Ihre HR, Gagne L, Frechet JM and Szoka FC Jr (2002) Polyester dendritic systems for drug delivery applications: *In vitro* and *in vivo* evaluation. *Bioconjug Chem* **13**:453.

127. Connor J and Huang L (1986) pH-sensitive immunoliposonies as an efficient and target-specific carrier for antitumor drugs. *Cancer Res* **46**:3431.
128. Simoes S, Moreira JN, Fonseca C, Duzgunes N and de Lima MC (2004) On the formulation of pH-sensitive liposomes with long circulation times. *Adv Drug Deliv Rev* **56**:947.
129. Fattal E, Couvreur P and Dubernet C (2004) "Smart" delivery of antisense oligonucleotides by anionic pH-sensitive liposomes. *Adv Drug Deliv Rev* **56**:931.
130. Sudimack JJ, Guo W, Tjarks W and Lee RJ (2002) A novel pH-sensitive liposome formulation containing oleyl alcohol. *Biochim Biophys Acta* **1564**:31.
131. Asokan A and Cho MJ (2003) Cytosolic delivery of macromolecules. II. Mechanistic studies with pH-sensitive morpholine lipids. *Biochim Biophys Acta* **1611**:151.
132. Roux E, Passirani C, Scheffold S, Benoit JP and Leroux JC (2004) Serum-stable and long-circulating, PEGylated, pH-sensitive liposomes. *J Control Rel* **94**:447.
133. Reddy AJ and Low PS (2000) Enhanced folate receptor mediated gene therapy using a novel pH-sensitive lipid formulation. *J Control Rel* **64**:27.
134. Turk MJ, Reddy JA, Chmielewski JA and Low PS (2002) Characterization of a novel pH-sensitive peptide that enhances drug release from folate-targeted liposomes at endosomal pHs. *Biochim Biophys Acta* **1559**:56.
135. Kakudo T, Chaki S, Futaki S, Nakase I, Akaji K, Kawakami T, Maruyama K, Kamiya H and Harashima H (2004) Transferrin-modified liposomes equipped with a pH-sensitive fusogenic peptide: An artificial viral-like delivery system. *Biochemistry* **43**:5618.
136. Shi G, Guo W, Stephenson SM and Lee RJ (2002) Efficient intracellular drug and gene delivery using folate receptor-targeted pH-sensitive liposomes composed of cationic/anionic lipid combinations. *J Control Rel* **80**:309.
137. Vives E, Brodin P and Lebleu B (1997) A truncated HIV-1 Tat protein basic domain rapidly translocates through the plasma membrane and accumulates in the cell nucleus. *J Biol Chem* **272**:16010.
138. Wagner E (1999) Application of membrane-active peptides for nonviral gene delivery. *Adv Drug Deliv Rev* **38**:279.
139. Lewin M, Carlesso N, Tung CH, Tang XW, Cory D, Scadden DT and Weissleder R (2000) Tat peptide-derivatized magnetic nanoparticles allow in vivo tracking and recovery of progenitor cells. *Nat Biotechnol* **18**:410.
140. Bhorade R, Weissleder R, Nakakoshi T, Moore A and Tung CH (2000) Macrocyclic chelators with paramagnetic cations are internalized into mammalian cells via a HIV-tat derived membrane translocation peptide. *Bioconjug Chem* **11**:301.

141. Torchilin VP, Rammohan R, Weissig V, Levchenko TS (2001) TAT peptide on the surface of liposomes affords their efficient intracellular delivery even at low temperature and in the presence of metabolic inhibitors. *Proc Natl Acad USA Sci* **98**:8786.
142. Tseng YL, Liu JJ and Hong RL (2002) Translocation of liposomes into cancer cells by cell-penetrating peptides penetratin and tat: A kinetic and efficacy study. *Mol Pharmacol* **62**:864.
143. Derossi D, Joliot AH, Chassaing G and Prochiantz A (1994) The third helix of the Antennapedia homeodomain translocates through biological membranes. *J Biol Chem* **269**:10444.
144. Elliott G and O'Hare P (1997) Intercellular trafficking and protein delivery by a herpesvirus structural protein. *Cell* **88**:223.
145. Wadia JS, Stan RV and Dowdy SF (2004) Transducable TAT-HA fusogenic peptide enhances escape of TAT fusion proteins after lipid raft macropinocytosis. *Nat Med* **10**:310.
146. Gupta B, Levchenko TS and Torchilin VP (2005) Intracellular delivery of large molecules and small particles by cell-penetrating proteins and peptides. *Adv Drug Deliv Rev* **57**:637.
147. Wadia JS and Dowdy SF (2005) Transmembrane delivery of protein and peptide drugs by TAT-mediated transduction in the treatment of cancer. *Adv Drug Deliv Rev* **57**:579.
148. Rothbard JB, Jessop TC, Lewis RS, Murray BA and Wender PA (2004) Role of membrane potential and hydrogen bonding in the mechanism of translocation of guanidinium-rich peptides into cells. *J Am Chem Soc* **126**:9506.
149. Rothbard JB, Jessop TC and Wender PA (2005) Adaptive translocation: The role of hydrogen bonding and membrane potential in the uptake of guanidinium-rich transporters into cells. *Adv Drug Deliv Rev* **57**:495.
150. Vocero-Akbani A, Lissy ang NA and Dowdy SF (2000) Transduction of full-length Tat fusion proteins directly into mammalian cells: Analysis of T cell receptor activation-induced cell death. *Methods Enzymol* **322**:508.
151. Soga N, Namba N, McAllister S, Cornelius L, Teitelbaum SL, Dowdy SF, Kawamura J and Hruska KA (2001) Rho family GTPases regulate VEGF-stimulated endothelial cell motility. *Exp Cell Res* **269**:73.
152. Zezula J, Casaccia-Bonnefil P, Ezhevsky SA, Osterhout DJ, Levine JM, Dowdy SF, Chao MV and Koff A (2001) p21cip1 is required for the differentiation of oligodendrocytes independently of cell cycle withdrawal. *EMBO Rep* **2**:27.
153. Hsia CY, Cheng S, Owyang AM, Dowdy SF and Liou HC (2002) c-Rel regulation of the cell cycle in primary mouse B lymphocytes. *Int Immunol* **14**:905.

154. Schwarze SR, Ho A, Vocero-Akbani A and Dowdy SF (1999) *In vivo* protein transduction: Delivery of a biologically active protein into the mouse. *Science* **285**:1569.

155. Gorodetsky R, Levdansky L, Vexler A, Shimeliovich I, Kassis I, Ben-Moshe M, Magdassi S and Marx G (2004) Liposome transduction into cells enhanced by haptotactic peptides (Haptides) homologous to fibrinogen C-termini. *J Control Rel* **95**:477.

156. Torchilin VP, Levchenko TS, Rammohan R, Volodina N, Papahadjopoulos-Sternberg B and D'Souza GG (2003) Cell transfection *in vitro* and *in vivo* with nontoxic TAT peptide-liposome-DNA complexes. *Proc Natl USA Acad Sci* **100**:1972.

# 9

# Folate-Mediated Delivery of Protein and Peptide Drugs into Tumors

*Joseph A. Reddy, Christopher P. Leamon and Philip S. Low*

The membrane-bound folate receptor (FR) is overexpressed on a wide range of human cancers, such as those originating in ovary, lung, breast, endometrium, kidney and brain. The vitamin folic acid (FA) is a high affinity ligand of the FR, which retains its receptor binding properties when conjugated to other molecules. Consequently, folate mediated targeting has been successfully applied for the delivery of a wide variety of payloads, including protein toxins, immunotherapeutic agents, adenoviral vectors and radio-imaging agents to FR-positive cancers. These folate-bearing delivery systems have exhibited major enhancements in cancer cell specificity and selectivity over their non-targeted counterparts. Hence, it is hopeful that this targeting strategy should lead to improvements in the safety and efficacy of anti cancer agents, resulting in an increased therapeutic advantage in the clinic. This chapter reviews the growing literature on folate-targeted drugs, with specific emphasis on the delivery of proteins and peptides to cancers.

# 1. Introduction

For most cancer therapies, an increase in the fraction of cells killed requires an exponential increase in the dose of anti cancer drug that is administered. However, as most chemotherapeutic agents distribute indiscriminately into both malignant and non-malignant cells, which leads to toxic side effects on actively proliferating healthy tissues, only limited doses of most chemotherapeutic agents can be administered. Therefore, techniques are required to selectively target drug molecules to the tumor site without concomitant delivery into normal tissues. One solution to this problem is to selectively deliver the anti-proliferative agents to tumor cells by conjugating them to tumor-homing ligands. An example of this type of targeting system is the use of the FA ligand to deliver attached therapeutic and imaging agents to cells that express the FR.[1–3]

The FR displays high affinity for FA, with $K_d$ values ranging between 0.1 and 1 nM.[4] When FA is covalently linked to another molecule, the resulting conjugate retains its affinity for the cell surface FR. Following their binding to the FR, folates and their drug conjugates are taken up by cells via a receptor-mediated endocytic mechanism.[5] As depicted in Fig. 1, the endocytosis process begins with the invagination of the surrounding plasma

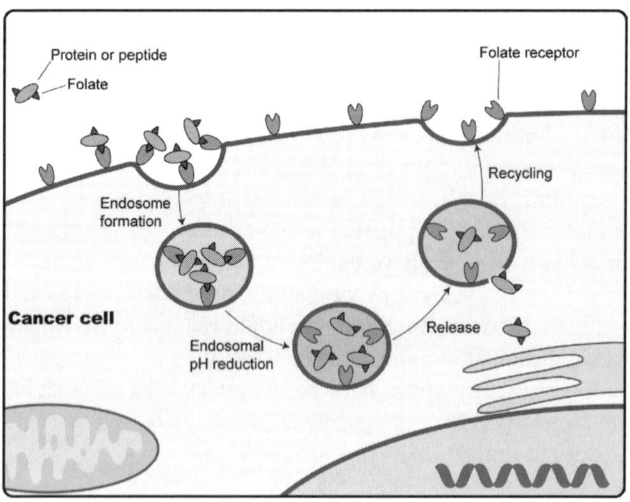


**Fig. 1** Schematic representation of the internalization of folate-protein/peptide conjugates by FR mediated endocytosis pathway.

membrane to form a distinct membrane-derived compartment (called an endosome) within the cytoplasm.

An internal proton gradient then acidifies the endosomes and promotes the release of FA (or the FA-drug conjugate) from its membrane-anchored receptor. The released FA conjugate is then free to enter the cytosol through membrane permeation, or by simple leakage of the vitamin during imperfect membrane fusion events.[6]

The FR protein is not evenly distributed among the various tissues and cell types of the body. For example, FR is expressed in limited quantities on certain normal epithelial cells, but it is significantly overexpressed on malignant tissues derived from the same cell types.[7–10] In fresh tumor specimens, the FR has been found in >90% of ovarian and other gynecological cancers, approximately 50% of lung carcinomas, and 25% of breast cancers.[9,11,12] Interestingly, when FR levels were compared among many ovarian cancer tissue samples, the highest levels of FR expression were measured in the most aggressive (late stage, high grade) forms of this disease.[13] As a consequence of this upregulation, the FR was deemed an appropriate tumor marker,[14] and it has therefore been exploited clinically to distinguish malignant from nonmalignant tissues.[7–10] FR has also been used pre-clinically and clinically as a targetable receptor for the delivery of various pharmaceutical agents to cancer cells.

While numerous applications of folate-targeted technology have been explored,[15] we will focus the remaining sections of this chapter on techniques related to the delivery of protein/peptide based therapeutic and imaging agents.

## 2. Methods for Coupling FA to Proteins and Peptides

FA is a light sensitive organic molecule with limited water solubility (10 mg/L at 0°C, 500 mg/L at 100°C) in its unionized form and significantly greater solubility (15 g/L at 0°C) as a disodium salt.[16] In aqueous solution, FA and some of its conjugates tend to associate into hydrogen bonded planar tetramers, and these tetramers may further stack to form octamers, dodecamers and so forth, that yield separate peaks when analyzed by HPLC.[17] FA is also poorly soluble in most organic solvents except for dimethylsulfoxide (DMSO) and dimethyl formamide (DMF). When dried in a vacuum desiccator under reduced pressure, FA retains

5 to 11% moisture, which corresponds to ~1 to 2.5 moles of water per mole of FA. However, this water can be removed by drying under vacuum at temperatures over 140°C.[16] Finally, FA can be destroyed by extended exposure to high temperatures or strongly alkaline solutions. All of these properties must be kept in mind when preparing and purifying folate-drug conjugates.

FA can be most easily conjugated to a ligand via either its $\alpha$ or $\gamma$ carboxylic acid moiety. The use of 1-ethyl-3-(3-dimethyl aminopropyl) carbodiimide (EDC) is the preferred method for coupling multiple FA ligands to the free amines of water soluble macromolecules such as proteins,[18] antibodies[19,20] and amine-containing polymers.[21] In this procedure, FA is dissolved in anhydrous DMSO and mixed with a fivefold molar excess of EDC for one hour. The macromolecule is then dissolved in a phosphate-boric acid buffer, pH 8.5, and incubated with a 10–20 fold molar excess of the activated vitamin for 1–16 hr. In the case of poly-lysine labeling, however, the reaction is successfully conducted in 20 mM sodium phosphate buffer, pH 4.5.[22] Unreacted FA can then be separated from the labeled macromolecule using a Sephadex G-25 column equilibrated in phosphate-buffered saline, pH 7.4. Adjustment of the reaction mixture to pH 9 has been found to be useful for complete removal of free FA from the FA-conjugate during gel separation.[21] To quantitatively eliminate unbound FA, the conjugate can also be extensively dialyzed against a neutral pH buffer. The extent of FA conjugation can be determined spectrophotometrically at 363 nm (FA: $E_m = 6197\ M^{-1}$ in phosphate buffered saline, pH 7.4).

Low molecular weight organic molecules like deferoxamine mesylate[23] or polyoxyethylene-bis-amine,[24] which are soluble in organic solvents like DMSO, can be conjugated to FA by employing the carboxylic acid activating agent, dicyclohexylcarbodiimide (DCC). Equimolar quantities of FA and DCC, plus trace amounts of pyridine are mixed with the ligand, and then stirred overnight. The insoluble by-product, dicyclohexylurea, is easily separated by centrifugation. If desired, the conjugate can often be precipitated with acetone and washed with diethyl ether. The product can then be purified by anion exchange chromatography on a DEAE-trisacryl Sepharose column using 50 mM ammonium bicarbonate buffer as an eluant.[25]

Both the EDC and DCC procedures yield a mixture of $\gamma$- and $\alpha$-folate derivatives, with attachment at the $\gamma$-carboxyl being favored. However, a $\gamma$-specific folate derivative can also be generated by using more involved procedures. In the first method (Fig. 2), FA (**a**) is first treated with excess trifluoroacetic anhydride and then hydrolyzed to give $N^{10}$-(trifluroacetyl) pyrofolic acid (**b**). Reaction of this intermediate (**b**) with hydrazine yields pteroyl hydrazide (**c**), that is then oxidized to pteroyl azide (**d**). Pteroyl azide is then reacted with methyl glutamate in the presence of tetramethyl

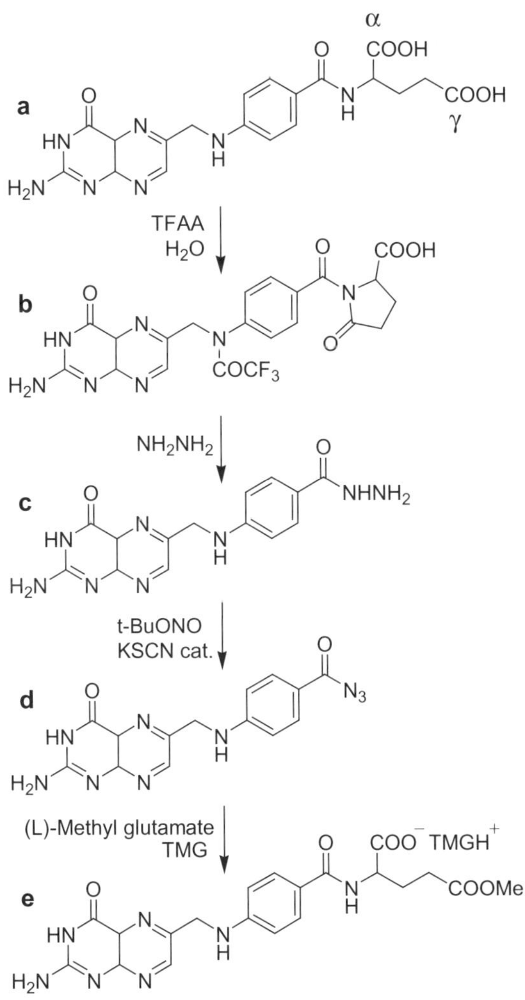


**Fig. 2** Synthetic scheme of the $\gamma$-methyl folate procedure. TFAA, trifluoroacetic anhydride; t-BuONO, tert-butyl nitrite; KSCN, potassium thiocyanate; TMG, tetramethyl guanidine.

guanidine to afford exclusively $\gamma$-methylfolate (**e**), a $\gamma$-activated FA. The $\gamma$-methylfolate can be reacted with a free amine-containing molecule to give a $\gamma$-specific conjugate. For example, reaction of $\gamma$-methylfolate with ethylenediamine generates folate-$\gamma$Glu-ethylenediamine in 87% yield.[26]

A second method relies on solid phase peptide chemistry.[27] With this approach, an appropriately modified p-hydroxymethyl phenoxymethyl polystyrene (HMP) resin is coupled sequentially with Fmoc-Glu-OtBu followed by terminal coupling with $N_{10}$-trifluoroacetylpteroic acid. The isomerically pure $\gamma$-Glu derivative is then recovered following standard cleavage and deprotection steps.

## 3. Cytotoxicity Studies of Folate Linked Toxins

Folate-targeted protein delivery was first introduced by Leamon and Low.[28] Here, FA conjugated proteins were found to be internalized within FR-positive cells via a FR-mediated endocytic process. The specific involvement of FR-mediated endocytosis in this novel delivery strategy was established with the following controls: (i) only FA-derivatized proteins were shown to concentrate inside FR-expressing cells, while identical proteins lacking the FA ligand displayed no affinity for the same cells[28]; (ii) pretreatment of FR-positive cells with an excess of free FA or anti-FR antibodies was found to completely inhibit the uptake of FA conjugates[29]; and (iii) selective release of FR's from their membrane anchors by cell surface digestion with a phosphatidylinositol-specific phospholipase C rendered the treated cells unable to internalize FA conjugates.[29] Taken together, these data demonstrated that the well-characterized FR was responsible for mediating the uptake of FA conjugates into FR-bearing cells.

To begin to explore possible functional applications of this novel targeting technique, conjugates of the highly toxic proteins, momordin[30] and a truncated version of pseudomonas exotoxin A (PE38)[25] were prepared and examined for toxicity. Momordin kills cells by inhibiting protein synthesis through modification of the 60S ribosomal subunit, whereas PE38 achieves the same result by ADP-ribosylating elongation factor 2. Importantly, unmodified forms of both momordin and the truncated recombinant PE are biologically inactive because, unlike ricin toxin, they lack a cellular binding domain that would normally facilitate their binding and entry into cells.

The FA-momordin conjugate was found to inhibit protein synthesis in a cell-free reticulocyte lysate to the same extent as the unmodified toxin ($IC_{50}$, ~1 nM).[30] To assess the contribution of folate ligation to FA-momordin's toxicity to whole cells, the cytotoxicity of FA-momordin was compared to that of unmodified momordin by determining the amount of radiolabeled leucine incorporated into KB (a human oral carcinoma), and HeLa cell (a human cervical carcinoma) protein after incubation with these toxins.[30] It was found that the FA-momordin conjugate displayed >1000-fold more cytotoxicity ($IC_{50}$~1 nM) against both cell types than the unconjugated protein. Further, this potency was shown to be completely abrogated when the cells were co-incubated with excess FA.[30] The amount of internalized FA-momordin neared saturation by 5 hr incubation, while inhibition of protein synthesis began within 8 hr and increased until the final time point of 48 hr. The latency period between toxin uptake and interruption of protein synthesis was hypothesized to derive from the trafficking time of the toxin along its endosomal pathway prior to entry into the cytoplasm, as seen with other antibody-toxin conjugates.[31]

To test whether FR mediated endocytosis might selectively target toxins to tumor cells, mixed cell cultures containing both normal and transformed cells were examined for their responses to the toxin conjugates. Thus, co-cultures of HeLa and WI38 cells (a normal human embryonic lung fibroblast), as well as co-cultures of KB and Hs67 cells (a normal human thymic fibroblast) were pulsed with FA-momordin or underivatized momordin for 20 min and then grown in the absence of toxin. In co-cultures pulsed with momordin, the more aggressive transformed cells dominated the co-culture, while in co-cultures pulsed with FA-momordin, the transformed cells were selectively eliminated, allowing the normal cells to achieve a near confluent state. Analysis of FR number showed that the KB and HeLa cells associated with 280 and 160 pmol FA-conjugate/mg of cellular protein, while the WI38 and Hs67 cells did not bind any detectable toxin. This correlation indicated that FR density might be a good predictor of FA-toxin sensitivity.[18]

Since FA-conjugates were observed to remain in endosomal compartments for extended periods after cellular uptake, bacterial toxins with and without their translocation domains were compared for relative toxicity. HeLa cells were pulsed with either FA-momordin lacking a translocation domain or FA-LysPE38, an engineered form of PE containing a translocation domain. FA-momordin required 19 hr to produce 50% inhibition of total cellular protein synthesis, whereas FA-LysPE38 caused 50% inhibition in only 3 hr. Further, the $IC_{50}$ value of the FA-LysPE38 conjugate was 10-fold higher ($10^{-10}$ M) than the FA-momordin conjugate. Since both toxins function by interrupting protein synthesis immediately upon entry into the cytosol, the accelerated rate of protein synthesis inhibition and the simultaneous 10-fold increase in potency of the FA-LysPE38, was attributed to the latter toxin's ability to facilitate escape from an intracellular compartment.

To confirm the importance of a mechanism of endosomal escape for toxin potency, a comparison of the cytotoxicities of translocation competent and translocation incompetent forms of the same truncated PE conjugate was made. CysPE35, a shorter form of PE, contains a single free cysteine ($Cys^{287}$) that must be reduced for translocation to occur. Consequently, conjugation of ligands (e.g. FA) to $Cys^{287}$ through a stable thioether linkage would be expected to inhibit the toxin's translocation ability, while attachment via a reducible disulfide bond should allow for restoration of the free cysteine, following exposure to the reducing milieu of the endosomal compartments.[32] When the toxicities of the $Cys^{287}$ thioether- and $Cys^{287}$ disulfide-linked folate conjugates towards HeLa cells were compared, the former exhibited an $IC_{50}$ of $>10^{-7}$ M, while the latter displayed a value of $\sim 4 \times 10^{-11}$ M (46). Clearly, the difference in potency of more than four orders of magnitude emphasizes the importance of incorporating some mechanism for transit from the endosome in the design of any folate targeted protein conjugate.

The value of a toxin translocation domain was also found to be heightened when the number of FR's on the targeted cancer cells was limited. Cells expressing large numbers of FR's (KB, HeLa and 2008-FBP cell lines) were found to be readily killed by both FA-momordin and FA-LysPE38. As anticipated, FA-LysPE38 was approximately 10-fold more toxic than FA-momordin in all cases. Importantly, cell lines FDXC, SKOV3 and Caco-2, which express lower levels of FR, were still susceptible to FA-LysPE38 and ricin (a toxin with similar translocation capability); however, they were insensitive to FA-momordin. Also, when co-cultures of Caco-2 cells ($FR^+$) and Hs67 cells ($FR^-$) were treated with FA-toxins, FA-momordin did not retard the growth of the tumor cells; whereas $Cys^{287}$ disulfide-FA selectively killed the Caco-2 cells without harming the Hs67 cells. Again, this suggested that in contrast to toxins bearing a translocation domain, FA-momordin cannot reach the cytosol in lethal quantities within cells expressing limited numbers of the FR. Therefore, it can be concluded that below a threshold level of FR expression, only toxins with translocation capabilities enter the cytosol in sufficient numbers to kill cells. It should be noted, however, that CaOV3 and OVCAR3 cells, which also express low levels of the FR receptor, are resistant to FA-LysPE38. This indicates that factors other than FR density and translocation capability may also be important for FA-toxin cytotoxicity.[25]

The free amino groups on most proteins can be conveniently derivatized with carboxyl-containing ligands via simple carbodiimide conjugation, without compromising the protein's biological activity. Since some toxins have been known to greatly reduce their activity when conjugated in this manner, an alternative conjugation technique has been developed. Here, the carbohydrate residues of the protein toxin, Gelonin, were oxidized, and FA was conjugated to the protein using the 3-(2-pyridyldithio)propionyl hydrazide carbohydrate-selective crosslinker. The activity of the resulting FA-ss-Gelonin (Fig. 3) conjugate was then compared to a related conjugate, prepared by linking FA to surface amino groups of Gelonin (FA-CO-Gelonin).[33] Although both conjugates were found to bind the FR with the same affinity, the FA-ss-Gelonin conjugate was approximately 225-fold more active at inhibiting protein synthesis in a cell free system than the FA-CO Gelonin conjugate. These results indicate that the conjugation methodology and linker design may be critical for preserving a protein's biological activity.

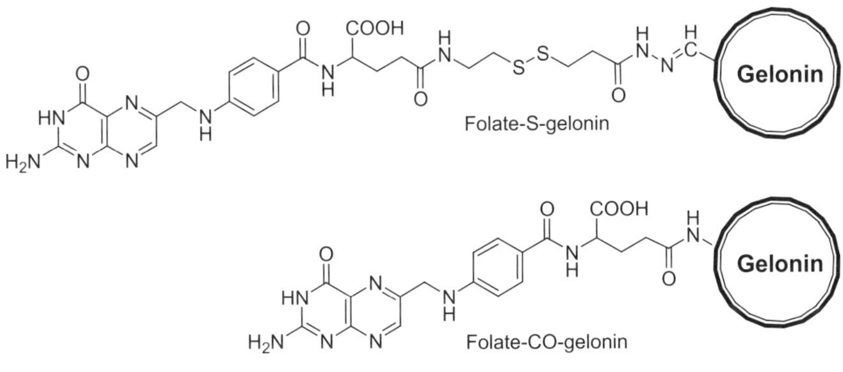


**Fig. 3** Chemical structures of FA-ss-Gelonin and FA-CO-Gelonin.

Protein toxin conjugates of tissue-specific antibodies have also been explored as tumor-selective killing agents in both clinical and preclinical studies.[34] Although several recent constructs have shown considerable promise, in many cases, the antibody-toxin conjugates have been compromised by (i) their large size and consequent reduced penetration of solid tumors; (ii) elicitation of a host immune response, except when humanized antibodies are used; (iii) reduction in antigen affinity due to steric hindrance from the attached toxin; (iv) expense of expression and purification of the monoclonal antibody; (v) inefficiency of site-specific conjugation of the toxin to the antibody; and (vi) poor tumor selectivity.[34–35] In contrast, except for the possible elicitation of a host immune response against the protein toxin, folic acid-toxin conjugates may suffer from few of the above handicaps. An increase in potency of $>10^4$ upon derivatization with FA simply emphasizes the merit of exploiting the folate uptake route for efficient intracellular delivery of toxins. When the benefits of tumor-specific targeting are considered, folate would seem to constitute a ligand worthy of consideration for toxin delivery in cancer therapy.

## 4. Folate Targeted Immunotherapy

It is widely accepted that although the immune system is capable of attacking and eliminating tumors, it frequently fails to do so because (1) insufficient tumor specific immune cells enter the tumor mass, or (2) the activity of those tumor-specific immune cells that enter the tumor mass has been suppressed.[36] To achieve maximal cytotoxicity, tumor reactive T cells must recognize a tumor antigen presented on the surface of the tumor cell within the context of a major histocompatibility complex (MHC), and then kill the cell by one of several pathways. Important in this process is the T cell receptor (TCR), which in association with the CD3 signaling complex, plays an important role in the recognition of the peptide/MHC complex on the tumor cell surface. In an effort to enhance the interaction between such cytotoxic T cells and the tumor, bispecific antibodies have been designed to simultaneously bind a unique antigen on the tumor cell and a classical receptor such as the TCR on a freely migrating T cell. These bispecific anti-TCR/anti-tumor cell antibodies have been found to stimulate and/or redirect the activity of T-cells against a tumor.

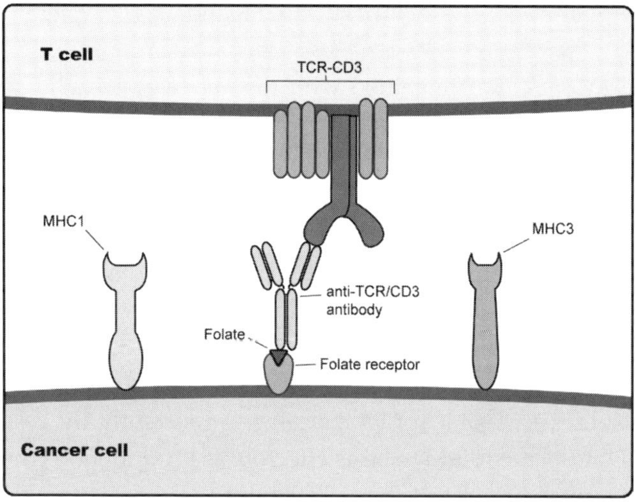


**Fig. 4** Schematic representation of T-cell activation by Folate-$\alpha$-TCR/CD3 antibody.

Knowing that the FR is a tumor-specific antigen,[14] Kranz *et al.* constructed an innovative set of "bispecific" antibodies that consisted of FA covalently linked to an antibody or antibody fragment directed against an epitope of the TCR/CD3 complex[19] (Fig. 4). Thus, these novel conjugates combined the specificities engineered into the bispecific antibodies into a much simpler complex, i.e. FA was used to mediate tumor cell binding, whereas the anti-TCR antibody was employed to mediate T-cell binding. Administration of the bridged homing agents was therefore predicted to direct the immune cells to kill the tumor cells.

Three different anti-TCR antibodies were used, including (i) a clonotypic antibody specific for the cytotoxic T cell (CTL) clone 2C, (ii) an anti-V$\beta$ antibody, which recognizes TCR epitopes in the V$\beta$ region, and (iii) an anti-CD3 antibody. The FA-antibodies were constructed with an average of five FA moieties per antibody, and the conjugates exhibited high affinity (50 to 90 nM Kd) for FR-positive tumor cells but did not bind to FR-negative cells. The activities of the FA-antibody constructs were later assessed on five different FR expressing mouse tumor cell lines using a $^{51}$Cr-release assay to quantitate cell lysis. T cell mediated cell lysis could be detected at FA-IgG concentrations as low as 1 pM, which was 1/1000th the concentration required to detect binding to the FR-positive cells. Furthermore, the extent of lysis correlated with the level of surface FR and was completely inhibited by free FA.

One drawback that remained with the above approach was that the large size of the anti-T cell antibodies (Mr ~160 000) greatly restricted their access to poorly perfused regions of solid tumors. As anticipated, biodistribution studies with IgG, $Fab_2$, Fab and single chain variable region antibodies (scFv) revealed that scFv molecules (Mr ~30 000) penetrated tumors more rapidly, to a greater depth, and more uniformly than all other forms of the antibody. Therefore, the Kranz group[37] elected to construct an FA-scfv conjugate directed against TCR. In this fusion protein, the $V_l$ and $V_h$ regions of the scFv were linked by a 26 amino acid flexible spacer that contained 8 lysine residues accessible for FA coupling. Importantly, the resulting FA-scFv conjugates were found to be as effective as FA-antibody conjugates in mediating lysis of tumor cells by CTL *in vitro*.[37]

To effectively evaluate the *in vivo* activity of the aforementioned constructs, one would normally inject human tumor cells, activated effector cells and the FA-antibody conjugates into an immunodeficient animal, and monitor tumor growth. However, since such a model would not accurately reflect a clinical scenario, Rund *et al.*[38] developed a mouse model by crossing a 2C TCR transgenic mouse with a RAG-1 knockout mouse. The resulting TCR/RAG-$1^{-/-}$ progeny, although lacking in a number of cell types (B cells and $CD4^+$ T cells), were able to grow FR-positive KB tumors and mobilize endogenous $CD8^+$ T cells. Thus, a day after receiving an injection of the antigenic peptide, SIYRYYGL (which is recognized by the transgenic TCR and causes CTL activation), these transgenic mice were subcutaneously injected with KB cells pre-mixed with PBS, unconjugated antibody or FA-antibody. In the animals treated with PBS or unconjugated antibody, tumors reached a target size of 20 mm in 35–38 days. In contrast, tumors in mice treated with a FA-Fab reached 20 mm in >114 days, while animals treated with a FA-scFv construct took 70–88 days. When tested in an alternate i.p. tumor model, mice receiving PBS or unconjugated Fab survived for 23–24 days, while mice treated with FA-Fab or FA-scFv survived for over 70 days. It was also observed that T-cell activation was an important requirement in this therapy, since survival of mice that did not receive the activating peptide was significantly reduced. Importantly, since the T-cells of the TCR/RAG-$1^{-/-}$ mice were limited to only a single clone, these FA-antibody conjugates need to be further tested in a syngeneic mouse model with a normal heterogenous immune system.

There are still other immunotherapy strategies for which applications of folate-mediated targeting have yet to be examined. One such strategy involves linking FA to a superantigen to concentrate the superantigen in the tumor tissue. Superantigens are unique in that they directly activate a substantial fraction of T cells, without the need for presentation in the context of a major histocompatibility complex. Consequently, T cells migrating through the cancer tissue would rapidly activate and proliferate, release chemokines to invite in other immune cells, and engage in a number of cytotoxic activities that can lead to tumor cell killing. By focusing these activities in the tumor, the chances of host rejection of the tumor are greatly increased. Other immune stimulants that might warrant similar evaluation include toll like receptor ligands such as muramyldipeptide and bacterial lipopolysaccharide fragments.

## 5. Folate Targeted Enzyme Prodrug Therapy

Another approach for improving the tumor selectivity of chemotherapy, termed antibody directed enzyme prodrug therapy (ADEPT), has traditionally combined antibody targeting with enzyme catalyzed prodrug activation. In this strategy, an enzyme-monoclonal antibody conjugate is administered and allowed to accumulate in the antigen-expressing tissue (e.g. tumor). A nontoxic prodrug is then injected which, in contact with the enzyme, is converted to its cytotoxic form killing the tumor.[39] Lu and Low have adapted this approach for use with FA by substituting the vitamin for the monoclonal antibody.[40] In these studies, penicillin-V amidase, a fungal enzyme known to hydrolyze the prodrug, doxorubicin-N-p-hydroxyphenoxyacetamide (DPO) to free doxorubicin, was conjugated to FA and tested for cytotoxicity *in vitro*. Although the potency of the DPO prodrug towards KB cells pretreated with FA-penicillin-V amidase was found to be comparable to that of active free doxorubicin, the FA-directed enzyme prodrug combination exhibited no toxicity towards A549 cells, a FR-negative cell line.

## 6. Folate-mediated Targeting of Adenovirus

Adenovirus particles are commonly used in gene therapy applications, because they can efficiently transfer genes into a wide range of mammalian cell types. Human adenovirus contains a 36 kb double strand of DNA that encodes more than 50 gene products required for various events in the viral life cycle. By eliminating the E1 region of the viral genome, space is made for inserting therapeutic genes into the vector, and in the absence of the transactivating E1a protein, the virus cannot replicate. Thus, after gene transfer, no viral spread is expected to occur and transgene expression should be facilitated. In contrast to retroviruses, adenoviral vectors (i) can be concentrated to high titer ($10^{12}$/mL); (ii) are capable of transducing nondividing cells; and (iii) do not integrate into host chromosomal DNA. However, since the coxsackie-Ad receptor (CAR) for the adenovirus capsid fiber protein is widely expressed on the surface of many cells, these adenoviral particles cannot be targeted to specific cell types. Therefore, to permit gene delivery strictly to certain target cell types, the endogenous viral tropism must be blocked and be replaced with the required novel tropism.

Adenovirus infection of cells is a two step process in which the adenovirus initially binds to the CAR receptor through the knob domain of the fiber capsid protein, after which the viral particle is internalized via an endocytic pathway mediated by the interaction of the penton base RGD sequence with an $\alpha v\beta$ integrin on the cell surface. Douglas *et al.*[20] blocked the endogenous viral tropism by employing a neutralizing Fab fragment of the anti-knob monoclonal antibody, and endowing the virus with a novel tropism by conjugating this Fab to FA. When mixed with adenoviral particles carrying a luciferase reporter gene, the resulting FA-Fab conjugate was shown to redirect adenoviral infection of KB cells via the FR at a level comparable to that achieved by native unmodified adenoviral infection. Without FA retargeting, the Fab-neutralized virus displayed only 1% of its normal transfection efficiency. FR-expressing KB cells could also be selectively transfected in the presence of FR-negative Jiyoye cells, confirming that the modified virus was dependent on FR for cell association. Evidence of this targeting strategy for cancer gene therapy was provided by complexing an adenoviral vector carrying the gene for herpes simplex virus thymidine kinase (HSV-TK) with the FA-Fab conjugate to achieve specific killing of KB cells on treatment with ganciclovir.

Along similar lines, Nyanguile *et al.*[41] used a novel intein-mediated protein ligation (IPL) technique to link FA to CAR D1, a 14 kDa protein that binds to the Ad fiber knob domain. When FA-CAR D1-modified adenovirus was incubated with KB cells, FR-specific transduction of the cells was observed. Therefore, modification of the first step of adenoviral infection, namely the binding of the knob domain, did not affect the ability of the virus to subsequently enter the cell and deliver its DNA into the nucleus. However, when FA derivatization of adenovirus was made at random sites on the virus, transduction of FR-positive cells was actually reduced.[42] It is conceivable that although the two steps of the adenovirus internalization process are not functionally linked, attachment of FA at a non-knob site may have severely compromised a critical surface component needed for transduction. Hence, the above studies suggested that retargeting of adenoviral vectors via a non-adenoviral cellular receptor is possible with a proper choice of ligand attachment site.

Although the above studies demonstrated that retargeting adenoviral vectors to FR-expressing cells might constitute a viable strategy for gene therapy of cancer, many questions still remain. For example, what measures must be taken to avoid elimination of the viral vector by an active immune system? Furthermore, what is the effect of such modifications on the efficiency and specificity of this method *in vivo*, especially since systemically administered unmodified virus leads to preferential virus accumulation in the liver?

## 7. *In vivo* Fate of Folate-protein Conjugates

The success of FR-targeted proteins *in vitro* prompted a similar investigation in tumor-bearing mice. When KB tumor-bearing nude mice were injected with an $^{111}$In-bovine serum albumin (BSA)-FA construct, the majority of the agent was found in the blood, liver and kidneys.[43] Unfortunately, tumor uptake of the $^{111}$In-BSA-FA construct was only 1.4-fold higher than the non-targeted $^{111}$In-BSA construct. We speculate that the low amount of tumor accumulation was the result of physical barriers to the penetration of a large protein, since low molecular weight folate-drug conjugates have been shown to readily reach virtually all malignant cells in a tumor mass. However, *in vivo* FA-protein targeting may still prove to be an effective approach if alternative dosing schemes are employed. For example, it was recently found that tumor cells from freshly isolated ascitic fluid of ovarian cancer patients bound and internalized a FA-BSA-FITC construct, and the net uptake was 22-fold higher than that measured in cultured FR-positive HeLa and Skov3.[44] These data suggest that FA-protein conjugates may still constitute highly effective therapeutic agents if applied to readily accessible tumors, such as cancers of the blood and peritoneal cavity.

## 8. Folate-targeted Peptidic Imaging Agent

An important practical problem in oncology today relates to improper disease management based on inadequate tools for detection/diagnosis of the patient's tumor. As a result, many cancer patients receive either insufficient or excessive treatment. More than one half of all cancer patients are known to have metastatic spread of cancer at the time of first detection. Although, three-dimensional radiological methods, such as high-resolution computed tomography (CT) and magnetic resonance imaging (MRI), are excellent for imaging normal tissues and for defining anatomy in the region of the tumor, these methods can only depend on gross enlargement of normal organs or an alteration in the signal to distinguish malignant from normal tissues. In comparison, targeted radiodetection can image a tumor at the nanomolar level by sensitive detection of a tumor associated antigen, thereby facilitating disease management at a much earlier stage and grade. FR is one among many tumor associated antigens that can be used for radiodiagnostic applications, as demonstrated by the progress made by a peptide-based folate-targeted radiopharmaceutical agent.

Gamma emitting technetium-99m ($^{99m}$Tc) based imaging agents are currently dominating the field of diagnostic nuclear medicine owing to $^{99m}$Tc's superior nuclear characteristics (6 hr half-life and 141 keV gamma rays), which yield high resolution images without the risk of hazardous radiation exposure.[45] The wide availability and cost effectiveness of $^{99m}$Tc are also of major importance for routine clinical applications. Furthermore, it is now generally recognized that the rapid blood clearance and fast target localization properties of small molecule based imaging agents are best utilized when short-lived radionuclides such as $^{99m}$Tc are employed.[45] Hence, our laboratory has synthesized a novel FA-peptide chelator, called EC20, which coordinates Tc (V) as a tetradentate ligand (Fig. 5).[46] This proprietary molecule is a folate containing peptide consisting in sequence of pteroic acid, D-$\gamma$-glutamic acid, $\beta$-L-diaminopropionic acid, L-aspartic acid and L-cysteine. EC20 was synthesized by a solid phase synthetic procedure.[46]

This new chelate was found to bind FR-positive tumor cells with high affinity ($K_d \sim 3$ nM). Following i.v. injection into Balb/c mice bearing

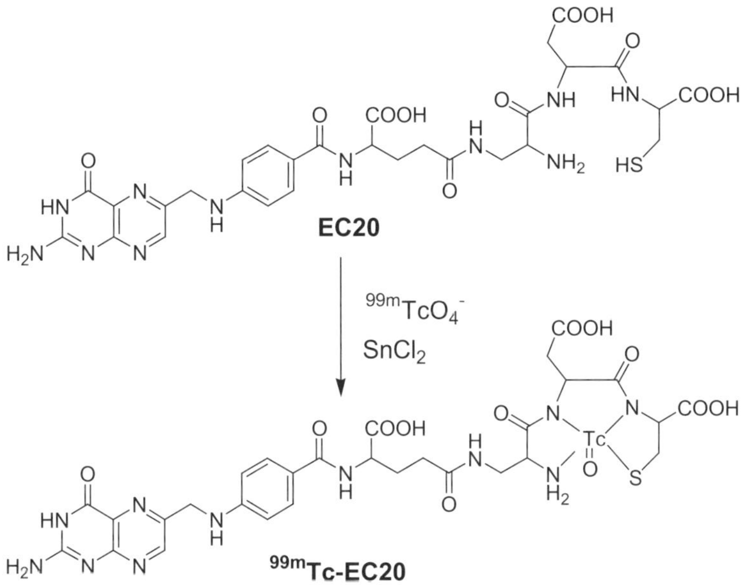


**Fig. 5** Formation of $^{99m}$Tc-EC20 chelation product.

M109 tumors, the net tumor accumulation of $^{99m}$Tc-EC20 (17% ID/g) was very similar to that of $^{111}$In-DTPA-FA (19% ID/g), a non-peptide based folate-radiopharmaceutical agent.[47] In addition, $^{99m}$Tc-EC20 was rapidly removed from circulation (plasma $t_{1/2}\sim$4 min) and excreted into the urine in a non-metabolized form. Additional *in vivo* studies performed using the M109-Balb/c animal model revealed that (i) the net tumor uptake of $^{99m}$Tc-EC20 was specific and proportional to FR expression levels in tumors derived from three cell lines[48]; (ii) the uptake of $^{99m}$Tc-EC20 increased proportionally as a function of tumor size over the tested range of 50–400 mm$^3$; (iii) the agent's uptake in an i.p. localized tumor was similar to that of a s.c. implanted tumor; (iv) following either i.p. or i.v. dose administration, no differences in $^{99m}$Tc-EC20 uptake within the kidneys, tumor and other abdominally-localized tissues is observed; (v) leucovorin (folinic acid) supplementation of the normally used FA-deficient chow had no effect on the agent's overall tumor uptake or tissue biodistribution pattern; and (vi) tumor to non-tumor ratios could be increased up to 2.7-fold when a single equivalent of free FA was co-administered with the $^{99m}$Tc-EC20. From this extensive analysis we concluded that the tumor size, position and the method of dosing did not affect the net uptake of $^{99m}$Tc-EC20 into tumors as long they expressed adequate amounts of FR. It was also realized that substituting FA in the chow with a compound like leucovorin should maintain similar folate pools as FA[49] and therefore satisfy the tissue folate needs in animals enduring long term FR-targeted therapy.

Phase 1 clinical trials of $^{99m}$Tc-EC20 (FolateScan) for the diagnostic imaging of ovarian, endometrial and renal cancers have successfully been completed. Radiation dosimetry data from eight adult female subjects, four normal volunteers and four subjects with known or suspected ovarian cancer were evaluated with a single injection of EC20 labeled with 15–20 mCi of $^{99m}$Tc. Results from these initial phase 1 clinical studies indicate that the optimal time for imaging is approximately 2–4 hrs after injection of the agent. In another phase 1 study designed to enroll patients with metastatic renal cell carcinoma, preliminary analysis of the data indicates that 33 of the 43 patients (77%) exhibit $^{99m}$Tc-EC20 uptake in their tumor masses. There are also no clinically significant trends observed, with regard to drug related changes in hematology, clinical chemistry, urinalysis or vital sign parameters in these patients. Importantly, a series of phase 2 clinical studies, each in a different tumor type, are currently in progress to determine which malignancies can best be imaged by $^{99m}$Tc-EC20.

## 9. Conclusions and Outlook

FR is significantly upregulated in a large number of solid and hematopoietic human cancers. Simple covalent attachment of FA to virtually any macromolecule produces a conjugate that can be internalized into FR-bearing cells in a similar fashion to free FA.[28] Although the low permeability of cell membranes to macromolecules limits their applications in intracellular drug delivery, FA conjugation offers a possible solution to this cell penetration problem. FA-drug conjugates also display significant selectivity for cancer cells *in vivo*, as evidenced by the fact that the same FA-drug conjugates do not bind to FR negative tumors. Since the magnitude of tumor cytotoxicity depends on the cumulative amount of therapeutic agent delivered to the cancer cell, the ability of FR to recycle and deliver multiple folate conjugates per cell represents an additional advantage.

FA constitutes an attractive alternative to antibodies and other proteins/peptides as a targeting ligand. FA is much smaller (Mr ~441) than monoclonal antibodies (Mr ~160 000) and hence can penetrate tumor tissues much more rapidly and completely. FA is also nonimmunogenic, whereas even humanized monoclonal antibodies can elicit a neutralizing immune response. FA is relatively easy to conjugate to a wide variety of molecules and molecular complexes, whereas antibody conjugation can be nonselective and relatively inefficient. FA is stable to mild acids/bases and a variety of solvents, temperatures, and storage conditions, whereas antibodies must be handled carefully to avoid their denaturation. FA is also inexpensive to procure, whereas monoclonal antibodies are costly to produce. Clearly, more animal and human clinical studies must be conducted before the usefulness of FA can be accurately evaluated; however, the results to date from many labs point to the fact that folate targeting will soon find an important niche in the diagnosis and/or treatment of cancer. Moreover, since many FR-positive cancers are associated with poor clinical outcomes,[50] FA-linked therapeutics may offer new options to patients with few other alternatives.

## References

1. Reddy JA and Low PS (1998) Folate-mediated targeting of therapeutic and imaging agents to cancers. *Crit Rev Ther Drug Carrier Syst* **15**:587–627.
2. Leamon CP and Low PS (2001) Folate-mediated targeting: From diagnostics to drug and gene delivery. *Drug Discov Today* **6**:44–51.
3. Lu Y and Low PS (2002) Folate-mediated delivery of macromolecular anticancer therapeutic agents. *Adv Drug Deliv Rev* **54**:675–693.
4. Shen F, Wu M, Ross JF, Miller D and Ratnam M (1995) Folate receptor type gamma is primarily a secretory protein due to lack of an efficient signal for glycosylphosphatidylinositol modification: Protein characterization and cell type specificity. *Biochemistry* **34**:5660–5665.
5. Kamen BA and Capdevila A (1986) Receptor-mediated folate accumulation is regulated by the cellular folate content. *Proc Natl Acad Sci USA* **83**:5983–5987.
6. Turek JJ, Leamon CP and Low PS (1993) Endocytosis of folate-protein conjugates: Ultrastructural localization in KB cells. *J Cell Sci* **106 (Pt 1)**:423–430.
7. Weitman SD, Lark RH, Coney LR, Fort DW, Frasca V, Zurawski VR Jr and Kamen BA (1992) Distribution of the folate receptor GP38 in normal and malignant cell lines and tissues. *Cancer Res* **52**:3396–3401.
8. Ross JF, Chaudhuri PK and Ratnam M (1994) Differential regulation of folate receptor isoforms in normal and malignant tissues *in vivo* and in established cell lines. Physiologic and clinical implications. *Cancer* **73**:2432–2443.

9. Garin-Chesa P, Campbell I, Saigo PE, Lewis JL, Jr, Old LJ and Rettig WJ (1993) Trophoblast and ovarian cancer antigen LK26. Sensitivity and specificity in immunopathology and molecular identification as a folate-binding protein. *Am J Pathol* **142**:557–567.
10. Coney LR, Tomassetti A, Carayannopoulos L, Frasca V, Kamen BA, Colnaghi MI and Zurawski VR Jr (1991) Cloning of a tumor-associated antigen: MOv18 and MOv19 antibodies recognize a folate-binding protein. *Cancer Res* **51**: 6125–6132.
11. Boerman OC, van Niekerk CC, Makkink K, Hanselaar TG, Kenemans P and Poels LG (1991) Comparative immunohistochemical study of four monoclonal antibodies directed against ovarian carcinoma-associated antigens. *Int J Gynecol Pathol* **10**:15–25.
12. Mattes MJ, Major PP, Goldenberg DM, Dion AS, Hutter R V and Klein KM (1990) Patterns of antigen distribution in human carcinomas. *Cancer Res* **50**:880s–884s.
13. Toffoli G, Cernigoi C, Russo A, Gallo A, Bagnoli M and Boiocchi M (1997) Overexpression of folate binding protein in ovarian cancers. *Int J Cancer* **74**:193–198.
14. Campbell IG, Jones TA, Foulkes WD and Trowsdale J (1991) Folate-binding protein is a marker for ovarian cancer. *Cancer Res* **51**:5329–5338.
15. Low PS and Antony AC (eds.) (2004) Folate receptor-targeted drugs for cancer and inflammatory diseases. *Adv Drug Deliv Rev* **56**:1055–1231.
16. Blakley RL, The Biochemistry of Folic Acid and Related Pteridines, Vol. 13. North Holland Publishing Company: Amsterdam, 1969.
17. Ciuchi F, Di Nicola G, Franz H, Gottarelli G, Mariani P, Bossi MGP and Spada GP (1994) Self-recognition and self-assembly of folic acid salts: Columnar liquid crystalline polymorphism and the column growth process. *J Am Chem Soc* **116**:7064.
18. Leamon CP and Low PS (1994) Selective targeting of malignant cells with cytotoxin-folate conjugates. *J Drug Target* **2**:101–112.
19. Kranz DM, Patrick TA, Brigle KE, Spinella MJ and Roy EJ (1995) Conjugates of folate and anti-T-cell-receptor antibodies specifically target folate-receptor-positive tumor cells for lysis. *Proc Natl Acad Sci USA* **92**:9057–9061.
20. Douglas JT, Rogers BE, Rosenfeld ME, Michael SI, Feng M and Curiel DT (1996) Targeted gene delivery by tropism-modified adenoviral vectors. *Nat Biotechnol* **14**:1574–1578.
21. Lee RJ, Wang S and Low PS (1996) Measurement of endosome pH following folate receptor-mediated endocytosis. *Biochim Biophys Acta* **1312**:237–242.
22. Citro G, Szczylik C, Ginobbi P, Zupi G and Calabretta B (1994) Inhibition of leukaemia cell proliferation by folic acid-polylysine-mediated introduction of c-myb antisense oligodeoxynucleotides into HL-60 cells. *Br J Cancer* **69**:463–467.

23. Wang S, Lee RJ, Mathias CJ, Green MA and Low PS (1996) Synthesis, purification, and tumor cell uptake of 67Ga-deferoxamine–folate, a potential radiopharmaceutical for tumor imaging. *Bioconjug Chem* **7**:56–62.
24. Lee RJ and Low PS (1995) Folate-mediated tumor cell targeting of liposome-entrapped doxorubicin in vitro. *Biochim Biophys Acta* **1233**:134–144.
25. Leamon CP, Pastan I and Low PS (1993) Cytotoxicity of folate-Pseudomonas exotoxin conjugates toward tumor cells. Contribution of translocation domain. *J Biol Chem* **268**:24847–24854.
26. Luo J, Smith MD, Lantrip DA, Wang S and Fuchs PL (1997) Efficient syntheses of pyrofolic acid and pteroyl azide, reagents for the production of carboxyl-differentiated derivatives of folic acid. *J Am Chem Soc* **119**:10004.
27. Leamon CP, Weigl D and Hendren RW (1999) Folate copolymer-mediated transfection of cultured cells. *Bioconjug Chem* **10**:947–957.
28. Leamon CP and Low PS (1991) Delivery of macromolecules into living cells: A method that exploits folate receptor endocytosis. *Proc Natl Acad Sci USA* **88**:5572–5576.
29. Leamon CP and Low PS (1993) Membrane folate-binding proteins are responsible for folate-protein conjugate endocytosis into cultured cells. *Biochem J* **291 (Pt 3)**:855–860.
30. Leamon CP and Low PS (1992) Cytotoxicity of momordin-folate conjugates in cultured human cells. *J Biol Chem* **267**:24966–24971.
31. Oeltmann TN and Heath EC (1979) A hybrid protein containing the toxic subunit of ricin and the cell-specific subunit of human chorionic gonadotropin. II. Biologic properties. *J Biol Chem* **254**:1028–1032.
32. Theuer CP, Kreitman RJ, FitzGerald DJ and Pastan I (1993) Immunotoxins made with a recombinant form of Pseudomonas exotoxin A that do not require proteolysis for activity. *Cancer Res* **53**:340–347.
33. Atkinson SF, Bettinger T, Seymour LW, Behr JP and Ward CM (2001) Conjugation of folate via gelonin carbohydrate residues retains ribosomal-inactivating properties of the toxin and permits targeting to folate receptor positive cells. *J Biol Chem* **276**:27930–27935.
34. Frankel AE, Neville DM, Bugge TA, Kreitman RJ and Leppla SH (2003) Immunotoxin therapy of hematologic malignancies. *Semin Oncol* **30**: 545–557.
35. Riethmuller G, Schneider-Gadicke E and Johnson JP (1993) Monoclonal antibodies in cancer therapy. *Curr Opin Immunol* **5**:732–739.
36. Davis ID (2000) An overview of cancer immunotherapy. *Immunol Cell Biol* **78**:179–195.
37. Cho BK, Roy EJ, Patrick TA and Kranz DM (1997) Single-chain Fv/folate conjugates mediate efficient lysis of folate-receptor-positive tumor cells. *Bioconjug Chem* **8**:338–346.

38. Rund LA, Cho BK, Manning TC, Holler PD, Roy EJ and Kranz DM (1999) Bispecific agents target endogenous murine T cells against human tumor xenografts. *Int J Cancer* **83**:141–149.
39. Bagshawe KD, Sharma SK and Begent RH (2004) Antibody-directed enzyme prodrug therapy (ADEPT) for cancer. *Exp Opin Biol Ther* **4**:1777–1789.
40. Lu JY, Lowe DA, Kennedy MD and Low PS (1999) Folate-targeted enzyme prodrug cancer therapy utilizing penicillin-V amidase and a doxorubicin prodrug. *J Drug Target* **7**:43–53.
41. Nyanguile O, Dancik C, Blakemore J, Mulgrew K, Kaleko M and Stevenson SC (2003) Synthesis of adenoviral targeting molecules by intein-mediated protein ligation. *Gene Ther* **10**:1362–1369.
42. Reddy JA, Clapp DW and Low PS (2001) Retargeting of viral vectors to the folate receptor endocytic pathway. *J Control Rel* **74**:77–82.
43. Shinoda T, Takagi A, Maeda A, Kagatani S, Konno Y and Hashida M (1998) *In vivo* fate of folate-BSA in non-tumor- and tumor-bearing mice. *J Pharm Sci* **87**:1521–1526.
44. Ward CM, Acheson N and Seymour LW (2000) Folic acid targeting of protein conjugates into ascites tumor cells from ovarian cancer patients. *J Drug Target* **8**:119–123.
45. Liu S and Edwards DS (1999) 99mTc-labeled small peptides as diagnostic radiopharmaceuticals. *Chem Rev* **99**:2235–2268.
46. Leamon CP, Parker MA, Vlahov IR, Xu LC, Reddy JA, Vetzel M and Douglas N (2002) Synthesis and biological evaluation of EC20: A new folate-derived, (99m)Tc-based radiopharmaceutical. *Bioconjug Chem* **13**:1200–1210.
47. Wang S, Luo J, Lantrip DA, Waters DJ, Mathias CJ, Green MA, Fuchs PL and Low PS (1997) Design and synthesis of [111In]DTPA-folate for use as a tumor-targeted radiopharmaceutical. *Bioconjug Chem* **8**:673–679.
48. Reddy JA, Xu LC, Parker N, Vetzel M and Leamon CP (2004) Preclinical evaluation of (99m)Tc-EC20 for imaging folate receptor-positive tumors. *J Nucl Med* **45**:857–866.
49. Schmitz JC, Stuart RK and Priest DG (1994) Disposition of folic acid and its metabolites: a comparison with leucovorin. *Clin Pharmacol Ther* **55**:501–508.
50. Toffoli G, Russo A, Gallo A, Cernigoi C, Miotti S, Sorio R, Tumolo S and Boiocchi M (1998) Expression of folate binding protein as a prognostic factor for response to platinum-containing chemotherapy and survival in human ovarian cancer. *Int J Cancer* **79**:121–126.

# 10

# Transferrin Receptor Mediated Delivery of Protein and Peptide Drugs into Tumors

*Julia Fahrmeir and Manfred Ogris*

## 1. Introduction

Crucial requirements in successful cancer therapy are the delivery of the therapeutic drug into target cells, a high level of selectivity, low toxicity, as well as the absence of side-effects and drug resistance. Therefore, researchers have been evaluating special drug delivery systems which are able to exploit natural pathways in a new setting such as transfer routes. One approach is the attachment of anticancer agents to the transport protein transferrin, creating a novel system for active targeting of drugs to proliferating malignant cells that overexpress transferrin receptors. In the last years, several studies have demonstrated that this pathway is highly effective in animal models and clinical studies for humans have already begun. This book chapter summarizes the structure and biological functions of transferrin and transferrin receptors, and provides an overview of the development of transferrin receptor-mediated delivery of protein and peptide drugs to cancer cells.

# 2. Transferrin (Tf)

## 2.1. *Structure*

Transferrin (Tf) belongs to a superfamily of iron-binding glycoproteins with a molecular weight of ~79 kDa and a single polypeptide chain of 679 amino acids.[1–3] They have been classified into three major sub-families: serum transferrin, ovotransferrin and lactotransferrin. The best known member is the serum Tf, which is found in blood and other mammalian fluids such as lymph and cerebrospinal fluid, while ovotransferrin is found in egg white, and lactoferrin in mammalian milks, as well as other secretions such as tears and saliva (for review see Refs. 2–4).

The Tf family has closely related amino acid sequences and three-dimensional structures: as x-ray structures show, the polypeptide chain is arranged in two homologous halves (with typically ca. 40% sequence identity) which are termed the N- (336 amino acids) and the C-lobe (343 amino acids), connected by a short linear peptide region.[5,6] 19 intra-chain disulfide bonds stabilize the protein and the three carbohydrate side chains protect the molecule to the outside. Each lobe is divided into two sub-domains comprising a series of $\alpha$-helices which overlay a central $\beta$-sheet backbone. They have one binding site for $Fe^{3+}$ as well as one anion-binding site, because strong iron binding by Tf is dependent on a synergistic anion, usually bicarbonate.[7] In the open, iron-free conformation, the two domains form a large cleft, easy to be accessed by $Fe^{3+}$ which can coordinate with the synergistic anion and four amino acids, suspected to form the iron binding pocket (two Tyr, a His and an Asp).[6] The bicarbonate anion is also needed for proper release of $Fe^{3+}$, once it is bound. The iron-loaded holo-Tf (with two iron atoms per Tf molecule) has an altered shape compared with iron-free apo-transferrin, exhibiting an increased stability to thermal or proteolytic degradation and an increased affinity to the Tf receptor.[3]

Iron release is triggered by a drop in pH during the endocytosis cycle. With protonation and dissociation of the synergistic anion, specifically the protonation of the responsive amino acids, Tf undergoes a conformational change with an opening of the binding cleft and a final release of iron.[8]

## 2.2. *Biological function*

Iron is distributed throughout the whole body and is a fundamental necessity for almost every living organism. Many proteins depend on iron as a co-factor for redox-reactions or ligand coordination, as it exists in two common oxidation states. However, under physiologic conditions, $Fe^{3+}$ forms a highly insoluble hydroxide complex which cannot be taken up by cells, so that despite its abundance, iron is not easily accessible. On the other hand, free $Fe^{2+}$ can lead to the formation of toxic hydroxyl radicals, thus resulting in oxidative damage to tissues. The fundamental role of Tf is to overcome these problems and to control the levels of free iron in the body fluids. By binding iron, serum Tf serves three purposes: it renders iron soluble, facilitates the transport through the body into the cells, and due to the redox-inactive form, it prevents iron-mediated free radical toxicity.

In the last few years, researchers found in many *in vitro* and clinical studies that high levels of free iron in the body are linked to an increased growth of pathogens.[9] The ability of apotransferrin to reduce bacterial infections was examined by the group of Parkkinen.[10] The antimicrobial activity of the iron-binding protein might not only be due to the reduction of free iron levels, but additionally, it could be due to the fact that apo-Tf is able to hinder the adhesion of bacteria to surfaces.[1] For ovotransferrin and lactoferrin, there is also a known antimicrobial activity, which apparently depends not only on simple iron deprivation, but also on direct contact with bacteria.[11]

# 3. Transferrin-Receptors (TfR)

## 3.1. *Structure*

The transferrin receptor 1 (TfR1) is the best-characterized receptor for human Tf. It is a homodimeric type II transmembrane glycoprotein, composed of two disulfide-bonded identical subunit monomers, each with ~90 kDa molecular mass. It consists of a small aminoterminal cytoplasmic domain (residues 1 to 67), a single-pass transmembrane region (residues 68 to 88), and a large extracellular portion (ectodomain, residues 89 to 760) which contains a binding site for the Tf molecule. The extracellular side has four glycosylation sites which are thought to be crucial for the binding and proper release of Tf.[2–4,12] The 3D crystal structure of the ectodomain has been determined by Lawrence *et al.*[13] One TfR monomer is butterfly-like shaped and has three distinct domains, identified as the protease-like, apical and helical domain which form a lateral cleft, thought to be responsible for getting in contact with Tf. The homodimeric receptor can bind up to two molecules of Tf. However, different TfR may have different affinities to Tf and generally, diferric holotransferrin binds with higher affinity than the mono-ferric and apo-forms. There also exists a homologous receptor (TfR2) which is predominantly expressed in the liver, but its function is not completely clarified thus far.[14]

## 3.2. *Clathrin-mediated endocytosis*

Cells have various mechanisms for taking up iron, however, higher organisms like almost every mammalian cell, are capable of internalizing Fe via receptor mediated endocytosis of Tf-bound iron, which is one of the best understood processes in cell biology (for review, see Refs. 15–17). There are basically two mechanisms for endocytosis, a clathrin independent and a clathrin dependent one. The second, which will be briefly described here, involves the binding of a ligand (e.g. Tf) to a specific cell surface receptor (e.g. TfR).

The Tf-mediated endocytosis pathway is triggered by the binding of diferric Tf (holo-Tf) to the receptor on the outer face of the plasma membrane, which leads to the assembly of cytosolic coat proteins to form "clathrin coated pits".[4] Those pits invaginate and pinch off to form endocytic vesicles, which are encapsulated by a clathrin coat. Upon maturation and the loss of clathrin coat, the vesicles merge into endosomes. The endosomal lumen can be rapidly acidified by proton pumps (to pH of ca. 5.5), which weakens the association between iron and Tf, leading to final iron release.[18] However, the whole mechanism through which iron is dissociated, is still not clear. A plasma oxidoreductase reduces free $Fe^{3+}$ to $Fe^{2+}$ and DMT1, an apical transmembrane iron transporter, actively transports $Fe^{2+}$ from the endosomal compartment into the cytosol. After the iron release, the intact receptor-bound apotransferrin is recycled through exocytic vesicles back to the cell surface, where neutral pH leads to the detachment of apo-Tf into the extracellular fluid.[2,19]

### 3.3. *Regulation of expression*

The abundance of TfR is dependent on the intracellular iron concentration. Several studies have shown that cells grown in the presence of iron salt showed a decrease in their Tf-binding capacity, associated with a reduction in the number of TfR which control iron uptake. This decrease is also associated with an enhanced intracellular level of ferritin. On the other hand, incubating these cells with chelators will increase the number of TfR, thereby reducing the content of ferritin.[20]

The molecular mechanisms of TfR expression are regulated largely at the posttranscriptional level, by interaction of specific proteins with TfR mRNA. The non-translated receptor mRNA contains a region with a series of five hairpin stem-loop structures known as iron responsive elements (IREs).[21,22] This specific structure of the mRNA is recognized and bound by specific trans-acting proteins known as iron-regulatory proteins (IRPs).[23,24] Those IRPs are able to stabilize the mRNA and control the rate of translation. Till now, there are two closely related IRPs known, i.e. IRP1 and IRP2, which respond via different mechanisms to the variations in iron concentrations.[25] IRP1 is considered a bifunctional sensor for iron with RNA binding affinity, when cellular iron levels are low, and subsequently switching to enzymatic activities in a similar fashion to aconitase under high intracellular iron concentrations.[26,27] To upregulate TfR expression, IRP1 binds to the 3′-untranslated region of TfR mRNA, stabilizing mRNA and increasing the synthesis of TfR.[28] On the other hand, IRP1 can bind to the 5′-untranslated mRNA of ferritin,[22] and due to the prevention of translation sterically, inhibit the synthesis of ferritin. Conversely, under high concentration of intracellular iron, IPR1 becomes enzymatically active and is no longer able to bind to the IRE, which leads to the degradation of TfR mRNA. In addition to iron concentration, other intracellular factors can regulate the TfR expression level.[29,30] Nitric oxide and $H_2O_2$ produced from oxidative stress can induce IRPs and modulate cellular iron metabolism (for review, see Refs. 16 and 31).

### 3.4. *Tissue distribution*

Iron is an essential micronutrient of all types of living cells. Therefore, TfR is expressed in many tissue types. However, most normal resting human cells only need little iron for their normal cellular functions and do not show over expression of TfR.[32] In contrast, cells which have a high requirement for iron due to increased metabolism rate during cell growth and development, express a large number of TfR.[33,34] These are mainly erythrocyte precursors together with all actively proliferating tissues. In particular, the brain tissue would greatly suffer from disturbed function if any iron deficiency or excess occurs.[35,36] Hence, the iron transport across the blood brain barrier must be well regulated, which is realized due to a high density of TfR in the brain capillary endothelial cells.

Also, cancer cells generally express higher levels of TfR and take up iron at a higher rate to sustain their rapid growth (for review see Refs. 34, 37, 38). This is also reflected by the ability of tumor cells to be radio localized using $^{67}Ga$, which binds like iron to Tf and is taken up via the same TfR mediated pathway.[39] There is also evidence that there is greater dependency in tumor cell proliferation on the metal ion, since some iron chelators like desferrioxamine are able to inhibit the growth of a variety of tumors.[40] The findings of elevated TfR levels in tumor tissues led to the proposal that anti-cancer drugs could be effectively delivered to tumor cells using TfR targeting. In the recent years, many studies have demonstrated that utilizing this specific uptake pathway is highly effective in animals, as well as in human anti cancer treatment.

## 4. *In vivo* Application of Proteins Drugs, General Considerations

Clearance of "foreign" material from the blood stream is mostly accompanied by cells of the reticulo endothelial system. After opsonization with antibodies, antibody-antigen complexes are recognized by macrophages and degraded. Also, the activation of the complement system can lead to rapid removal of peptidic and proteinous therapeutics. In order to circumvent the clearance of macromolecular drugs from the bloodstream, several strategies can be applied. Induction of antibody mediated responses is of lower importance when a toxin-conjugate is applied only once, but after repeated administration, a rapid immunological response can occur. Polyethyleneglycol (PEG) has been extensively used to increase the circulation time in the blood of various peptides, proteins, liposomes and other macromolecular drug carriers.

Targeted toxins modified with PEG have shown increased half lives in the blood and reduced immunogenicity by fully retained binding capacity.[41] Using Tf as a targeting domain, there is less probability for immunological reactions against the targeting domain, whereas when using antiTfR immunity against the antibody can occur. For clinical applications, antibodies which are usually derived from murine or other non-human sources are humanized. A humanized antibody contains the amino acid sequences from the six complementary-determining regions of the murine monoclonal antibody, which are grafted onto a human antibody framework. Humanized antibodies usually show reduced immunogenicity and prolonged half lives in the blood stream.

## 5. Generation of Conjugates

Coupling of toxins to the transferrin molecule or to the TfR targeting antibodies can either be achieved by direct chemical coupling, e.g. hetero-bifunctional crosslikers, or by expression of a recombinant fusion protein. In any case, it is important to bear in mind that there is the need for an intracellular cleavage of targeting domain (Tf or anti-TfR) and effector domain (the toxin).

### 5.1. *Chemical coupling*

Chemical coupling to the Tf protein can be directed to specific domains of the molecule by modifying the glycosylation sites of Tf.[42] Within the carbohydrate chains, the two terminal carbon atoms of the sialic acids are selectively removed by periodate oxidation, and the oxidized aldehyde form of Tf is used for coupling to primary amino groups within the peptide or protein to be coupled. The junction resulting from aldimide formation is stabilized by reduction with sodium cyanoborohydride to the corresponding amine linkage. This coupling procedure results in a conjugate with defined binding site, without interfering with the receptor binding domain of the Tf molecule. Nevertheless, for the delivery of certain toxins, it is necessary to obtain intracellular cleavage from the targeting domain for the toxin to achieve full functionality. Chemical coupling with the heterobifunctional crosslinker SPDP (N-Hydroxysuccinimidyl 3-[2-pyridyldithio]-propionic acid) results in the formation of a reducible disulfide bond. The primary amino groups of the protein (N-terminus or epsilon-amino groups in lysines) react with the activated acid group of SPDP, resulting in protein modified with dithiopyridine groups. The reaction partner can be modified in a similar way and subsequently reduced with excess of DTT to obtain a free sulfhydryl group. Alternatively, modification with iminothiolane (Trauts reagent) is possible, which also reacts with primary amino groups and introduces a free thiol group into the protein. After reacting the two modified proteins under oxygen-free condition, a disulfide bond forms between them. Scott *et al.*[43] have used this coupling method for the generation of a Tf-gelonin conjugate, and Citores and colleagues[44] have also described such a procedure to couple different toxins to Tf. Although certain toxin conjugates are only active when coupled via a cleavable bond, premature degradation of the disulfide bond can impair the targeting effect by releasing free toxin.[43]

Alternative crosslinking molecules have been used where the disulfide bond is protected against attack by thiolate anions. Thorpe *et al.*[45] synthesized sodium S-4-succinimidyloxycarbonyl-alpha-methyl benzyl thiosulfate (SMBT) and 4-succinimidyloxycarbonyl-alpha-methyl-alpha(2-pyridyldithio)toluene (SMPT). Both reagents generated a hindered disulfide linkage, in which a methyl group and a benzene ring were attached to the carbon atom adjacent to the disulfide bond. Conjugates of ricin A chain and a targeting antibody were far more stable in the blood stream, compared with conjugates crosslinked with iminothiolane. Arpicco *et al.*[46] synthesized a whole series of new crosslinkers with hindered disulfide bonds, based on thioimidate group linked to an S-acetyl thiol or a substituted aryldithio group. As Tf is internalized into endosomes which are subsequently acidified to release the iron from Tf, an acid labile linker will allow a triggered release of the effector domain from the targeting domain.[47]

## 5.2. *Recombinant production of conjugates*

Recombinant production of functional Tf has been carried out in different expression systems, including bacteria, insect cells and mammalian cells. Bacterial produced Tf lacks glycosylation,[48] but the lack of glycosylation does not seem to impair the binding properties of the Tf to the TfR.[49] Some general considerations have to be kept in mind when producing fusion proteins with Tf or antiTfR antibodies in certain expression systems. Recombinant production of fusion proteins in bacteria leads to the formation of inclusion bodies, where the recombinant proteins are stored in an aggregated, insoluble form. After denaturation and renaturation steps, purification is usually carried out by methods based on chromatography. Folding of the individual protein components must not be impaired, and the cleavage of targeting domain and effector domain after cellular internalization is also important for the functionality of certain toxins (see above). The inclusion of a spacer domain within the fusion construct promotes independent folding of the two protein components; additionally, this spacer region can be designed in a way for specific enzymatic cleavage to release the effector domain inside the cell. Enzymatic activity of angiogenin (a ribonuclease) fused to an antiTfR antibody and the antitumor activity of the fusion protein was optimal, when introducing a five amino acid spacer into the fusion protein.[50] A 12-residue spacer designed for cleavage by furin (TRHRQPRGWEQL) was incorporated into a single chain antiTfR(sFv)-restrictocin construct, which led to a significant increase in cell killing activity.[51]

## 6. Tf and AntiTFR Targeted Toxins

A broad range of toxins have been coupled to Tf or antiTfR for targeted delivery to solid tumors. Trowbridge and Domingo[52] already described the usefulness of targeted delivery of toxins to TfR expressing cells in 1981. Ricin or diphtheria toxin subunits coupled to antiTfR inhibited growth of melanoma cells in a nude mouse model.

### 6.1. *Ribosomal inhibitors*

Ribosome inactivating proteins (RIP's) are mainly found in plants, but are also found in some bacteria and fungi.[53] Their common feature is a specific N-glycosidase activity, which can cleave an invariant adenine from the 28S rRNA of eukaryotic ribosomes, and subsequently inhibiting their function. Type II RIP's consist of two polypeptide chains: the A chain carries the catalytic glycosidase domain, whereas cell binding and internalization are mediated by the B-chain targeting certain carbohydrates on the cell surface. Ricin is a type II RIP found in the seed of *Ricinus Communis*. Replacing the B chain by Tf or antiTfR, selective delivery to the tumor cells was obtained concomitant with reduced toxicity towards TfR negative cells.[54] Immunotoxins targeted to the TfR are internalized into endosomes, followed by acidification. Subsequent release of the toxin into the cytoplasm can be rather poor. In order to potentiate the endosomal release of the immunotoxin, Chignola *et al.*[55] have utilized a fusion protein of ricin A chain and the N-terminus of protein-G of vesicular stomatitis virus. Conjugates of this fusion protein with Tf showed a significant increase in specific toxicity, mostly due to facilitated release into the cytoplasm. Tf-ricin conjugates have also been shown to be effective against astrozytoma *in vitro* and *in vivo*.[56] Compared with ricin, both type II RIP's, Nigrin b and ebulin I, are 200–50 000-fold less toxic to cultured cells and mice, even though their anti-ribosomal activities are similar.[57] Coupling them to Tf results in immunotoxins that are highly active against cancer cells.[44] Another plant derived inhibitor of mammalian ribosomes is saporin from *Saponaria officinalis*; due to lack of internalization, saporin alone is nontoxic to cells. After coupling the peptide to Tf, rapid internalization into leukemia cells and cellular toxicity was observed.[58] The internalization of saporin-Tf conjugates was highly dependent on the iron saturation of Tf; only iron loaded Tf led to the internalization of different Tf targeted immunotoxins, whereas iron depletion abolished the uptake.[59] Gelonin is a single chain known as type I ribosome inactivating protein. Members of this family possess the catalytic A-chain necessary for protein synthesis inhibition, but lack the B-chain that is characteristic of the type II toxins, e.g. ricin. Due to the lack of internalization, gelonin is virtually nontoxic, after coupling the molecule to Tf or antiTfR cellular internalization, followed by potent inhibition of protein synthesis in tumor cells.[60]

## 6.2. *Ribonucleases*

The natural function of ribonucleases is the posttranscriptional control of gene expression by regulating the RNA population in the cytoplasm. However, several plant and animal derived toxins also show nuclease activity, and their potential use as diagnostics and therapeutics have been evaluated for several diseases including cancer.[61] Cytotoxic ribonucleases in bacteria, fungi or plants contribute to host defense pathways, and have been evaluated for therapeutic purposes.[62] Secreted ribonucleases, e.g. pancreatic RNAse A, are internalized by endocytosis into acidic vesicles, but the mechanism is different from the Tf internalization pathway.[63] Onconase, a secreted amphibian RNAse, isolated from *Rana pipiens*, exhibits strong cytotoxicity towards tumor cells, and is in the clinical phase III for the treatment of malignant mesothelioma in combination with doxorubicine at the moment (http://www.clinicaltrials.gov/ct/show/NCT00003034). RNAses which do not carry an efficient cell internalizing function have been utilized for targeted delivery into cancer cells, by fusing them to internalizing moieties including an anti TfR sFv.[64] Restrictocin, a fungal ribonuclease from *Aspergillus restrictus*, expressed as a fusion protein with sFv directed to the human TfR, was selectively toxic to tumor cells.[65] By using protease and metabolic inhibitors, it could be shown that the proteolytic cleavage of toxin and targeting domain was necessary to obtain full functionality of the recombinant protein. In a similar way, alpha sarcin, a ribonuclease derived from *Aspergillus giganteus*, was expressed as a fusion protein and shown to be functionally active on tumor cells overexpressing TfR.[66] Human ribonucleases belonging to the ribonuclease A superfamily have the advantage of being non-immunogenic, which is an important feature in respect to further clinical use. Eosinophil-derived neurotoxin, a ribonuclease found in cytotoxic granules of eosinophiles, is thought to play a role in their anti-tumor and anti-parasite functions. A fusion protein with a sFv directed against the TfR mediated a cytotoxic response against leukemia cells, whereas cells lacking the TfR were not affected.[67] Zewe and colleagues[68] generated recombinant fusion proteins of either eosinphil-derived neurotoxin or human pancreatic nuclease with a sFv directed against TfR; both constructs inhibited protein synthesis in different human tumor cell lines.

## 6.3. *Diphtheria toxin and Pseudomonas exotoxin*

Diphtheria toxin from *Corynebacterium diphtheriae* (DT), and pseudomonas exotoxin (PE) from *Pseudomonas aeruginosa*, both irreversibly inhibit protein synthesis by catalyzing ADP-ribosylation of elongation factor 2, leading to cell lysis or apoptosis.[69] The gene for DT, located on a bacteriophage present in the bacteria, is translated into a single chain protein consisting of three domains. Binding of DT to heparin-binding epidermal growth factor receptors is mediated by the binding domain B. After endosomal internalization into clathrin coated vesicles, DT undergoes furin cleavage at the arginine rich loop. The lowered endosomal pH induces different conformational changes, leading to the insertion of the translocation domain into the endosomal membrane. Additional conformational changes and the reduction of disulfide bridges within the molecule unfold the catalytic domain A, which after subsequent translocation into the cytoplasm, catalyzes the ADP-ribosylation of elongation factor 2 (for review, refer to Ref. 70). To redirect DT to certain tumors, the B domain has to be removed or inactivated by specific mutations, and a tumor specific ligand attached without negatively influencing the toxins catalytic and the translocation domain.[70] Early work by O'Keefe and Draper[71] described the covalent attachment of Tf to DT, with SPDP leading to a disulfide linkage. Uptake of the conjugate into murine fibroblasts was Tf specific and resulted in a 50% inhibition of protein synthesis, already at concentrations of 1 ng/ml. As mentioned above, binding of DT to its receptor can also be inhibited by the mutations in the binding domain of the toxin. A point mutation in the B domain inactivates binding to the mammalian cells, while the covalent attachment of Tf via a thioether bond reactivated the mutated DT, leading to TfR mediated internalization into tumor cells and potent cytotoxicity.[72] The conjugate named Tf-CRM107 was highly toxic on a broad range of tumor cells, whereas CMR107 lacking Tf had to be applied at 10 000-fold higher concentrations to achieve similar toxicities (for review, see Ref. 73). Preclinical studies on orthotopically implanted human brain tumor models in immunodeficient mice, led to complete tumor regression after local administration of the conjugate.[74] After further studies in primates, a clinical phase I trial was started: in 9 out of 15 patients, at least 50% reduction of tumor volume was achieved including two complete responses.[75] The encouraging outcome led to a clinical phase II trial with 44 patients evaluating the efficiency of Tf-CRM107, after local administration of the conjugate into the brain of glioblastoma and astrocytoma patients, where the safety and efficiency of the drug was confirmed.[73] Meanwhile, a randomized phase III clinical trial is ongoing, where Tf-CRM107 is compared with the best practice chemotherapy in treating glioblastoma multiforme (http://www.clinicaltrials.gov/ct/show/NCT00087230).

Similar as DT, pseudomonas exotoxin (PE) is a single chain peptide with three domains for binding the cellular PE receptor (domain I), the translocation across membranes (domain II) and the ADP ribosylating activity (domain III). In the first few experiments to achieve the targeted delivery of PE, a truncated version of the toxin lacking domain I was chemically coupled to an anti TfR antibody.[76] As this conjugate only showed low cytotoxic activity due to poor reactivity of lysines near the C-terminus of the protein, another mutated form of PE, with lysine residues near the N-terminus, was coupled to different antibodies including antiTfR.[77] This conjugate was highly active of TfR expression tumor cells *in vitro*, and it also caused tumor regression in a xenografted mouse tumor model after intraperitoneal application.

## 7. Conclusions

The development of new biological drug entities is fostered by the high specificity and the usually low toxicity of biological molecules. In contrast to the development of new chemical drugs, a critical aspect of medium- to large-sized new biological drugs is the restricted extravasation and diffusion to the target tissue, as well as uptake into the target cells. Such an apparent weakness, however, can be converted into a strength, by strategies as reviewed in this chapter, the targeting of tumors with TfR targeting molecules. The unique properties of the tumor vasculature enable the enhanced extravasation and accumulation of macromolecules in the tumor; incorporation of TfR targeting domains such as Tf, antiTfR antibodies or scFv increases tumor specificity, since TfR are overexpressed in many tumor types. The exceptional efficacy of the TfR mediated uptake pathway makes the strategy especially useful for the intracellular delivery of toxin molecules. The first phase III clinical trial with Tf-CRM107 illustrates the progress in this field.

## References


1. Gomme PT, McCann KB and Bertolini J (2005) Transferrin: Structure, function and potential therapeutic actions. *Drug Discov Today* **10**:267–273.
2. Qian ZM, Li H, Sun H and Ho K (2002) Targeted drug delivery via the transferrin receptor-mediated endocytosis pathway. *Pharmacol Rev* **54**:561–587.
3. Widera A, Norouziyan F and Shen WC (2003) Mechanisms of TfR-mediated transcytosis and sorting in epithelial cells and applications toward drug delivery. *Adv Drug Deliv Rev* **55**:1439–1466.
4. Li H and Qian ZM (2002) Transferrin/transferrin receptor-mediated drug delivery. *Med Res Rev* **22**:225–250.
5. Sun H, Li H and Sadler PJ (1999) Transferrin as a metal ion mediator. *Chem Rev* **99**:2817–2842.

6. Cheng Y, Zak O, Aisen P, Harrison SC and Walz T (2004) Structure of the human transferrin receptor-transferrin complex. *Cell* **116**:565–576.
7. Hirose M (2000) The structural mechanism for iron uptake and release by transferrins. *Biosci Biotechnol Biochem* **64**:1328–1336.
8. He QY, Mason AB, Nguyen V, MacGillivray RT and Woodworth RC (2000) The chloride effect is related to anion binding in determining the rate of iron release from the human transferrin N-lobe. *Biochem J* **350 (Pt 3)**:909–915.
9. Teehan GS, Bahdouch D, Ruthazer R, Balakrishnan VS, Snydman DR and Jaber BL (2004) Iron storage indices: Novel predictors of bacteremia in hemodialysis patients initiating intravenous iron therapy. *Clin Infect Dis* **38**:1090–1094.
10. Parkkinen J, von Bonsdorff L, Ebeling F and Sahlstedt L (2002) Function and therapeutic development of apotransferrin. *Vox Sang* **83** (Suppl 1):321–326.
11. Dalmastri C, Valenti P, Visca P, Vittorioso P and Orsi N (1988) Enhanced antimicrobial activity of lactoferrin by binding to the bacterial surface. *Microbiologica* **11**:225–230.
12. Huebers HA and Finch CA (1987) The physiology of transferrin and transferrin receptors. *Physiol Rev* **67**:520–582.
13. Lawrence CM, Ray S, Babyonyshev M, Galluser R, Borhani DW and Harrison SC (1999) Crystal structure of the ectodomain of human transferrin receptor. *Science* **286**:779–782.
14. Kawabata H, Yang R, Hirama T, Vuong PT, Kawano S, Gombart AF and Koeffler HP (1999) Molecular cloning of transferrin receptor 2. A new member of the transferrin receptor-like family. *J Biol Chem* **274**:20826–20832.
15. Conner SD and Schmid SL (2003) Regulated portals of entry into the cell. *Nature* **422**:37–44.
16. Lieu PT, Heiskala M, Peterson PA and Yang Y (2001) The roles of iron in health and disease. *Mol Aspects Med* **22**:1–87.
17. Morgan EH (1996) Cellular iron processing. *J Gastroenterol Hepatol* **11**:1027–1030.
18. Nunez MT, Gaete V, Watkins JA and Glass J (1990) Mobilization of iron from endocytic vesicles: The effects of acidification and reduction. *J Biol Chem* **265**:6688–6692.
19. Morgan EH (1996) Cellular iron processing. *J Gastroenterol Hepatol* **11**: 1027–1030.
20. Testa EP, Testa U, Samoggia P, Salvo G, Camagna A and Peschle C (1986) Expression of transferrin receptors in human erythroleukemic lines: Regulation in the plateau and exponential phase of growth. *Cancer Res* **46**:5330–5334.
21. Casey JL, Hentze MW, Koeller DM, Caughman SW, Rouault TA, Klausner RD and Harford JB (1988) Iron-responsive elements: Regulatory RNA sequences that control mRNA levels and translation. *Science* **240**:924–928.

22. Hentze MW, Caughman SW, Rouault TA, Barriocanal JG, Dancis A, Harford JB and Klausner RD (1987) Identification of the iron-responsive element for the translational regulation of human ferritin mRNA. *Science* **238**:1570–1573.
23. Basilion JP, Kennedy MC, Beinert H, Massinople CM, Klausner RD and Rouault TA (1994) Overexpression of iron-responsive element-binding protein and its analytical characterization as the RNA-binding form, devoid of an iron-sulfur cluster. *Arch Biochem Biophys* **311**:517–522.
24. Leibold EA and Munro HN (1988) Cytoplasmic protein binds in vitro to a highly conserved sequence in the 5′ untranslated region of ferritin heavy- and light-subunit mRNAs. *Proc Natl Acad Sci USA* **85**:2171–2175.
25. Henderson BR, Sciser C and Kuhn LC (1993) Characterization of a second RNA-binding protein in rodents with specificity for iron-responsive elements. *J Biol Chem* **268**:27327–27334.
26. Constable A, Quick S, Gray NK and Hentze MW (1992) Modulation of the RNA-binding activity of a regulatory protein by iron *in vitro*: Switching between enzymatic and genetic function? *Proc Natl Acad Sci USA* **89**:4554–4558.
27. Basilion JP, Rouault TA, Massinople CM, Klausner RD and Burgess WH (1994) The iron-responsive element-binding protein: Localization of the RNA-binding site to the aconitase active-site cleft. *Proc Natl Acad Sci USA* **91**:574–578.
28. Erlitzki R, Long JC and Theil EC (2002) Multiple, conserved iron-responsive elements in the 3′-untranslated region of transferrin receptor mRNA enhance binding of iron regulatory protein 2. *J Biol Chem* **277**:42579–42587.
29. Koeller DM, Casey JL, Hentze MW, Gerhardt EM, Chan LN, Klausner RD and Harford JB (1989) A cytosolic protein binds to structural elements within the iron regulatory region of the transferrin receptor mRNA. *Proc Natl Acad Sci USA* **86**:3574–3578
30. Mullner EW, Neupert B and Kuhn LC (1989) A specific mRNA binding factor regulates the iron-dependent stability of cytoplasmic transferrin receptor mRNA. *Cell* **58**:373–382.
31. Hentze MW and Kuhn LC (1996) Molecular control of vertebrate iron metabolism: mRNA-based regulatory circuits operated by iron, nitric oxide, and oxidative stress. *Proc Natl Acad Sci USA* **93**:8175–8182.
32. Gatter KC, Brown G, Trowbridge IS, Woolston RE and Mason DY (1983) Transferrin receptors in human tissues: Their distribution and possible clinical relevance. *J Clin Pathol* **36**:539–545.
33. Iacopetta BJ, Morgan EH and Yeoh GC (1982) Transferrin receptors and iron uptake during erythroid cell development. *Biochim Biophys Acta* **687**:204–210.
34. Richardson DR and Ponka P (1997) The molecular mechanisms of the metabolism and transport of iron in normal and neoplastic cells. *Biochim Biophys Acta* **1331**:1–40.

35. Jefferies WA, Brandon MR, Hunt SV, Williams AF, Gatter KC and Mason DY (1984) Transferrin receptor on endothelium of brain capillaries. *Nature* **312**: 162–163.
36. Moos T and Morgan EH (2000) Transferrin and transferrin receptor function in brain barrier systems. *Cell Mol Neurobiol* **20**:77–95.
37. Kwok JC and Richardson DR (2002) The iron metabolism of neoplastic cells: Alterations that facilitate proliferation? *Crit Rev Oncol Hematol* **42**:65–78.
38. Le NT and Richardson DR (2002) The role of iron in cell cycle progression and the proliferation of neoplastic cells. *Biochim Biophys Acta* **1603**:31–46.
39. Chan SM, Hoffer PB, Maric N and Duray P (1987) Inhibition of gallium-67 uptake in melanoma by an anti-human transferrin receptor monoclonal antibody. *J Nucl Med* **28**:1303–1307.
40. Richardson DR (2002) Iron chelators as therapeutic agents for the treatment of cancer. *Crit Rev Oncol Hematol* **42**:267–281.
41. Tsutsumi Y, Onda M, Nagata S, Lee B, Kreitman RJ and Pastan I (2000) Site-specific chemical modification with polyethylene glycol of recombinant immunotoxin anti-Tac(Fv)-PE38 (LMB-2) improves antitumor activity and reduces animal toxicity and immunogenicity. *Proc Natl Acad Sci USA* **97**: 8548–8553.
42. Wagner E, Cotten M, Mechtler K, Kirlappos H and Birnstiel ML (1991) DNA-binding transferrin conjugates as functional gene-delivery agents: Synthesis by linkage of polylysine or ethidium homodimer to the transferrin carbohydrate moiety. *Bioconjug Chem* **2**:226–231.
43. Scott CF, Jr, Goldmacher VS, Lambert JM, Jackson JV and McIntyre GD (1987) An immunotoxin composed of a monoclonal antitransferrin receptor antibody linked by a disulfide bond to the ribosome-inactivating protein gelonin: Potent *in vitro* and *in vivo* effects against human tumors. *J Natl Cancer Inst* **79**: 1163–1172.
44. Citores L, Ferreras JM, Munoz R, Benitez J, Jimenez P and Girbes T (2002) Targeting cancer cells with transferrin conjugates containing the non-toxic type 2 ribosome-inactivating proteins nigrin b or ebulin l. *Cancer Lett* **184**:29–35.
45. Thorpe PE, Wallace PM, Knowles PP, Relf MG, Brown AN, Watson GJ, Knyba RE, Wawrzynczak EJ and Blakey DC (1987) New coupling agents for the synthesis of immunotoxins containing a hindered disulfide bond with improved stability *in vivo*. *Cancer Res* **47**:5924–5931.
46. Arpicco S, Dosio F, Brusa P, Crosasso P and Cattel L (1997) New coupling reagents for the preparation of disulfide cross-linked conjugates with increased stability. *Bioconjug Chem* **8**:327–337.
47. Braslawsky GR, Kadow K, Knipe J, McGoff K, Edson M, Kaneko T and Greenfield RS (1991) Adriamycin(hydrazone)-antibody conjugates require

internalization and intracellular acid hydrolysis for antitumor activity. *Cancer Immunol Immunother* **33**:367–374.

48. de Smit MH, Hoefkens P, de Jong G, van Duin J, van Knippenberg PH and van Eijk HG (1995) Optimized bacterial production of nonglycosylated human transferrin and its half-molecules. *Int J Biochem Cell Biol* **27**:839–850.
49. Mason AB, Miller MK, Funk WD, Banfield DK, Savage KJ, Oliver RW, Green BN, MacGillivray RT and Woodworth RC (1993) Expression of glycosylated and nonglycosylated human transferrin in mammalian cells. Characterization of the recombinant proteins with comparison to three commercially available transferrins. *Biochemistry* **32**:5472–5479.
50. Newton DL, Xue Y, Olson KA, Fett JW and Rybak SM (1996) Angiogenin single-chain immunofusions: Influence of peptide linkers and spacers between fusion protein domains. *Biochemistry* **35**:545–553.
51. Goyal A and Batra JK (2000) Inclusion of a furin-sensitive spacer enhances the cytotoxicity of ribotoxin restrictocin containing recombinant single-chain immunotoxins. *Biochem J* **345 (Pt 2)**:247–254.
52. Trowbridge IS and Domingo DL (1981) Anti-transferrin receptor monoclonal antibody and toxin-antibody conjugates affect growth of human tumor cells. *Nature* **294**:171–173.
53. Stirpe F (2004) Ribosome-inactivating proteins. *Toxicon* **44**:371–383.
54. Zovickian J, Johnson VG and Youle RJ (1987) Potent and specific killing of human malignant brain tumor cells by an anti-transferrin receptor antibody-ricin immunotoxin. *J Neurosurg* **66**:850–861.
55. Chignola R, Anselmi C, Dalla SM, Franceschi A, Fracasso G, Pasti M, Chiesa E, Lord JM, Tridente G and Colombatti M (1995) Self-potentiation of ligand-toxin conjugates containing ricin A chain fused with viral structures. *J Biol Chem* **270**:23345–23351.
56. Hall WA (2000) Targeted toxin therapy for malignant astrocytoma. *Neurosurgery* **46**:544–551.
57. Munoz R, Arias Y, Ferreras JM, Jimenez P, Rojo MA and Girbes T (2001) Sensitivity of cancer cell lines to the novel non-toxic type 2 ribosome-inactivating protein nigrin b. *Cancer Lett* **167**:163–169.
58. Bergamaschi G, Cazzola M, Dezza L, Savino E, Consonni L and Lappi D (1988) Killing of K562 cells with conjugates between human transferrin and a ribosome-inactivating protein (SO-6). *Br J Haematol* **68**:379–384.
59. Gosselaar PH, van Dijk AJ, de Gast GC, Polito L, Bolognesi A, Vooijs WC, Verheul AF, Krouwer HG and Marx JJ (2002) Transferrin toxin but not transferrin receptor immunotoxin is influenced by free transferrin and iron saturation. *Eur J Clin Invest* **32** (Suppl 1):61–69.

60. Yazdi PT, Wenning LA and Murphy RM (1995) Influence of cellular trafficking on protein synthesis inhibition of immunotoxins directed against the transferrin receptor. *Cancer Res* **55**:3763–3771.
61. Schein CH (1997) From housekeeper to microsurgeon: The diagnostic and therapeutic potential of ribonucleases. *Nat Biotechnol* **15**:529–536.
62. Rybak SM and Newton DL (1999) Natural and engineered cytotoxic ribonucleases: Therapeutic potential. *Exp Cell Res* **253**:325–335.
63. Haigis MC and Raines RT (2003) Secretory ribonucleases are internalized by a dynamin-independent endocytic pathway. *J Cell Sci* **116**:313–324.
64. Rybak SM, Hoogenboom HR, Newton DL, Raus JC and Youle RJ (1992) Rational immunotherapy with ribonuclease chimeras. An approach toward humanizing immunotoxins. *Cell Biophys* **21**:121–138.
65. Rathore D, Nayak SK and Batra JK (1997) Overproduction of fungal ribotoxin alpha-sarcin in Escherichia coli: Generation of an active immunotoxin. *Gene* **190**:31–35.
66. Rathore D and Batra JK (1997) Construction, expression and characterization of chimaeric toxins containing the ribonucleolytic toxin restrictocin: Intracellular mechanism of action. *Biochem J* **324 (Pt 3)**:815–822.
67. Newton DL, Nicholls PJ, Rybak SM and Youle RJ (1994) Expression and characterization of recombinant human eosinophil-derived neurotoxin and eosinophil-derived neurotoxin-anti-transferrin receptor sFv. *J Biol Chem* **269**:26739–26745.
68. Zewe M, Rybak SM, Dubel S, Coy JF, Welschof M, Newton DL and Little M (1997) Cloning and cytotoxicity of a human pancreatic RNase immunofusion. *Immunotechnology* **3**:127–136.
69. Michl P and Gress TM (2004) Bacteria and bacterial toxins as therapeutic agents for solid tumors. *Curr Cancer Drug Targets* **4**:689–702.
70. Frankel AE, Rossi P, Kuzel TM and Foss F (2002) Diphtheria fusion protein therapy of chemoresistant malignancies. *Curr Cancer Drug Targets* **2**:19–36.
71. O'Keefe DO and Draper RK (1985) Characterization of a transferrin-diphtheria toxin conjugate. *J Biol Chem* **260**:932–937.
72. Greenfield L, Johnson VG and Youle RJ (1987) Mutations in diphtheria toxin separate binding from entry and amplify immunotoxin selectivity. *Science* **238**:536–539.
73. Weaver M and Laske DW (2003) Transferrin receptor ligand-targeted toxin conjugate (Tf-CRM107) for therapy of malignant gliomas. *J Neurooncol* **65**:3–13.
74. Laske DW, Ilercil O, Akbasak A, Youle RJ and Oldfield EH (1994) Efficacy of direct intratumoral therapy with targeted protein toxins for solid human gliomas in nude mice. *J Neurosurg* **80**:520–526.

75. Laske DW, Youle RJ and Oldfield EH (1997) Tumor regression with regional distribution of the targeted toxin TF-CRM107 in patients with malignant brain tumors. *Nat Med* **3**:1362–1368.
76. Kondo T, FitzGerald D, Chaudhary VK, Adhya S and Pastan I (1988) Activity of immunotoxins constructed with modified Pseudomonas exotoxin A lacking the cell recognition domain. *J Biol Chem* **263**:9470–9475.
77. Batra JK, Jinno Y, Chaudhary VK, Kondo T, Willingham MC, FitzGerald DJ and Pastan I (1989) Antitumor activity in mice of an immunotoxin made with anti-transferrin receptor and a recombinant form of Pseudomonas exotoxin. *Proc Natl Acad Sci USA* **86**:8545–8549.

# 11

# Transmembrane Delivery of Protein and Peptide Drugs into Cancer Cells

*Cheryl C. Saenz and Steven F. Dowdy*

## Abstract

Progress in the treatment of cancer has been slow over the past few decades, with the development of new therapeutics being hampered by the poor target specificity of small molecules and the impermeability of the plasma membrane to any molecule greater than 500 Da. Adenoviral, retroviral, and lipid-based delivery systems have all met with only limited clinical success secondary to problems with immunogenicity, prolonged gene expression, and significant long-term, occasionally lethal, side effects. The recent identification of a particular group of proteins with enhanced ability to cross the plasma membrane has led to the discovery of a class of protein domains with cell penetrating properties. The fusion of these protein transduction domain sequences with large, biologically active, macromolecular cargo is sufficient to induce their rapid transduction into virtually every type of mammalian cell. Moreover, this nascent technology appears to circumvent many of the problems associated with DNA and viral-based methods. Protein transduction-mediated delivery of previously unavailable large molecules, peptides and proteins, offers a unique opportunity to specifically modulate tumor cell biology *in vivo*.

## 1. Introduction

One of the dictums of the Hippocratic Oath is to "first do no harm". Likewise, the major goal of anticancer therapeutics is to specifically kill cancer cells while leaving surrounding normal tissues unharmed. As traditional therapies like cytotoxic chemotherapy are nonspecific for cancer cells and the need to develop more targeted treatments is clearly necessary. For example, certain proteins and/or protein–protein interactions are specifically altered or deregulated in cancer cells. One attractive strategy for anticancer therapy is to develop molecules that target and modulate tumor-specific proteins and protein–protein interactions. Currently these approaches are being developed using DNA and viral therapy, however, these efforts suffer from potentially limiting side-effects which have contributed to poor outcomes in several clinical trials.[1–3]

An alternate effort to modulate the biology of cancers involves the direct introduction of peptides, full-length proteins, and/or protein functional domains into tumor cells. In general, the plasma membrane of eukaryotic cells is impermeable to the vast majority of macromolecules, limiting intracellular uptake to those which are sufficiently nonpolar and less than 500 Da in size. However, with the identification of several protein transduction domains (PTD), numerous recent studies have now revealed the potential ability of PTDs to modulate the biology of living organisms and to treat diseases such as cancer by the direct intracellular delivery of proteins and peptides.[4,5] The three most widely studied PTDs are from the *Drosophila* homeotic transcription protein, antennapedia (Antp),[6–8] the Herpes Simplex Virus structural protein VP22[9] and the HIV-1 transcriptional activator TAT protein.[10,11] Unlike traditional based techniques, the transduction of these proteins does not appear to be affected by cell type and they can be efficiently transduced into most, if not all, mammalian cells both *in vitro* and *in vivo* with no apparent toxicity.[12] In addition to full-length proteins, protein transduction domains have also been used successfully to induce the intracellular uptake of DNA,[13] antisense oligonucleotides,[14] small molecule drugs[15] and even inorganic 40 nm iron particles[16–19] suggesting that there is no apparent size restriction to this process.

The direct delivery and efficient cellular uptake of transducing proteins and peptides is an exciting new scientific tool that offers several advantages over traditional DNA-based methods for manipulating the cellular phenotype. This chapter will examine the discovery and development of

PTD-mediated transduction technology and will illustrate recent therapeutic applications in the treatment of cancer.

## 2. Functional Characteristics of Protein Transduction Domains

### 2.1. *TAT peptide*

The HIV-1 TAT protein is an essential viral regulatory factor which is involved in the trans-activation of genes involved in the HIV long terminal repeat and therefore plays a critical role in viral replication.[20] Full length TAT protein is encoded by two exons and is between 86 and 102 amino acids in length, depending on the strain of virus. In the late 1980s, two groups independently reported that TAT protein was rapidly internalized into cells in culture and could induce the expression of reporter activity at low concentrations.[10,11] The first examples of the possible therapeutic usefulness offered by TAT-mediated protein transduction were shown by Fawell *et al.* in 1994 when they demonstrated that large proteins such as $\beta$-galactosidase, horseradish peroxidase and RNAse A could be transduced into cells by chemically conjugating them to peptides corresponding to amino acids 1-72 or 37-72 of TAT.[21] Significantly, all the cells in culture showed uptake of the TAT protein and transduction of TAT-$\beta$-galactosidase could be achieved in all cell types that were tested including HeLa, COS-1, CHO, H9, NIH 3T3, primary human keratinocytes, and umbilical endothelial cells.

While it had been shown that chemical conjugates of heterologous full length proteins with the TAT 37-72 peptide could be effectively delivered through the plasma membrane of cells, in 1997, Vives *et al.* began to further characterize shorter domains of the TAT protein to determine the critical region for cell internalization in an effort to improve the cellular uptake and activity of conjugated proteins.[22] Their research demonstrated that the arginine-rich segment in the TAT protein (positions 48–60) (GRKKRRQRRRPPQ) appears to be more efficiently transduced than any of the other active peptide sequences.

### 2.2. *TAT-related peptides*

Although TAT was the first PTD characterized, subsequent studies have identified several other novel transduction domains. Transduction properties have been identified in the *Drosophila* Antennapedia (Antp) homeodomain transcription factor.[23] The minimal PTD region was characterized down to16 residues (RQIKIWFQNRRMKWKK). In addition, VP22 protein,[24–26] as well as synthetic peptoid carriers such as poly-arginine,[27–29] have been shown to promote efficient transduction of organic compounds and peptides. In general, transduction efficiency appears to correspond to the number and location of arginine residues in the protein transduction domain sequence.

While these different protein transduction domains show similar characteristics for cellular uptake, they vary in their efficacy for transporting protein cargo into cells. To date, fusion proteins created with the transduction domain consisting of TAT-(47-57) have shown markedly better cellular uptake than similar fusions using the 16 amino acid sequence from antennapedia or VP22, although recently devised peptide sequences such as the retro-inverso form of TAT-(57-47) or homeopolymers of arginine appear to increase cellular uptake several-fold.[30,31]

## 3. Mechanism of PTD-Mediated Transduction

As the popularity of TAT mediated protein transduction increased over the last decade a growing schism began to form between descriptive studies which demonstrated a phenotypic effect and the mechanistic studies which could not uncover how, if at all, these transducing motifs were able to cross the cell membrane. Clearly, there were cases in which the cargo covalently attached to the TAT peptide was able to elicit intracellular events strongly suggesting that the protein transduction domain was able to enter into the cell. In spite of this, simultaneous studies aimed at determining the method of uptake suggested disparate mechanisms.

Early observations regarding the nature of TAT-mediated cellular uptake relied on visualization and were strongly influenced by the cell surface binding of these domains. Mechanistic studies on transduction suggested that PTDs penetrated cells directly across the cell membrane in a temperature- and energy-independent process. However, these proposed mechanisms appear to reflect strong ionic interactions between the PTD and the cell surface, as it was subsequently discovered that following fixation extracellular bound, TAT proteins were redistributed within the cytoplasm and nucleus of the cell, resulting in apparent internalization.[32,33]

In order to determine the true mechanism of TAT-mediated transduction while avoiding these potential pitfalls Wadia *et al.* used a phenotypic assay for cellular uptake based on a *LoxP*-Stop-*LoxP* GFP gene reporter assay and the transduction of Cre recombinase.[34] In this system, exogenous TAT-Cre protein must enter the cell, translocate to the nucleus, and excise the transcriptional STOP DNA segment in live cells in a nontoxic fashion before scoring positive for eGFP expression. Treatment of reporter T cells for as little as 5 min was sufficient to induce recombination confirming that the cellular uptake was a rapid process.

Endocytosis is an essential cellular process for the uptake of a wide variety of extracellular factors that occurs through functionally distinct mechanisms.[35] Recent work has demonstrated the co-localization of fluorescently labeled TAT-Cre with endosomal markers in live cells.[34] By treating the cells with a variety of cholesterol depleting agents and endocytosis inhibitors, it was determined that the internalization of TAT-fusion proteins and peptides occurs by lipid-raft dependent macropinocytosis,[34] a specialized type of fluid-phase endocytosis.[35] Although macropinosomes are thought to be inherently leaky vesicles, compared with other types of endosomes,[36,37] the majority of the TAT proteins remain trapped within these intracellular compartments, functioning as internal reservoirs up to 24 hrs following treatment.

## 4. Delivery of Anti-Cancer Therapies Using Protein Transduction Domains

The field of transduction has expanded remarkably over the last decade from cell culture observations to the delivery of biologically-active macromolecules in animal model systems.[38–40] Although the eventual role of protein transduction in cancer therapy has yet to be completely defined, it is clear that transduction has tremendous potential to deliver therapeutic biologically active cargo to specifically kill tumor cells. One of the most impressive features of transduction technology is its tremendous versatility. A broad spectrum of molecular cargo has been successfully delivered, representing a wide range of sizes and biophysical properties, including small molecules, peptides, protein functional domains, full-length proteins, DNA, siRNA, phage particles, magnetic nanoparticles, and liposomes. In the context of cancer therapy, transduction overcomes several problems encountered using standard techniques. Conventional chemotherapeutic treatment regimens show low specificity of target and largely affect normal cells as well as malignant cells. Furthermore, these agents are often poorly transported into certain tissues or cells, or are actively exported from cells, as in the case of tumor cells expressing multidrug resistance protein. In contrast, transduction as a delivery modality permits transport independent of these factors and conjugation to protein transduction domains allows the development of novel agents which specifically target cancer cells.

Loss of function of tumor suppressor and DNA damage repair genes, genetic alterations of proto-oncogenes, and loss of sensitivity to cytotoxic chemotherapeutic agents are all involved in the progression of normal cells into the malignant phenotype. In recent years, a growing number of researchers have reported on the use of protein transduction domains as a novel method to modulate these variant aberrant processes and kill malignant cells in cell culture and preclinical animal models. The remainder of this review will focus on the application of protein transduction technology to deliver proteins and peptides in cancer therapy.

## 4.1. *Loss of tumor suppressor function*

The epigenetic reconstitution of tumor suppressor function represents a specific goal in cancer therapy by which it should be possible to selectively target tumor cells while leaving surrounding normal cells unaffected.

### 4.1.1. *p53 tumor suppressor*

The p53 tumor suppressor gene, a DNA damage sensor, is mutated in half of all human tumors, and loss of normal p53 function increases resistance of cancer cells to therapy.[41] Due to mutations and alterations in the DNA damage repair machinery, tumor cells are thought to undergo continuous DNA damage, whereas normal cells maintain the ability to repair damaged DNA. Consequently, reconstitution of p53 function restores the link between pro-apoptotic stimuli and the apoptotic execution machinery, resulting in cell death in a DNA damage-dependent manner. In 1997, Selivanova *et al.* linked a C-terminal p53 peptide, which previously was shown to activate wild type p53 and several DNA-contact mutant forms of p53,[42] to the antennapedia transduction domain and found that the conjugate peptide induced p53-dependent apoptosis in several tumor cells lines.[43]

In a similar study, Harbour *et al.* found that an N-terminal p53 peptide fused to the TAT transduction domain was able to accumulate within cells and disrupt p53 binding to its negative regulator HDM2.[44] By binding to the N-terminal transactivation domain of p53, HDM2 inhibits transcriptional activity of p53, and promotes p53 degradation.[45] Not surprisingly, HDM2 is overexpressed in many tumors that contain wild type p53, resulting in a functional inactivation of p53. Using a rabbit xenograph model of retinoblastoma, intravitreal injection of TAT-N-terminal p53 peptide circumvented HMD2 regulation of p53 and induced rapid accumulation of the tumor suppressor protein, activation of apoptotic genes, and preferential killing of the retinoblastoma cells.[44]

The full length p53 protein has also been conjugated to TAT peptide and transduced into human cancer cell lines, successfully restoring wild-type p53 function. By transducing recombinant TAT-p53 protein in cells, Ryu *et al.* were able to demonstrate a dose- and time-dependent increase in the activation of the p53-dependent gene p21.[46] Importantly, transduction of TAT-p53 resulted in the transcriptional activation of p53-dependent genes not only in p53 positive human cancer cell lines, but also in p53 null cell lines. In the cells treated with TAT-p53, the biological activity of this protein was further confirmed by a decrease in the levels of pro-caspase-3 and Bcl-2 protein and the appearance of activated caspase-3 and PARP cleavage, all consistent with the induction of apoptosis. One drawback of such an approach is the nonspecific delivery of transducible TAT-p53 peptide to any and all cell types, normal as well as tumorigenic. As a result, normal cells are as vulnerable to this treatment as malignant.

By structurally modifying a TAT-C-terminal p53-activating peptide to a retro-inverso form (RI-TATp53C′), Snyder *et al.* were able to synthesize a transducible peptide that was resistant to protease degradation.[47] Although RI-TATp53C′ was able to enter into all cells, p53-specific genes were only activated within cancer cells, not normal cells, as normal cells do not accumulate p53, lacking any significant DNA damage. Snyder *et al.* demonstrated the *in vivo* efficacy of this peptide in a mouse model of ovarian carcinomatosis, where like the human equivalent, there was widespread tumor growth scattered throughout the peritoneal cavity.[47] Intraperitoneal injection of RI-TATp53C′ for 12 days resulted in a significant reduction in tumor growth and a greater than six fold increase in lifespan over vehicle-treated mice.

### 4.1.2. *p16 tumor suppressor*

The p16$^{INK4}$ tumor suppressor gene is frequently altered in human cancers by point mutation, deletion, or silencing due to promoter methylation, each of which can result in its functional inactivation.[48,49] The p16 tumor suppressor protein binds to monomeric Cdk4 or Cdk6 and prevents the formation of active cyclinD:Cdk4/6 complexes. Inhibition of cyclinD:Cdk4/6 activity leads to a G0/G1 phase cell cycle arrest.[50] As a result, restoration of p16 function in tumor cells is an attractive target for therapeutic PTDs.

Lane and co-workers have previously shown that the third ankyrin-like repeat of p16 is responsible for Cdk4/6 binding.[51] A 20 amino acid p16 peptide derived from this domain is sufficient to bind to Cdk4/6 and inhibit cyclin D:Cdk4/6-dependent pRb phosphorylation *in vitro*. Linkage of the p16 peptide to the Antp PTD blocked S-phase entry in asynchronous human keratinocytes.[51,52] Similarly, Gius *et al.* have demonstrated that covalent linkage of this p16 peptide to the TAT transduction domain inhibits cyclinD:Cdk4/6 activity and results in a G1 cell cycle arrest.[53] In addition, Ezhevsky *et al.* have shown that the full-length p16 protein fused to the TAT protein transduction domain blocked cyclin D:Cdk4/6 kinase activity and elicited a G1 cell cycle arrest at a concentration of 300 nm, approximately 100 times lower than that required for the p16 peptide.[54]

These observations provide experimental evidence that the Antp and TAT protein transduction domains may be functionally equivalent in the delivery of p16 peptide. In addition, larger peptides and full-length proteins may be more effective at lower concentrations secondary to an increased specificity for their intracellular targets.

### 4.1.3. *p21 and p27 tumor suppressors*

The cell cycle proteins p21 and p27 bind to and inhibit preexisting, active cyclin D:Cdk4/6, cyclin E:Cdk2, and cyclin A:2 complexes.[55] Overexpression of p21 or p27 leads to an early G1 phase cell cycle arrest. In addition, the cyclin-dependent kinase inhibitor p21 is a major mediator of the p53-dependent growth-arrest pathway.[56]

Recognition of Cdk/cyclin complexes by p21 is thought to occur through a protein–protein interaction with a binding groove on the cyclin subunit. This groove has also been shown to be involved in the recruitment of Cdk substrates, including pRb and E2F. Blocking of this docking site prevents recognition and subsequent phosphorylation of Cdk substrates, restoring p21-like tumor suppressor activity.[57] Treatment of tumor cells with a 20 amino-acid peptide based on the carboxy-terminal Cdk binding domain of p21 coupled to the antennapedia transduction domain was able to inhibit pRb phosphorylation and induce a strong G1 phase cell cycle arrest.[58] Through linkage of the Antp PTD to two peptides corresponding to p21 Cdk-binding domains, (residues 17–33 and 63–77), Bonfanti *et al.* were able to prevent cell growth in two human ovarian cancer cell lines.[59] Chen *et al.* have proposed that E2F deregulation (which frequently occurs during transformation) and Cdk2/cyclin A inactivation are synthetically lethal. They offered this hypothesis based on the demonstration that fusion of the cyclin A recognition site on E2F1 to the TAT transduction domain results in apoptosis in transformed cells, but not normal cells.[60] These observations have been further extended *in vivo* using peptides with similar recognition motifs fused to an antennapedia sequence to induce apoptosis and inhibit tumor growth in nude mice with SVT2 tumor grafts and mammary tumors derived from HER2 transgenic mice.[61]

Overexpression of the p27 tumor suppressor protein in tumor cells leads to cell cycle arrest and apoptosis.[62] In order to reconstitute tumor suppressor function, p27 was synthesized with the TAT transduction domain and used to treat human Jurkat T cells in culture.[12] Transduced TAT-p27 was found bound to Cdk2 within the cells and antiCdk2 immunoprecipitation kinase assays from cells transduced with TAT-p27 showed loss of Cdk2 kinase activity compared to controls. Moreover, treatment of cells with TAT-p27 resulted in a substantial, dose-dependant, $G_1$ phase cell cycle arrest. In a similar study Parada *et al.* showed that overexpression of p27 *via* TAT fusion resulted in cell cycle arrest and apoptosis in human B cell lymphomas.[63] Notably, only the wild type p27 and not the mutant form, which cannot interact with Cdk2, had any biological effect.

The dominant-negative form of Cdk2, that sequesters cyclin from endogenous wild type Cdk, is a functional analog of p21 and p27, exclusively targeting cyclin E and A. TAT-dn-Cdk2 fusion proteins efficiently and reversibly block the proliferation of human fibroblasts and maintain pRb in its active, growth inhibitory state.[64] TAT-dn-Cdk2 transduced into 100% of human diploid fibroblasts, sequestered cyclins A and E, and elicited an early $G_1$ cell cycle arrest by effectively blocking cyclin E/Cdk2 activity.

Based upon these observations, a theoretical use of transducible cell cycle inhibitors would be to temporarily restrain normal cell proliferation during the course of cytotoxic cancer chemotherapy. In this case, pretreatment of normal cells (epithelial, bone marrow, etc.) with cell cycle inhibitors would impose a cell cycle arrest and hypothetically protect normal tissues from chemotherapy-induced damage.[65]

### 4.1.4. *VHL tumor suppressor*

The von Hippel-Lindau (VHL) tumor suppressor gene is functionally inactivated in many renal cell carcinomas (RCC).[66] A small region of VHL normally binds to the cytoplasmic domain of the insulin-like growth factor 1 (IGF-1) receptor and interrupts IGF-1 signaling.[67] As RCC growth depends on IGF-1 receptor activation, mutations in the IGF-1 binding region of VHL permit unrestricted IGF-1 receptor signaling resulting in unrestricted cell proliferation and RCC growth.[68] Epigenetic compensation is a reasonable strategy to restore VHL tumor suppressor function. Datta *et al.* treated RCC cells with TAT-VHL peptide and measured the incorporation of tritiated thymidine and found that TAT-VHL was able to reduce cell proliferation by 80% and inhibit tumor invasiveness through a Matrigel barrier by 81% compared to control TAT-FLAG.[68] Moreover, TAT-VHL peptide inhibited tyrosine phosphorylation of MAP kinase, an essential downstream signal transduction molecule involved in cell proliferation. Impressively, when tested *in vivo* in nude mice with subcutaneous RCC tumors, intraperitoneal injection of TAT-VHL peptide retarded tumor growth and dramatically reduced tumor invasiveness into adjacent muscle wall.

### 4.1.5. *NF2/merlin tumor suppressor*

Schwannomas occur sporadically in patients with neurofibromatosis and account for 8% of all intracranial tumors. Presently, the only treatment available is surgical removal. The majority of schwannomas have a biallelic mutation in the NF2 or merlin tumor suppressor gene.[69] Loss of merlin function causes an alteration in cell shape, aberrant cell–cell communication, and contributes to tumor formation. The addition of wild-type TAT-merlin protein to primary schwannoma cells *in vitro* reverses the cytoskeletal defects and uncontrolled tumor growth due to loss of merlin, and restores these cells to a near normal phenotype.[70] These effects were time and concentration dependent and were reversible after washing the cells. TAT-merlin isoform 2, an inactive isoform, had no effect on the cytoskeletal or growth defects in these transformed cell lines. Since TAT-linked proteins have been shown to transduce across the blood-brain barrier, these observations suggest that, in principle, it should be possible to use TAT-merlin to reconstitute merlin function and reverse aberrant cell growth and tumor formation *in vivo*.

Collectively, the above observations demonstrate that both transducible peptides and proteins are capable of restoring tumor suppressor function. Since tumor suppressor genes are deregulated in the vast majority of human tumors,[41] targeting of these genes and/or their products is now under investigation as an important adjunct for the treatment of cancer.

## 4.2. *Resistance to apoptosis*

Tumor cell populations expand not only as a result of disordered cell cycle regulation but also from aberrations in the rate of programmed cell death (apoptosis). Avoidance of apoptosis is a hallmark of most and perhaps all cancers.[71] Induction of apoptosis is an ideal target for the specificity of PTD-mediated delivery of anticancer peptides.

### 4.2.1. *Caspase activation*

Inhibitors of apoptosis proteins (IAPs) regulate intracellular caspase activity by blocking caspase-active sites.[72] IAPs are often overexpressed in malignant cells and contribute to resistance to chemotherapy and ionizing radiation. Consequently, selective sequestration of IAPs in tumor cells is a promising strategy to overcome resistance to pro-apoptotic therapy. The interaction between caspases and IAPs can be inhibited by a mitochondrial polypeptide called Smac/Diablo.[73] During apoptosis, Smac is released from the mitochondria into the cytosol where it binds to IAPs, disrupts IAP-sequestration of caspases, and facilitates caspase activation.[74] One strategy to induce tumor-specific apoptosis, therefore, is to introduce Smac peptide directly into the cytoplasm and circumvent the necessity of mitochondrial release.

To explore this possibility, Arnt *et al.* fused the IAP binding site on Smac (N-terminal 4–8 residues) to the Antp PTD.[75] Treatment of human breast cancer cells with the Smac$^{WT}$-Antp peptide resulted in binding of the peptide to XIAP and cIAP1 and displacement of caspase 3. Moreover, while sub-therapeutic concentrations of antineoplastic agents including paclitaxel, etoposide, and doxorubicin failed to induce apoptosis in these cells on their own, combined treatment with Smac$^{WT}$-Antp significantly enhanced the induction of apoptosis and long-term antiproliferative effects.

These observations were extended to demonstrate the ability of transducing Smac peptide to improve the efficacy of anticancer therapy in a mouse model of cancer. In the first example, Smac-derived peptides covalently attached to TAT enhanced the ability of TRAIL, a death receptor ligand with specificity for tumor cells, to reduce the size of malignant gliomas.[76] Fulda *et al.* showed that local administration of TRAIL with Smac-TAT peptide reduced tumor volume and extended the life of nude mice bearing established intracranial U87MG tumors. The co-administration of TRAIL and Smac-TAT resulted in a synergistic effect with a reduction in tumor volume and increased survival, with mice living greater than 70 days from the inception of the experiment. In contrast, local treatment with vehicle, Smac-TAT alone, or TRAIL alone had no effect on tumor growth, with control mice succumbing to their tumor burden by 30 days. In addition, by tethering the N-terminal seven amino acids of Smac to a TAT-like arginine-rich transduction sequence, Yang *et al.* were able to demonstrate enhanced chemotherapy-induced suppression of H460 nonsmall cell lung cancers grown as flank xenografts.[77]

Solid tumors often contain large hypoxic areas which are resistant to conventional chemotherapy and radiation. The HIF-1$\alpha$ transcription factor regulates multiple genes involved in a cell's response to intracellular oxygen levels, including the up-regulation of VEGF.[78] The half-life of HIF-1$\alpha$ is regulated through a decrease in its degradation rate in response to hypoxia.[79] A 200 amino acid oxygen dependent degradation (ODD) domain within HIF-1$\alpha$ promotes ubiquitination and proteosomal degradation of the protein under normoxic conditions.[80] Low oxygen partial pressure in solid tumors results in the inactivation of ODD-dependent degradation, stabilization of HIF-1$\alpha$, and transcriptional induction of target genes to stimulate angiogenesis.[81] By utilizing the "oxygen-sensing" properties of the ODD domain, Harada *et al.*, devised a novel anticancer therapy based on a TAT-ODD-caspase 3 fusion protein to induce cell death within the hypoxic regions of tumors.[82] NIH3T3 or A549 cells treated with

TAT-ODD-caspase 3 under hypoxic conditions resulted in cell death by apoptosis. When TAT-ODD-caspase 3 was injected intraperitoneally into mice bearing solid tumors, the active protein was found to be preferentially stabilized in the tumors and not present throughout the normal tissues. Significantly, the administration of TAT-ODD-caspase-3 wild type, but not an inactive mutant form of TAT-ODD-caspase-3, was able to suppress tumor growth, reduce the tumor mass, and induce tumor-specific apoptosis.

### 4.2.2. *Bcl-2 family proteins*

The Bcl-2 family of proteins plays a critical role in the regulation of apoptosis. Members of the Bcl-2 family include pro-apoptotic proteins (e.g. Bax and Bak) and antiapoptotic proteins (e.g. Bcl-2 and Bcl-XL).[83] The Bcl-2 related antiapoptotic proteins have been found to play a role in cancer development by inhibiting programmed cell death. BH1-4 are Bcl-2 homology domains involved in protein–protein interactions and specific mutations in these regions can result in either an increase or a decrease in antiapoptotic activity.[84]

The BH3 domain has been found to antagonize the antiapoptotic function of Bcl-XL. Conjugation of the BH3 domain peptide to the transduction domain from antennapedia (Antp-BH3) resulted in efficient transduction into HeLa cells and induced apoptosis.[85] In contrast, Antp PTD or BH3 peptide added to cells alone was ineffective. Importantly, the mutant Antp-BH3-L78A peptide with a single amino acid substitution and reduced binding activity to Bcl-XL failed to initiate apoptosis. Letai *et al.* further refined these observations by utilizing the poly-arginine transduction domain to deliver BH3 domain peptides derived from multiple pro-apoptotic Bcl-2 family members to induce synergistic killing of leukemia cells in culture.[86]

### 4.2.3. *S100 family proteins*

The S100 family of proteins is involved in a wide variety of cellular functions including cell growth and motility, cell cycle regulation, transcription, and differentiation. Recent evidence also suggests that S100 proteins are associated with a number of human diseases including cancer.[87] Among these proteins, S100A4 appears to play an important role in the acquisition of the metastatic phenotype and S100B has been found to be upregulated in most human melanomas. Recently, another member, S100C/A11 has been found to be involved in growth inhibition following translocation from the cytoplasm to the nucleus.[88] This regulation appears to be disordered in malignant cells. Further characterization revealed that the N-terminal 19 amino acids (MAK19) constituted a domain which inhibits growth of human keratinocytes by a mechanism similar to intact S100C/A11. When MAK19 was transduced into cells by fusion with the TAT (TAT-MAK19) there was a dose-dependent decrease in DNA synthesis followed by cell death. This cytotoxic effect occurred in various human cancer cell lines including squamous cell carcinoma, melanoma, colon cancer, mammary carcinoma, and lung cancer. One drawback of this approach, however, was the nonspecific induction of apoptosis in normal as well as tumorigenic cells. Future refinements in selective tumor transduction, however, may aid with the further development of this strategy.

### 4.2.4. *Protein kinase 2 (casein kinase 2)*

Phosphorylation of substrates by casein kinase 2 (CK2) is frequently dysregulated in human oncogenesis.[89] Increased CK2 enzymatic activity has been associated with aberrations in cell growth and proliferation, cell viability and apoptosis in various solid tumors. Inhibition of CK2 has been shown to induce apoptosis *in vitro*. Using a phage library, Perea *et al.* identified a cyclic peptide (P15) that selectively blocks CK2 phosphorylation by blocking substrate interaction.[89] P15 was then conjugated to the TAT PTD and various human cancer cell lines were transduced with the P15-TAT fusion peptide. Induction of apoptosis was demonstrated through caspase activation and cytotoxicity in cells treated with P15-TAT. These observations were extended *in vivo* where direct injection of P15-TAT into solid-tumor bearing mice resulted in a significant reduction or complete inhibition of tumor growth compared with the control groups.

### 4.2.5. *Activating transcription factor 2*

Activating Transcription Factor 2 (ATF2) is a member of the bZIP family of transcription factors that requires phosphorylation by JNK and p38 to elicit its transcriptional activities.[90] A 50 amino acid peptide derived from the N-terminal of ATF2 has been shown to inhibit ATF2 and result in the sensitization of human cancer cells to apoptosis after treatment with either ionizing radiation or chemotherapy that by themselves fail to affect these tumors.[91] Bhoumi *et al.* fused the $ATF2^{51-100}$ peptide to the TAT transduction domain and demonstrated a significant decrease in tumor mass in mice bearing SW1 melanomas with complete regression in 9 of 20 tumors.[92] In addition, none of the mice treated with TAT–$ATF2^{51-100}$ developed lung metastases, whereas the control mice treated with TAT alone developed multiple metastatic pulmonary lesions. This significant inhibition of tumorigenesis and metastasis is believed to be mediated through activation of the apoptotic cascade as cells expressing $ATF2^{51-100}$ were shown to induce caspases 3 and 9, and PARP cleavage, reflecting cell commitment for apoptosis.

Two basic strategies have emerged to subvert the evasion of apoptosis in tumor cells: either restore the missing sensor function at the beginning of the apoptotic pathway or directly trigger the downstream execution machinery within the tumor cell. The above observations demonstrate that PTD-mediated delivery of peptides and proteins can be used to re-sensitize tumor cells to induce apoptosis regardless of which class of components has been altered.

## 4.3. *Activation of oncogenes*

Oncogenes are altered versions of proto-oncogenes that are directly responsible for cancer progression through deregulation of cell cycle control and differentiation. Mutations in a proto-oncogene can result in either constitutive activation or excess amounts of protein product thereby "converting" the gene into an oncogene. Protein transduction domains have been used to target protein–protein interactions and interrupt the aberrations of cell signaling set into play in malignant cells through activation of oncogenes.

### 4.3.1. *Ras signaling pathways*

The importance of targeting the Ras signaling pathway in cancer cannot be overstated. Oncogenic mutations in Ras occur in approximately 25% of human malignancies leading to constitutively active Ras and promoting growth factor-independent cell proliferation.[93] Grb2 is a small adaptor protein that is essential in the Ras signaling pathway. Grb2, complexed with

Sos, the exchange factor of Ras, recruits Sos to the membrane, activating Ras under its GTP form.[94] Activated Ras initiates the MAPK cascade by recruiting Raf, which triggers an extracellular signal-regulated kinase (ERK). ERK then translocates to the nucleus and stimulates transcription of early genes.[95] The blocking of Grb/Sos complexes has great potential to avert a vast downstream signaling cascade.

Blocking the interaction of Grb2 and Sos relies on peptides that mimic the proline-enriched protein–protein contact domains of Sos that bind the two SH3 domains of Grb2. Sos contains four proline-enriched regions and each of them shows low affinity for the SH3 domains of Grb2.[96] However, linking of two of Sos's proline-enriched domains together increases the affinity for Grb2 by 400-fold.[94] When linked to the Antp PTD, this peptide dimer disrupts Grb2/Sos complex formation and inhibits MAP kinase phosphorylation induced by the addition of EGF.[96] Importantly, the Antp peptide dimer conjugates did not alter proliferation in nonmalignant cells in culture.

One well-defined anticancer strategy is to block Ras-mediated signaling by introduction of a dominant-negative form of this small GTPase.[97] Dominant-negative Ras (dn-Ras) competes with wild-type Ras for binding to the GTP exchange factor, forming unproductive complexes and inhibiting activation of wild-type Ras. Fused to the TAT PTD, dn-Ras efficiently transduced into isolated human eosinophils and prevented IL-5-dependent activation of ERK 1 and 2.[98] Transduction of TAT-dn-Ras chimeric protein also blocked all Ras-mediated cellular responses from stimulation of cytokine-, chemokine-, and G-protein-coupled receptors.[99]

Another approach to reversing the Ras-dependent phenotype involves abolishing function of downstream intermediates such as the MEK/ERK interaction. The final effector of this pathway, ERK, has been targeted by cell-permeable inhibitors.[100] A MEK1-derived peptide, fused to the membrane-translocating sequence HBV, has been shown to inhibit ERK activation in TPA-stimulated 3T3 cells in a concentration-dependent manner. In addition, cell-permeable peptides inhibited ERK-mediated activation of the transcriptional activity of ELK1.

### 4.3.2. *HER-2 pathway*

HER-2, the oncogenic analog of epidermal growth factor, is associated with aggressive, highly metastatic breast cancers with increased resistance to chemotherapy. Thus, modulation of the HER-2 signal transduction pathway is a logical strategy for treating HER-2 overexpressing cancers. The transcriptional factor ESX, in complex with nuclear cofactor DRIP130, binds and strongly activates the HER-2 promoter.[101] Disruption of ESX/DRIP130 interactions impairs HER-2 gene expression and reduces proliferation and the viability of HER-2-expressing breast cancer cells. Asada *et al.* determined the critical region for the DRIP130-ESX interaction, and designed a TAT-ESX peptide with competitive binding capabilities.[102] When added in culture to HER-2 over-expressing breast cancer cells, the TAT-ESX peptide reduced HER-2 protein levels, retarded cell growth, and induced apoptosis. Control TAT transduction domain alone and an irrelevant TAT-VP16 chimeric protein had no effect on the breast cancer cells. Importantly, cells with low levels of endogenous HER-2 expression or with ectopic expression of HER-2 under the control of a heterologous promoter (hence not dependent on the DRIP130/ESX interaction), remained insensitive to the TAT-ESX peptide.

The interaction of protein–protein contact domains constitutes a target for the design of therapeutic anticancer peptides. Historically, these peptides have had limited bioavailability due to their inability to cross the cell membrane. With the discovery of PTDs, however, molecules greater than 500 Da are able to be transported to both the cytoplasm and the nucleus. Targeting signal-mediated cascades with PTDs confirms the precision that is achievable by protein transduction.

## 4.4. *Resistance to conventional cancer therapeutics*

The development of resistance to cytotoxic chemotherapy represents a major clinical obstacle in the treatment of malignant disease. In response to previous treatments, tumor cells often develop mutations in genes that allow them to escape apoptosis cascades or activate the transcription of genes involved in the active export of these agents out of cells. Given that these genetic changes usually occur in malignant cells and not the surrounding normal cells, they are an ideal target for PTD-mediated delivery of anticancer peptides.

### 4.4.1. *RasGAP-derived peptide*

RasGAP, a regulator of Ras and Rho GTP-binding proteins, is a paradoxical caspase substrate as it can induce both anti and proapoptotic signals, depending on the extent of its cleavage by caspases.[103] At low levels of caspase activity RasGAP is cleaved at position 455, generating an N-terminal fragment that is antiapoptotic.[104] At higher levels of caspase activity, however, the N-terminal fragment is cleaved, suppressing the antiapoptotic activity and generating two fragments, N1 and N2. Further, N2 has been shown to potently sensitize HeLa cells toward cisplatinum-induced apoptosis.[103] Michod *et al.* fused TAT to a peptide derived from the N2 fragment of RasGAP and treated several different cancer cell lines with TAT-$RasGAP_{317-326}$ and sublethal concentrations of cytotoxic chemotherapy.[105] TAT- $RasGAP_{317-326}$ efficiently transduced into cancer cells and rendered them sensitive to doses of cisplatinum, adriamycin, and mitoxantrone that did not, or only marginally, induced apoptosis in control conditions. Importantly, TAT- $RasGAP_{317-326}$ also readily transduced into noncancer cell lines but did not modulate their sensitivity to cytotoxic chemotherapy.

### 4.4.2. *Repair of mitochondrial DNA*

Recently a growing number of studies have reported on the role of mitochondrial DNA (mtDNA) in modulating cancer cell sensitivity to ionizing radiation, resistance to oxidative stress, and cytotoxic chemotherapy.[106–108] The manipulation of mtDNA repair can have a profound effect on cell survival under conditions of cytotoxic stress.[109] Targeting of specific repair enzymes into mitochondria can lead to an imbalance in base excision repair and enhanced cell killing. Specifically, there is a negative effect on mtDNA repair when Exonuclease III (ExoIII), a DNA repair enzyme from *E. coli*, is stably expressed and targeted to mitochondria in breast cancer cells.[110] By fusing a mitochondrial targeting sequence to ExoIII and then conjugating that construct to TAT (MTS-ExoIII-TAT), Shokolenko *et al.* were able to generate a transducible protein that trafficked to the mitochondrial matrix in breast cancer cells.[111] Repair of mtDNA was reduced by almost 20% following oxidative stress in the cells that were transduced with MTS-ExoIII-TAT. The results from long-term survival assays demonstrated a significant decrement in cellular survival after even a moderate decrease in the ability of mitochondria to repair itself, further supporting the hypothesis that disruption of mtDNA repair can sensitize cells to antitumor agents which act through reactive oxygen species.

The efficacy of the agents used to treat cancer cells relies on their ability to kill malignant cells, however, these agents are often severely limited by the fact that they can also adversely affect normal tissues. Protein transduction-mediated delivery of peptides and proteins can improve and/or restore the sensitivity of cancer cells to various anticancer therapeutics thereby allowing a lower effective dose to be used in treatment and minimizing side effects to surrounding normal cells.

## 4.5. *Additional applications of PTD-mediated delivery in the treatment of cancer cells*

### 4.5.1. *Cancer cell vaccines*

Tumor-associated antigens (TAA) have been shown to elicit cytotoxic responses that result in the development of immune memory cells that could safeguard against the development of recurrent tumorigenesis.[112] One of the main obstacles to the success of cancer vaccines, however, is the low immunogenicity of natural epitopes expressed by tumor cells which impairs the capabilities of the host T-cells to induce a sufficient immune response.[113] One strategy for overriding the immunogenicity barrier is to utilize tumor antigen-loaded dendritic cells (DC) as enhancers of an anti-tumor response. DCs proteolytically degrade antigens into peptides and present them in complexes with MHC classes I and II to initiate an immune response in naive T cells. DCs loaded with tumor antigens *in vitro* have been shown to induce immune responses *in vivo* after transplantation into the host.[114] Tumor–associated antigens can be loaded into DC cells either *via* transfection with cDNAs or infection with various viruses, neither process being ideal for clinical practice.

Direct attachment of the TAA to a PTD, however, can facilitate delivery of the antigen to the cytosolic compartment of DCs, while circumventing the limitations of transfection and the concerns surrounding the use of viral vectors in patients. Linkage of the Antp PTD to a nonimmunogenic peptide promotes internalization, processing, and presenting of the epitope to immature DCs in an efficient enough manner to activate Ag-specific cytotoxic T lymphocytes (CTL).[115] These observations demonstrate the successful conversion of an otherwise nonimmunogenic peptide into an immunogenic antigen by PTD-mediated delivery and are a suitable model for the development of cancer vaccines.

Tyrosinase-related protein 2 (TRP2) is a tumor rejection antigen for the human B16 melanoma cell line.[116] DCs loaded with TRP2 peptide can initiate a CTL response and promote B16 tumor rejection. However, the efficiency of TRP2 pulsed DCs is unacceptably low. In contrast, TRP2 peptide covalently linked to TAT dramatically potentiated the DCs ability to present MHC loaded TRP2 to T cells and increased antigen immunogenicity. Vaccination of mice with DCs pretreated with TAT-TRP2 peptides resulted in protection from a B16 tumor challenge as measured by a decrease in the number of lung metastases. Moreover, the survival of mice vaccinated with DCs transduced with TAT-TRP2 was significantly increased over control mice.

Due to the rapid MHC turnover and peptide degradation, the half-life of peptide-MHC class I complexes is limited. Consequently, DCs loaded with peptides *in vitro* will have a limited time to present antigen to CTLs *in vivo*. Fusion of Ag-peptides to PTDs, however, increases the intracellular reservoirs of antigen and may prolong the time of DC-dependent antigen presentation. Thus, the loading of DCs with PTD-Ag peptides remains an attractive strategy to improve the immunogenicity of tumor antigens and serves as a starting point to improve anticancer vaccines.

### 4.5.2. *Cancer cell migration*

The majority of patients that die from cancer do so secondary to the formation of metastases at distant organ sites and not from the primary tumor growth. A pivotal event in the development of metastatic tumors is the ability of individual tumor cells to detach from the primary tumor and migrate through the extracellular matrix. The adaptor protein phospholipase C-$\gamma$1(PLC-$\gamma$1) plays a key role in growth factor mediated tumor migration and therefore is an appropriate target for limiting the spread of metastasis. To create a specific inhibitor of tumor cell migration, the SH2 domains of PLC-$\gamma$1 were fused to the TAT transduction domain (PS2-TAT) and used to prevent PLC-$\gamma$1 tyrosine phosphorylation, thereby inhibiting EGF induced migration.[117] Treatment of MDA-HER2 breast cancer cells with PS2-TAT resulted in the competition between native PLC-$\gamma$1 for binding the activated kinase, thereby reducing PLC-$\gamma$1 phosphorylation and activation by 30%. Long-term PS2-TAT treatment was sufficient to reduce EGF-mediated migration by 75%, as well as decrease the overall proliferation of MDA-HER2 cells by 50%. Based on these results, PS2-TAT therapy represents a novel anti-tumor strategy which could be useful in minimizing metastasis formation and cell proliferation *in vivo*.

## 5. Conclusions

Over the past few years, research in the field of protein transduction has opened a new opportunity to develop novel anticancer therapeutic strategies. By overcoming the problem of delivering molecules greater than 500 Da, PTD-mediated delivery has greatly expanded the spectrum of potential intracellular targets. PTD-mediated transduction represents a novel means of developing new cancer therapies through the targeting of aberrant protein–protein interactions that occur in malignant cells, and not in surrounding normal tissues. Although these strategies are only beginning to be explored, they have already been used in several models of cancer to reconstitute tumor suppressor function, induce the apoptosis cascade, suppress the activation of oncogenes, and increase sensitivity to conventional cancer therapeutics. If PTD-mediated transduction can be adapted into humans with the ease and efficacy observed *in vitro* and in animal models, then within our grasp is the capability to devise entirely new therapeutic compounds, peptides and proteins, to combat the initiation and progression of cancer.

## References


1. Therapy ESoG (2003) French gene therapy group reports on the adverse event in a clinical trial of gene therapy for X-linked severe combined immune deficiency (X-SCID). *J Gene Med* **5**:82–84.
2. Reid T, Warren R and Kirn D (2002) Intravascular adenoviral agents in cancer patients: Lessons from clinical trials. *Cancer Gene Ther* **9**:979–986.
3. Vile R, Ando D and Kirn D (2002) The oncolytic virotherapy treatment platform for cancer: Unique biological and biosafety points to consider. *Cancer Gene Ther* **9**:1062–1067.
4. Wadia JS and Dowdy SF (2002) Protein transduction technology. *Curr Opin Biotechnol* **13**:52–56.
5. Wadia JS and Dowdy SF (2003) Modulation of cellular function by TAT mediated transduction of full length proteins. *Curr Protein Pept Sci* **4**:97–104.
6. Joliot A, Pernelle C, Deagostini-Bazin H and Prochiantz A (1991) Antennapedia homeobox peptide regulates neural morphogenesis. *Proc Natl Acad Sci USA* **88**:1864–1868.
7. Joliot A, Triller A, Volovitch M, Pernelle C and Prochiantz A (1991) Alpha-2-8-polysialic acid is the neuronal surface receptor of antennapedia homeobox peptide. *New Biol* **3**:1121–1134.

8. Le Roux I, Joliot AH, Bloch-Gallego E, Prochiantz A and Volovitch M (1993) Neurotrophic activity of the Antennapedia homeodomain depends on its specific DNA-binding properties. *Proc Natl Acad Sci USA* **90**:9120–9124.
9. Elliott G and O'Hare P (1997) Intercellular trafficking and protein delivery by a herpesvirus structural protein. *Cell* **88**:223–233.
10. Green M and Loewenstein P (1988) Autonomous functional domains of chemically synthesized human immunodeficiency virus Tat trans-activator protein. *Cell* **55**:1179–1188.
11. Frankel A and Pabo C (1988) Cellular uptake of the Tat protein from human immunodeficiency virus. *Cell* **55**:1189–1193.
12. Nagahara H, Vocero-Akbani AM, Snyder EL, Ho A, Latham DG, Lissy NA, Becker-Hapak M, Ezhevsky SA and Dowdy SF (1998) Transduction of full-length TAT fusion proteins into mammalian cells: TAT-p27Kip1 induces cell migration. *Nat Med* **4**:1449–1452.
13. Abu-Amer Y, Dowdy SF, Ross FP, Clohisy JC and Teitelbaum SL (2001) TAT fusion proteins containing tyrosine 42-deleted IkappaBalpha arrest osteoclastogenesis. *J Biol Chem* **276**:30499–30503.
14. Astriab-Fisher A, Sergueev D, Fisher M, Shaw BR and Juliano RL (2002) Conjugates of antisense oligonucleotides with the Tat and antennapedia cell-penetrating peptides: Effects on cellular uptake, binding to target sequences, and biologic actions. *Pharm Res* **19**:744–754.
15. Polyakov V, Sharma V, Dahlheimer JL, Pica CM, Luker GD and Piwnica-Worms D (2000) Novel Tat-peptide chelates for direct transduction of technetium-99m and rhenium into human cells for imaging and radiotherapy. *Bioconjug Chem* **11**:762–771.
16. Dodd CH, Hsu HC, Chu WJ, Yang P, Zhang HG, Mountz JD, Jr, Zinn K, Forder J, Josephson L, Weissleder R, Mountz JM and Mountz JD (2001) Normal T-cell response and *in vivo* magnetic resonance imaging of T cells loaded with HIV transactivator-peptide-derived superparamagnetic nanoparticles. *J Immunol Methods* **256**:89–105.
17. Wunderbaldinger P, Josephson L and Weissleder R (2002) Tat peptide directs enhanced clearance and hepatic permeability of magnetic nanoparticles. *Bioconjug Chem* **13**:264–268.
18. Lewin M, Carlesso N, Tung CH, Tang XW, Cory D, Scadden DT and Weissleder R (2000) Tat peptide-derivatized magnetic nanoparticles allow in vivo tracking and recovery of progenitor cells. *Nat Biotechnol* **18**:410–414.
19. Josephson L, Tung CH, Moore A and Weissleder R (1999) High-efficiency intracellular magnetic labeling with novel superparamagnetic-Tat peptide conjugates. *Bioconjug Chem* **10**:186–191.

20. Sodroski J, Patarca R, Rosen C, Wong-Staal F and Haseltine W (1985) Location of the trans-activating region on the genome of human T-cell lymphotrophic virus type III. *Science* **229**:74–77.
21. Fawell S, Seery J, Daikh Y, Moore C, Chen L, Pepinsky B and J Barsoum (1994) Tat-mediated delivery of heterologous proteins into cells. *Proc Natl Acad Sci USA* **91**:664–668.
22. Vives E, Brodin P and Lebleu B (1997) A truncated HIV-1 Tat protein basic domain rapidly translocates through the plasma membrane and accumulates in the cell nucleus. *J Biol Chem* **272**:16010–16017.
23. DeRossi D, Joliot AH, Chassaing G and Prochiantz A (1994) The third helix of the Antennapedia homeodomain translocates through biological membranes. *J Biol Chem* **269**:10444–10450.
24. Phelan A, Elliot G and O'Hare P (1998) Intercellular delivery of functional p53 by the herpesvirus protein VP22. *Nat Biotechnol* **16**:440–443.
25. Elliott G and O'Hare P (1999) Intercellular trafficking of VP22-GFP fusion proteins. *Gene Ther* **6**:149–151.
26. Brewis N, Phelan A, Webb J, Drew J, Elliott G and O'Hare P (2000) Evaluation of VP22 spread in tissue culture. *J Virol* **74**:1051–1056.
27. Uemura S, Rothbard JB, Matsushita H, Tsao PS, Fathman CG and Cooke JP (2002) Short polymers of arginine rapidly translocate into vascular cells. *Circ J* **66**:1155–1160.
28. Wender PA, Rothbard JB, Jessop TC, Kreider EL and Wylie BL (2002) Oligocarbamate molecular transporters: Design, synthesis and biological evaluation of a new class of transporters for drug delivery. *J Am Chem Soc* **124**:13382–13383.
29. Rothbard JB, Kreider E, VanDeusen CL, Wright L, Wylie BL and Wender PA (2002) Arginine-rich molecular transporters for drug delivery: Role of backbone spacing in cellular uptake. *J Med Chem* **45**:3612–3618.
30. Futaki S, Suzuki T, Ohashi W, Yagami T, Tanaka S, Ueda K and Sugiura Y (2001) Arginine-rich peptides. An abundant source of membrane-permeable peptides having potential as carriers for intracellular protein delivery. *J Biol Chem* **276**:5836–5840.
31. Wender PA, Mitchell DJ, Pattabiraman K, Pelkey ET, Steinman L and Rothbard JB (2000) The design, synthesis and evaluation of molecules that enable or enhance cellular uptake: Peptoid molecular transporters. *Proc Natl Acad Sci USA* **97**:13003–13008.
32. Lundberg M and Johansson M (2002) Positively charged DNA-binding proteins cause apparent cell membrane translocation. *Biochem Biophys Res Commun* **291**:367–371.

33. Vives E, Richard JP, Rispal C and Lebleu B (2003) TAT peptide internalization: Seeking the mechanism of entry. *Curr Protein Pept Sci* **4**:125–132.
34. Wadia JS, Stan RV and Dowdy SF (2004) Transducible TAT-HA fusogenic peptide enhances escape of TAT-fusion proteins after lipid raft macropinocytosis. *Nat Med* **10**:310–315.
35. Conner SD and Schmid SL (2003) Regulated portals of entry into the cell. *Nature* **422**:37–44.
36. Meier O, Boucke K, Hammer SV, Keller S, Stidwill RP, Hemmi S and Greber UF (2002) Adenovirus triggers macropinocytosis and endosomal leakage together with its clathrin-mediated uptake. *J Cell Biol* **158**:1119–1131.
37. Norbury CC, Hewlett LJ, Prescott AR, Shastri N and Watts C (1995) Class I MHC presentation of exogenous soluble antigen via macropinocytosis in bone marrow macrophages. *Immunity* **3**:783–791.
38. Schwarze SR, Ho A, Vocero-Akbani A and Dowdy SF (1999) *In vivo* protein transduction: Delivery of a biologically active protein into the mouse. *Science* **285**:1569–1572.
39. Jo D, Nashabi A, Doxsee C, Lin Q, Unutmaz D, Chen J and Ruley HE (2001) Epigenetic regulation of gene structure and function with a cell-permeable Cre recombinase. *Nat Biotechnol* **19**:929–933.
40. Xia H, Mao Q and Davidson BL (2001) The HIV Tat protein transduction domain improves the biodistribution of beta-glucuronidase expressed from recombinant viral vectors. *Nat Biotechnol* **19**:640–644.
41. Sherr CJ and McCormick F (2002) The RB and p53 pathways in cancer. *Cancer Cell* **2**:103–112.
42. Hupp TR, Sparks A and Lane DP (1995) Small peptides activate the latent sequence-specific DNA binding function of p53. *Cell* **83**:237–245.
43. Selivanova G, Iotsova V, Okan I, Fritsche M, Strom M, Groner B, Grafstrom RC and Wiman KG (1997) Restoration of the growth suppression function of mutant p53 by a synthetic peptide derived from the p53 C-terminal domain. *Nat Med* **3**:632–638.
44. Harbour JW, Worley L, Ma D and Cohen M (2002) Transducible peptide therapy for uveal melanoma and retinoblastoma. *Arch Ophthalmol* **120**:1341–1346.
45. Wu X, Bayle JH, Olson D and Levine AJ (1993) The p53-mdm-2 autoregulatory feedback loop. *Genes Dev* **7**:1126–1132.
46. Ryu J, Lee HJ, Kim K-A, Lee JY, Lee KS, Park J and Choi SY (2004) Intracellular delivery of p53 fused to the basic domain of HIV-1 Tat. *Mol Cells* **17**:353–359.
47. Snyder EL, Meade BR, Saenz CC and Dowdy SF (2004) Treatment of terminal peritoneal carcinomatosis by a transducible p53-activating peptide. *PLoS Biol* **2**:186–193.

48. Ruas M, Brookes S, McDonald NQ and Peters G (1999) Functional evaluation of tumour-specific variants of p16INK4a/CDKN2A: Correlation with protein structure information. *Oncogene* **18**:5423–5434.
49. Rocco JW and Sidransky D (2001) p16(MTS-1/CDKN2/INK4a) in cancer progression. *Exp Cell Res* **264**:42–55.
50. Shapiro GI, Edwards CD and Rollins BJ (2000) The physiology of p16 (INK4A)-mediated G1 proliferative arrest. *Cell Biochem Biophys* **33**:189–197.
51. Fahraeus R, Paramio JM, Ball KL, Lain S and Lane DP (1996) Inhibition of pRb phosphorylation and cell-cycle progression by a 20-residue peptide derived from p16CDKN2/INK4A. *Curr Biol* **6**:84–91.
52. Fahraeus R, Lain S, Ball KL and Lane DP (1998) Characterization of the cyclin-dependent kinase inhibitory domain of the INK4 family as a model for a synthetic tumour suppressor molecule. *Oncogene* **16**:587–596.
53. Gius DR, Ezhevsky SA, Becker-Hapak M, Nagahara H, Wei MC and Dowdy SF (1999) Transduced p16INK4a peptides inhibit hypophosphorylation of the retinoblastoma protein and cell cycle progression prior to activation of Cdk2 complexes in late G1. *Cancer Res* **59**:2577–2580.
54. Ezhevsky SA, Nagahara H, Vocero-Akbani AM, Gius DR, Wei MC and Dowdy SF (1997) Hypo-phosphorylation of the retinoblastoma protein (pRb) by cyclin D:Cdk4/6 complexes results in active pRb. *Proc Natl Acad Sci USA* **94**:10699–10704.
55. Sherr CJ and Roberts JM (1999) CDK inhibitors: positive and negative regulators of G1-phase progression. *Genes Dev* **13**:1501–1512.
56. el-Deiry WS, Harper JW, O'Connor PM, Velculescu VE, Canman CE, Jackman J, Pietenpol JA, Burrell M, Hill DE, Wang Y *et al.* (1994) WAF1/CIP1 is induced in p53-mediated G1 arrest and apoptosis. *Cancer Res* **54**: 1169–1174.
57. Zheleva DI, McInnes C, Gavine A-L, Zhelev NZ, Fischer PM and Lane DP (2002) Highly potent $p21^{WAF1}$-derived peptide inhibitors of CDK-mediated pRb phosphorylation: delineation and structural insight into their interactions with cyclin A. *J Peptide Res* **60**:257–270.
58. Ball KL, Lain S, Fahraeus R, Smythe C and Lane DP (1997) Cell-cycle arrest and inhibition of Cdk4 activity by small peptides based on the carboxy-terminal domain of p21WAF1. *Curr Biol* **7**:71–80.
59. Bonfanti M, Taverna S, Salmona M, D'Incalci M and Broggini M (1997) p21WAF1-derived peptides linked to an internalization peptide inhibit human cancer cell growth. *Cancer Res* **57**:1442–1446.
60. Chen YN, Sharma SK, Ramsey TM, Jiang L, Martin MS, Baker K, Adams PD, Bair KW and Kaelin WG Jr (1999) Selective killing of transformed cells

by cyclin/cyclin-dependent kinase 2 antagonists. *Proc Natl Acad Sci USA* **96**:4325–4329.

61. Mendoza N, Fong S, MArsters J, Koeppen H, Schwall R and Wickramasinghe D (2003) Selective cyclin-dependent kinase 2/cyclin A antagonists that differ from ATP site inhibitors block tumor growth. *Cancer Res* **63**:1020–1024.
62. Snyder EL, Meade BR and Dowdy SF (2003) Anti-cancer protein transduction strategies: Reconstitution of p27 tumor suppressor function. *J Control Rel* **91**:45–51.
63. Parada Y, Banerji L, Glassford J, Lea NC, Collado M, Rivas C, Lewis JL, Gordon MY, Thomas NS and Lam EW (2001) BCR-ABL and interleukin 3 promote haematopoietic cell proliferation and survival through modulation of cyclin D2 and p27Kip1 expression. *J Biol Chem* **276**:23572–23580.
64. Ezhevsky SA, Ho A, Becker-Hapak M, Davis PK and Dowdy SF (2001) Differential regulation of retinoblastoma tumor suppressor protein by G(1) cyclin-dependent kinase complexes in vivo. *Mol Cell Biol* **21**:4773–4784.
65. Blagosklonny MV and Pardee AB (2001) Exploiting cancer cell cycling for selective protection of normal cells. *Cancer Res* **61**:4301–4305.
66. Zbar B, Glenn G, Lubensky I, Choyke P, Walther MM, Magnusson G, Bergerheim US, Pettersson S, Amin M and Hurley K (1995) Hereditary papillary renal cell carcinoma: Clinical studies in 10 families. *J Urol* **153**:907–912.
67. Datta K, Nambudripad R, Pal S, Zhou M, Cohen HT and Mukhopadhyay D (2000) Inhibition of insulin-like growth factor-I-mediated cell signaling by the von Hippel-Lindau gene product in renal cancer. *J Biol Chem* **275**:20700–20706.
68. Datta K, Sundberg C, Karumanchi SA and Mukhopadhyay D (2001) The 104-123 amino acid sequence of the beta-domain of von Hippel-Lindau gene product is sufficient to inhibit renal tumor growth and invasion. *Cancer Res* **61**:1768–1775.
69. Merel P, Hoang-Xuan K, Sanson M, Moreau-Aubry A, Bijlsma EK, Lazaro C, Moisan JP, Resche F, Nishisho I, Estivill X *et al.* (1995) Predominant occurrence of somatic mutations of the NF2 gene in meningiomas and schwannomas. *Genes Chromosomes Cancer* **13**:211–216.
70. Bashour AM, Meng JJ, Ip W, MacCollin M and Ratner N (2002) The neurofibromatosis type 2 gene product, merlin, reverses the F-actin cytoskeletal defects in primary human Schwannoma cells. *Mol Cell Biol* **22**:1150–1157.
71. Hanahan D and Weinberg RA (2001) The hallmarks of cancer. *Cell* **100**:57–70.
72. Yang YL and Li XM (2000) The IAP family: Endogenous caspase inhibitors with multiple biological activities. *Cell Res* **10**:169–177.
73. Holcik M, Gibson H and Korneluk RG (2001) XIAP: Apoptotic brake and promising therapeutic target. *Apoptosis* **6**:253–261.

74. Shi Y (2002) Mechanisms of caspase activation and inhibition during apoptosis. *Mol Cell* **9**:459–470.
75. Arnt CR, Chiorean MV, Heldebrant MP, Gores GJ and Kaufmann SH (2002) Synthetic Smac/DIABLO peptides enhance the effects of chemotherapeutic agents by binding XIAP and cIAP1 *in situ*. *J Biol Chem* **277**:44236–44243.
76. Fulda S, Wick W, Weller M and Debatin KM (2002) Smac agonists sensitize for Apo2L/TRAIL- or anticancer drug-induced apoptosis and induce regression of malignant glioma *in vivo*. *Nat Med* **8**:808–815.
77. Yang L, Mashima T, Sato S, Mochizuki M, Sakamoto H, Yamori T, Oh-Hara T and Tsuruo T (2003) Predominant suppression of apoptosome by inhibitor of apoptosis protein in non-small cell lung cancer H460 cells: Therapeutic effect of a novel polyarginine-conjugated Smac peptide. *Cancer Res* **63**:831–837.
78. Semenza GL (2000) HIF-1: Using two hands to flip the angiogenic switch. *Cancer Metastasis Rev* **19**:59–65.
79. Yu AY, Frid MG, Shimoda LA, Wiener CM, Stenmark K and Semenza GL (1998) Temporal, spatial and oxygen-regulated expression of hypoxia-inducible factor-1 in the lung. *Am J Physiol* **275**:L818–L826.
80. Huang LE, Gu J, Schau M and Bunn HF (1998) Regulation of hypoxia-inducible factor 1alpha is mediated by an O2-dependent degradation domain via the ubiquitin-proteasome pathway. *Proc Natl Acad Sci USA* **95**:7987–7992.
81. Maxwell PH and Ratcliffe PJ (2002) Oxygen sensors and angiogenesis. *Semin Cell Dev Biol* **13**:29–37.
82. Harada H, Hiraoka M and Kizaka-Kondoh S (2002) Antitumor effect of TAT-oxygen-dependent degradation-caspase-3 fusion protein specifically stabilized and activated in hypoxic tumor cells. *Cancer Res* **62**:2013–2018.
83. Gross A, McDonnell JM and Korsmeyer SJ (1999) BCL-2 family members and the mitochondria in apoptosis. *Genes Dev* **13**:1899–1911.
84. Lutz RJ (2000) Role of the BH3 (Bcl-2 homology 3) domain in the regulation of apoptosis and Bcl-2-related proteins. *Biochem Soc Trans* **28**:51–56.
85. Holinger EP, Chittenden T and Lutz RJ (1999) Bak BH3 peptides antagonize Bcl-xL function and induce apoptosis through cytochrome c-independent activation of caspases. *J Biol Chem* **274**:13298–13304.
86. Letai A, Bassik MC, Walensky LD, Sorcinelli MD, Weiler S and Korsmeyer SJ (2002) Distinct BH3 domains either sensitize or activate mitochondrial apoptosis, serving as prototype cancer therapeutics. *Cancer Cell* **2**:183–192.
87. Donato R (1999) Functional roles of S100 proteins, calcium-binding proteins of the EF-hand type. *Biochim Biophys Acta* **1450**:191–231.
88. Makino E, Sakaguchi M, Iwatsuki K and Huh NH (2004) Introduction of an N-terminal peptide of S100C/A11 into human cells induces apoptotic cell death. *J Mol Med* **82**:612–620.

89. Perea SE, Reyes O, Puchades Y, Mendoza O, Vispo NS, Torrens I, Santos A, Silva R, Acevedo B, Lopez E, Falcon V and Alonso DF (2004) Antitumor effect of a novel proapoptotic peptide that impairs the phosphorylation by the protein kinase 2 (casein kinase 2). *Cancer Res* **64**:7127–7129.
90. Gupta S, Campbell D, Derijard B and Davis RJ (1995) Transcription factor ATF2 regulation by the JNK signal transduction pathway. *Science* **267**:389–393.
91. Bhoumik A, Ivanov V and Ronai Z (2001) Activating transcription factor 2-derived peptides alter resistance of human tumor cell lines to ultraviolet irradiation and chemical treatment. *Clin Cancer Res* **2**:331–342.
92. Bhoumik A, Gangi L and Ronai Z (2004) Inhibition of melanoma growth and metastasis by ATF2-derived peptides. *Cancer Res* **64**:8222–8230.
93. Adjei AA (2001) Blocking oncogenic Ras signaling for cancer therapy. *J Natl Cancer Inst* **93**:1062–1074.
94. Garbay C, Liu WQ, Vidal M and Roques BP (2000) Inhibitors of Ras signal transduction as antitumor agents. *Biochem Pharmacol* **60**:1165–1169.
95. Chang F, Steelman LS, Shelton JG, Lee JT, Navolanic NM, Blalock WL, Franklin R and McCubrey JA (2003) Regulation of cell cycle progression and apoptosis by the Ras/Raf/MEK/ERK pathway. *Int J Oncol* **22**:469–480.
96. Cussac D, Frech M and Chardin P (1994) Binding of the Grb2 SH2 domain to phosphotyrosine motifs does not change the affinity of its SH3 domains for Sos proline-rich motifs. *Embo J* **13**:4011–4021.
97. Baldari CT, Heguy A and Telford JL (1993) Ras protein activity is essential for T-cell antigen receptor signal transduction. *J Biol Chem* **268**:2693–2698.
98. Hall DJ, Cui J, Bates ME, Stout BA, Koenderman L, Coffer PJ and Bertics PJ (2001) Transduction of a dominant-negative H-Ras into human eosinophils attenuates extracellular signal-regulated kinase activation and interleukin-5-mediated cell viability. *Blood* **98**:2014–2021.
99. Myou S, Zhou XD, Boetticher E, Myo S, Meliton A, Lambertino A, Munoz NM and Leff AR (2002) Blockade of focal clustering and active conformation in beta(2)-integrin-mediated adhesion of eosinophils to intercellular adhesion molecule-1 caused by transduction of HIV TAT-dominant negative Ras. *J Immun* **169**:2670–2676.
100. Kelemen BR, Hsiao K and Goueli SA (2002) Selective *in vivo* inhibition of mitogen-activated protein kinase activation using cell-permeable peptides. *J Biol Chem* **277**:8741–8748.
101. Chang HC, Samaniego F, Nair BC, Buonaguro L and Ensoli B (1997) HIV-1 Tat protein exits from cells via a leaderless secretory pathway and binds to extracellular matrix-associated heparan sulfate proteoglycans through its basic region. *Aids* **11**:1421–1431.

102. Asada S, Choi Y, Yamada M, Wang SC, Hung MC, Qin J and Uesugi M (2002) External control of Her2 expression and cancer cell growth by targeting a Ras-linked coactivator. *Proc Natl Acad Sci USA* **99**:12747–12752.
103. Yang J-Y and Widmann C (2001) Antiapoptotic signaling generated by caspase-induced cleavage of RasGAP. *Mol Cell Biol* **21**:5346–5358.
104. Yang J-Y and Widmann C (2002) The RasGAP N-terminal fragment generated by caspase cleavage protects cells in a Ras/PI3K/Akt-dependent manner that does not rely on NFkappa B activation. *J Biol Chem* **277**:14641–14646.
105. Michod D, Yang J-Y, Chen J, Bonny C and Widmann C (2004) A RasGAP-derived cell permeable peptide potently enhances genotoxin-induced cytotoxicity in tumor cells. *Oncogene* **23**:8971–8978.
106. Yoneda M, Katsumata K, Hayakawa M, Tanaka M and Ozawa T (1995) Oxygen stress induces an apoptotic cell death associated with fragmentation of mitochondrial genome. *Biochem Biophys Res Commun* **209**:723–729.
107. Singh KK, Russell J, Sigala B, Zhang Y, Williams J and Keshav KF (1999) Mitochondrial DNA determines the cellular response to cancer therapeutic agents. *Oncogene* **18**:6641–6646.
108. Tang JT, Yamazaki H, Inoue T, Koizumi M, Yoshida K and Ozeki S (1999) Mitochondrial DNA influences radiation sensitivity and induction of apoptosis in human fibroblasts. *Anticancer Res* **19**:4959–4964.
109. Fishel ML, Seo YR, Smith ML and Kelley MR (2003) Imbalancing the DNA base excision repair pathway in the mitochondria; targeting and overexpressing *N*-methylpurine DNA glycosylase in mitochondria leads to enhanced cell killing. *Cancer Res* **63**:608–615.
110. Shokolenko IN, Alexeyev MF, Robertson FM, LeDoux SP and Wilson GL (2003) The expression of Exonuclease III from *E. coli* in mitochondria of breast cancer cells diminishes mitochondrial DNA repair capacity and cell survival after oxidative stress. *DNA Repair* **2**:471–472.
111. Shokolenko IN, Alexeyev MF, Le Doux SP and Wilson GL (2005) TAT-mediated protein transduction and targeted delivery of fusion proteins into mitochondria of breast cancer cells. *DNA Repair* **4**:511–518.
112. Eisenbach L, Bar-Haim E and El-Shami K (2000) Antitumor vaccination using peptide based vaccines. *Immunol Lett* **74**:27–34.
113. Meng WS and Butterfield LH (2002) Rational design of peptide-based tumor vaccines. *Pharm Res* **19**:926–932.
114. Tarte K and Klein B (1999) Dendritic cell-based vaccine: a promising approach for cancer immunotherapy. *Leukemia* **13**:653–663.
115. Chikh GG, Kong S, Bally MB, Meunier JC and Schutze-Redelmeier MPM (2001) Efficient delivery of antennapedia homeodomain fused to CTL epitope

with liposomes into dendritic cells results in activation of CD8(+) T cells. *J Immunol* **167**:6462–6470.

116. Wang HY, Fu T, Wang G, Zeng G, Perry-Lalley DM, Yang JC, Restifo NP, Hwu P and Wang RF (2002) Induction of CD4(+) T cell-dependent antitumor immunity by TAT-mediated tumor antigen delivery into dendritic cells. *J Clin Invest* **109**:1463–1470.
117. Katterle Y, Brandt BH, Dowdy SF, Niggemann B, Zanker KS and Dittmar T (2004) Antitumour effects of PLC-gamma1-(SH2)2-TAT fusion proteins on EGFR/c-erbB-2-positive breast cancer cells. *Br J Cancer* **90**:230–235.

# 12

# Protein and Peptide Drugs to Suppress Tumor Angiogenesis

*Curzio Rüegg*

## 1. Introduction

Over the past decade, significant advances have been made in the understanding of the cellular and molecular events that regulate and mediate physiological and pathological angiogenesis, including tumor angiogenesis.[1,2] Many extracellular, cell surface and intracellular molecules modulating angiogenesis have been identified and characterized. They include: (1) Growth factors and growth factor receptors, such as vascular endothelial growth factors (VEGFs) and VEGF-Receptors; (2) Adhesion molecules of the integrin, cadherin, and immunoglobulin families; (3) Extracellular matrix proteins, such as fibronectin, collagens, and laminins; (4) Remodeling and guidance molecules including Ephrin/Eph, Delta/Notch, Slit/Robo, Netrin/UNC5B, and Semaphorin/Plexin ligand/receptor families; (5) Matrix-degrading proteinases, in particular matrix metalloproteinases (MMPs) and their inhibitors (i.e. TIMPs); (6) Signaling molecules (e.g. Raf, MAPK, PKA, Rac, PKB, mTOR); (7) Transcription factors (e.g. HIF-1$\alpha$, NF-$\kappa$B) and homeobox gene products (e.g. HoxD3 and HoxB3).[3–6] Many of these molecules represent putative therapeutic targets of antiangiogenic strategies.

At present, there are over 250 angiogenesis inhibitors described, and over 80 entered clinical testing in cancer.[7,8] Many of these molecules are peptides, polypeptides, or proteins. They include cytokines with multiple antiangiogenic effects (e.g. VEGI, PF-4), antibodies inhibiting specific targets (e.g. antiVEGF-A; antiVEGF-R2), small fragments and polypeptides derived from larger precursor molecules (e.g. collagen XVIII-derived endostatin) and synthetic peptides (e.g. Cilengitide).[7–9] In this chapter we will describe the features of these classes of compounds and report more in details the leading compounds of each class. Table 1 gives an overview of proteins and peptides with antiangiogenic and antitumor activities.

## 2. Natural Cyokines

Tumor necrosis factor (TNF) is the paradigm example of a cytokine capable of selectively targeting and disrupting the tumor vasculature. TNF was originally identified as an endotoxin-induced factor causing hemorrhagic tumor necrosis in mice.[10] Subsequently, TNF was shown to modulate physiological and pathological effects, including tissue homeostasis, normal inflammation and immune responses, septic shock, insulin resistance and autoimmune diseases.[11] TNF is synthesized as a type II transmembrane protein and cleaved extracellularly close to the cell membrane by the TNF-$\alpha$-converting enzyme (TACE), giving rise to a soluble complex consisting of three identical polypeptides of 17 kDa each.[12,13] Both transmembrane and secreted forms are biologically active. TNF signals through two transmembrane receptors: TNF-R1 (55 kDa) and TNF-R2 (75 kDa).[14] TNF-R1 is constitutively expressed in a large variety of tissues, while TNF-R2 expression is restricted to lymphoid tissues and the endothelium. The availability of recombinant TNF in the mid-1980s made it possible to test TNF in cancer patients. Systemic administration, however, induced unacceptable side effects such as systemic inflammatory response syndrome and multiple organ failures and was quickly abandoned. To circumvent this problem, TNF was subsequently delivered regionally using the isolated limb perfusion (ILP) technique. ILP consists of the surgical isolation of the limb vasculature connected to a heart-lung machine allowing

**Table 1** Nonexhaustive list of proteins, antibodies, an peptides with antiangiogenic activities.

| Class | Molecule or drug | Molecular target | Comments | References |
|---|---|---|---|---|
| Monoclonal antibodies | Bevacizumab (Avastin®) | VEGF-A | Approved for the treatment of advanced colorectal cancer; extends survival also in advanced NSLC, breast, and kidney cancer (Genentech, Inc.) | 63 |
| | HuMV833 | VEGF-A | Tested in clinical trials with some evidence of activity in breast cancer | 68 |
| | IMC-1C11 | VEGF-R2/KDR | Tested in clinical trials (ImClone, Inc.) | 73 |
| | Vitaxin® (MEDI 522) | Integrin $\alpha V\beta 3$ | Tested in clinical trials, but no definitive evidence of activity (Medimmum) | 79 |
| | M200 | Integrin $\alpha 5\beta 1$ | Entering clinical testing (Protein Design Labs. Inc.) | www.pdl.com |
| | 17E6 | Integrin $\alpha V$ | Antitumor and antiangiogenic effects. Preclinical data only (Merck KGaA) | 119 |
| | BV13 | VE Cadherin | Inhibits tumor angiogenesis without side effects on endothelial permeability | 132 |
| Single chain antibodies | L19 | EDB + Fibronectin | High-affinity scFv tested in the clinic for tumor targeting. Antitumor agents have been coupled to L19 resulting in enhanced therapeutic activity in experimental models | 88 |
| | V65 | VEGF | Blocks angiogenesis in the chorioallantoic membrane and reduces tumor growth in mice | 95 |

**Table 1** (*Continued*)

| Class | Molecule or drug | Molecular target | Comments | References |
|---|---|---|---|---|
| | scFv-Ang2 | Angiopoietin-2 | Inhibits VEGF-induced endothelial cell proliferation and migration *in vitro* | 96 |
| | c-p1C11 | Flk-1/KDR | Blocks VEGF-KDR interaction and inhibits VEGF-mediated endothelial cell proliferation *in vitro* | 97 |
| | L36 | Laminin | Antilaminin scFv that blocks the formation of capillary-like structures *in vitro* | 98 |
| | VN18 | Vitronectin | Recognizes the activated conformation of vitronectin present in tumor tissues but not in normal tissues | 99 |
| Soluble receptors | VEGF-R1/2-trap | PlGF, VEGF-A, -B | VEGF-R1/2 Ig fusion protein. Potent inhibitor of angiogenesis in preclinical models. In clinical testing | 101 |
| | VEGF-R3-trap | VEGF-C, -D | VEGF-R3-Ig fusion protein. Inhibits tumor lymphangiogenesis and lymph node metastasis | 104 |
| Natural cyokines | IFN$\alpha$, $\beta$ | IFN$\alpha$/$\beta$-Rs | Interferon alpha was introduced to the clinic for the treatment of life-threatening infant angioma | 29, 30 |
| | IFN$\gamma$ | IFN$\gamma$-R | Interferon gamma enhances TNF-anti tumor activity | 31 |
| | TNF | TNF-R1 and -R2 | Tumor Necrosis Factor. Approved for sarcoma and widely used for melanoma of the limbs. Administered by ILP | 27 |

**Table 1** *(Continued)*

| Class | Molecule or drug | Molecular target | Comments | References |
|---|---|---|---|---|
| | PF4 | CXCR3B, EGF, FGF2 | Platelets Factor-4. Binds to and neutralize VEGF and FGF2. Elicits antiangiogenc signals via CXCR3B | 35 |
| | IL-10 | IL-10R | Suppresses VEGF-mediated angiogenesis and tumor growth | 32 |
| | IL-12 | IL-12R | Pleiotropic immunomodulatory, antiangiogenic, and antitumor effects | 33 |
| | VEGI (TL1A) | TRAMP (DR3) | A cytokine of the TNF family, that inhibits angiogenesis and suppresses tumor growth | 34 |
| | CXCL10 (IP10) | CXCR3 | Interferon-Inducible Protein 10 | 36 |
| | CXCL14 (BRAK) | Unknown | Potent inhibitor of angiogenesis and a chemotactic factor for immature dendritic cells | 37 |
| Cryptic protein fragments | Angiostatin | Integrins? | 38 kDa internal fragment of plasminogen, consisting of the first four Kringle domains (K1-4) | 38 |
| | Endostatin | Integrin $\alpha 5\beta 1$ | 20 kDa carboxyl-terminal fragment of collagen XVIII | 43 |
| | Tumstatin | Integrin $\alpha V\beta 3$ | Carboxyl-terminal noncollagenous 1 (NC1) domain of the alpha3 chain of collagen IV | 46 |
| | Vasculostatin | CD36 | 120 kDa fragment derived by proteolytic cleavage from brain angiogenesis inhibitor 1 | 48 |

**Table 1** *(Continued)*

| Class | Molecule or drug | Molecular target | Comments | References |
|---|---|---|---|---|
| | ABT-526; ABT510 | Unknown | Peptides derived from the amino-terminal globular region containing the type 1 repeats of thrombospondin | 54 |
| Small peptides | EMD121974 (Cilengitide) | Integrin $\alpha V\beta 3$ and $\alpha V\beta 5$ | High affinity inhibitor of $\alpha V\beta 3$ and $\alpha V\beta 5$ integrin. Enhanced antitumor effects of radiotherapy in preclinical models. In phase I clinical studies was well tolerated and some responses observed. Currently in phase II trials | 117, 118 |
| | ATN-161 | Integrin $\alpha 5\beta 1$ | Inhibits the fibronectin synergy-site on $\alpha 5\beta 1$. Enhances antitumor effects of 5FU chemotherapy in tumor models | 126 |
| | Anginex | Unknown | $\beta$-sheet-forming 33-meric peptides containing short sequences from the $\beta$-sheet domains of antiangiogenic proteins. Entered clinical testing | 129 |
| | cRGD-4C | $\alpha V$ integrins | Used to target cytotoxic drugs to tumors | 108 |
| | NGR | Aminopeptidase N (CD13) | Used to target TNF, doxorubicin, melphalan, and liposomes to tumors | 108 |
| | CGKRK/CDTRL | Unknown | Motif binding to tumor endothelial cells | 107 |
| | cCGNKRTRC | Unknown | Motif binding to tumor lymphatics | 107 |

the continuous circulation of high doses of cytotoxic agents over a short period of time (60–120 min). After perfusion, the vascular space is extensively rinsed and the vasculature re-established. Administration of high dose TNF (3–4 mg/limb), Interferon gamma (IFN$\gamma$) (0.2 mg) and melphalan (10–13 mg/L limb volume) under mild hyperthermia (38°C) in advanced melanoma and soft tissue sarcoma resulted in extensive tumor necrosis in the absence of significant local and systemic toxicity.[15,16] In this treatment, the tumor vasculature is selectively disrupted and complete tumor regression occurs in over 80% of the patients.[17] The antitumor vasculature activity of TNF involves two events. First, a rapid increase (within minutes) in the permeability of the tumor vasculature,[18] which favors the selective accumulation of the chemotherapeutic agent in the tumor tissue.[19] The molecular mechanisms leading to enhanced permeability involve remodeling of the cytoskeleton,[20,21] redistribution of the cell–cell adhesion molecules PECAM-1[22] and VE-cadherin.[23] Second, TNF elicits a late (within hours) cytotoxic damage to the tumor endothelial cells. This effect is associated with the suppression of vascular integrin $\alpha V\beta 3$ function and endothelial cell apoptosis.[17,24] Because of the clear antitumor effects of TNF, efforts have been undertaken to develop strategies allowing systemic administration. Among the several approaches chosen antibody- and peptide-mediated TNF targeting to the tumor vasculature have been successfully tested in preclinical models. Targeted TNF retains is ability to increase vascular permeability and to synergize with chemotherapy to induce tumor regression.[25,26] Clinical trials are planned and if successful, they may open the road to the systemic administration of TNF to treat human cancer in combination with chemotherapy.[27]

Interferon alpha (IFN$\alpha$) is another example of a cytokine being used as an antiangiogenic agent in the clinic. Interferons are natural polypeptides expressed by most cells in response to viral infections (IFN$\alpha/\beta$) or by immune and inflammatory cells upon activation (IFN$\gamma$). IFN$\alpha$ was used to treat life-threatening infant angiomas.[28] Treatment of tumor-bearing animals with IFN$\alpha$ or $\beta$ decreased blood vessel density within the tumors and inhibited tumor growth.[29,30] IFN$\gamma$ has only weak if any antiangiogenic activity, but it greatly enhances the antitumor activity of TNF,[31] and is used in the clinic in combination with TNF and chemotherapy in ILP.[17]

Additional natural polypeptides targeting and inhibiting angiogenesis include IL-10,[32] IL-12,[33] VEGI,[34] PF4,[35] IP10/CXCL10,[36] CXCL14.[37]

## 3. Cryptic Protein Fragments

An interesting group of polypeptides that selectively suppress tumor angiogenesis, are proteolytically-derived terminal and internal fragments derived from larger precursor proteins. The precursor molecules are in general extracellular matrix proteins (e.g. thrombospondins, collagens) or components of the coagulation system (e.g. plasminogen, antithrombin).

Angiostatin, the paradigm example of this class of antiangiogenic polypeptides is a 38 kDa internal fragment of plasminogen, consisting of the first four Kringle domains (K1-4).[38] Angiostatin was identified while studying the phenomenon of inhibition of tumor metastasis growth by the primary tumor. These pioneering studies demonstrated that angiostatin is produced within the environment of the primary tumor and is released into the systemic circulation. When it reaches the metastatic site, angiostatin inhibits angiogenesis and suppresses growth of small metastases. When the primary tumor is surgically removed, the source of angiostatin is removed with it and the metastases will resume growth.[39] Exogenously administered angiostatin can induce regression of established angiogenic tumor vessels, leading to tumor regression, or it can induce sustained tumor dormancy of primary tumors and metastasis by inhibiting de-novo angiogenesis.[40] For clinical testing angiostatin was produced as a recombinant molecule consisting of the first four (K1-4), or the first three (K1-3) Kringle domains of plasminogen. Approaches have also been developed to directly convert plasminogen into angiostatin *in vivo*.[41] Long-term patient treatment with recombinant human angiostatin was well tolerated and systemic levels were within the range of drug exposure that has biological activity in preclinical models.[42]

Endostatin is a 20 kDa carboxyl-terminal fragment of collagen XVIII, a proteoglycan/collagen found in vessel walls and basement membranes. The endostatin fragment was originally identified in conditioned media from a murine endothelial tumor cell line based on its ability to inhibit endothelial cell proliferation *in vitro* and angiogenesis and tumor growth *in vivo*.[43] The generation of endostatin is mediated by a variety of proteases (e.g. cathepsin L and MMPs) that cleave peptide bonds within the protease-sensitive hinge region of the carboxyl-terminal domain of collagen XVIII. Recombinant human endostatin retaining antiangiogenic activity in animal models was produced in *E. coli*, yeast, and mammalian cells.[44] Recombinant endostatin entered clinical testing and was found to be safe and well tolerated. Pharmacokinetic profiles (concentration-time curves) associated with activity in preclinical models were observed in patients.[45]

Tumstatin consists of the carboxyl-terminal noncollagenous 1 (NC1) domain of the alpha3 chain of collagen IV. It inhibited *in vivo* neovascularization in Matrigel plug assays, suppressed tumor growth in various xenograft models and induced endothelial cell-specific apoptosis.[46] Functionally active recombinant tumstatin was produced in *E. coli* and yeast but as of today, no testing in humans has been reported.[47]

Vasculostatin. Vasculostatin is a 120 kDa fragment derived by proteolytic cleavage from brain angiogenesis inhibitor 1, a transmembrane protein of unknown function expressed in normal brain. It was reported to inhibit migration of endothelial cells *in vitro* and angiogenesis *in vivo*.[48]

Thrombospondins (TSPs) and TSP fragments. TSPs are a family of large (>600 kDa) modular glycoprotein with both adhesive and antiadhesive function, which play a role in blood coagulation and tissue remodeling.[49,50] At least three homologous TSP genes (TSP-1, -2, and -3) exist in both human and mouse. TSPs have been reported to suppress angiogenesis by inhibiting endothelial cell proliferation, migration and capillary tubes formation.[51,52] TSP-1 and -2 can be proteolytically processed into well-defined short fragments with antiangiogenic activity. Recombinant fragments of human TSP-1 and -2, encompassing the amino-terminal globular region containing the type 1 repeats were shown to inhibit tumor growth and to reduce tumor vascularization *in vivo*, and to modulate endothelial cell functions and to induce endothelial cell apoptosis *in vitro*.[51–53] Synthetic antiangiogenic peptides derived from the second type-1 repeat of TSP-1 were generated, including the heptapeptide GVITRIR, the nonapeptides 5 (ABT-526) and 6 (ABT-510) and peptides containing the consensus sequence WSPW.[54] When tested *in vitro*, ABT-526 was found more potent than ABT-510 to inhibit endothelial cell migration while ABT-510 was more effective in inhibiting tube formation. ABT-510 and ABT-526 significantly increased apoptosis of cultured endothelial cells and ABT-510 was effective in blocking neovascularization in the mouse Matrigel plug model and in suppressing the growth of experimental tumors in mice and spontaneous tumors in dogs. While full-length thrombospondin molecules present poor bioavailability and are highly susceptible to proteolytic degradation, TSP-derived angiostatic peptides have favorable pharmacokinetic profiles and are stable. ABT-510 has improved endothelial cell binding (low nM range) and improved pharmacokinetic compared to ABT-526. ABT-510 appear as potent and promising therapeutic agents in antiangiogenic therapy, and is currently in phase II clinical studies.[54]

The processing of these matrix-derived endogenous inhibitors of angiogenesis is likely to represent a local control mechanism for the regulation of angiogenesis in physiological conditions other than cancer (e.g. inflammation, tissue remodeling, wound healing). The molecular mechanism by which these fragments inhibit angiogenesis is still a matter of investigation. A common emerging theme, however, is the interference with integrin-mediated endothelial cell adhesion events (in particular integrin $\alpha V\beta 3$ and $\alpha 5\beta 1$-dependent) and perturbation of the actin cytoskeleton and intracellular signaling events, resulting in reduced endothelial cell proliferation, migration and survival.[55–58] Recombinant angiostatin and endostatin entered clinical testing already several years ago. In general these fragments were well tolerated, and some cases of transient disease stabilizations were observed. A clear demonstration of efficacy, however, is still missing.[59] The production of functional angiostatin and endostatin for clinical testing proved to be technically difficult due to problems related to folding, degradation, pH sensitivity, solubility, precipitation, and aggregation. *In vivo* expression of recombinant angiostatin or endostatin through gene transfer ("gene therapy") may represent an alternative method of delivery.[60]

## 4. Antibodies

### 4.1. *Monoclonal antibodies (mAb)*

The establishment of monoclonal antibody (mAb) technology in the early 80s, made large-scale production of monospecific antibodies possible, and raised great expectations for therapeutic applications, in particular in cancer. This promise, however, was not realized until the late 1990s due to a combination of factors, including unfocused and inappropriate target selection, low antitumor activity when used alone, limited penetration into solid tumors, and development of neutralizing human antimouse antibodies. Subsequently, relevant molecular targets were identified and clinical trials demonstrated efficacy of some antibodies in selected patient populations. There are today eight monoclonal antibodies approved by the FDA for anticancer treatment.[61]

Anti-VEGF-A. Bevacizumab (Avastin®, Genentech, Inc.) is a humanized version of a murine monoclonal antibody (mAb A.4.6.1) raised against VEGF-A and blocking its biological activity.[62,63] The residues of the six complementarity-determining regions (CDR) of mAb A.4.6.1 were introduced into a human antibody framework by site-directed mutagenesis resulting in a 93% humanized antibody which binds VEGF-A with affinity close to that of the original murine antibody (Kd = 0.5 nM). Bevacizumab binds to and neutralizes all human VEGF-A isoforms and bioactive proteolytic fragments but not PlGF, VEGF-B, -C, nor -D. The half-life of bevacizumab in humans is $19 \pm 2$ days. First evidence that bevacizumab could have potent anti-tumor activity in human cancer came from a phase II trial in metastatic renal cancer. Patients treated with high dose bevacizumab had a significant prolongation of time to progression, compared to placebo-treated patients.[64] In a phase II trial, administration of bevacizumab in combination with fluorouracil and leucovorin, to patients with metastatic colorectal cancer, resulted in a higher response rates, longer median time to progression, and longer median survival, compared to chemotherapy alone.[65] In a phase III trial bevacizumab (5 mg/kg every 2 weeks), in combination with irinotecan, fluorouracil and leucovorin, resulted in improved objective tumor response, extended duration of the response, prolonged time to progression and significantly extended overall survival in patients with advanced colorectal cancer compared to chemotherapy alone.[66] Based on this study, bevacizumab was approved by the FDA as the first systemic antiangiogenic drug for human cancer treatment and became standard first-line treatment of metastatic colon cancer. This achievement validated the notion that VEGF-A is a relevant target to inhibit tumor angiogenesis and, in turn, that blocking angiogenesis is an effective strategy to treat human cancer.[63] Additional clinical trials with bevacizumab in advanced breast, nonsmall lung, and kidney cancers have shown significant improvement of time to progression and overall survival (for more information: www.asco.org and www.gene.com). Bevacizumab was generally well tolerated, although some serious side effects and toxicities were observed, such as gastrointestinal perforations and wound healing complications, in about 2% of the patients. Additional side effects included hypertension, thromboembolic events, bleedings, and proteinuria.[67] No evidence of antibody response to bevacizumab has been found in any of the clinical trials performed so far. At time of writing (spring 2005) there are over 70 open clinical trials with bevacizumab, including in neo-adjuvant settings.

A second anti-VEGF mAb (HuMV833) was tested in breast cancer patients and produced complete and partial responses in a subset of the patients.[68]

Anti-VEGF-R2. High affinity anti-VEGF-R2 blocking antibodies have been developed by ImClone Inc. (DC-101, anti-mouse Flk-1; IMC-IC11 anti-human Flk-1/KDR) ($Kd = 4.9 \times 10^{-10}$–$1.1 \times 10^{-9}$ M). The DC-101 antibody showed significant antiangiogenic activity in several experimental tumor models (e.g. Lewis lung, 4T1 mammary, B16 melanoma, CT-26 colon).[69,70] Combined administration of DC101 and chemotherapy was very efficient in inhibiting growth and metastasis of human tumor xenografts.[71,72] The anti-KDR antibodies competed on an equimolar basis with VEGF for binding to KDR and potently inhibited VEGF-induced signaling in human endothelial cells.[73,74] When tested in cancer patients IMC-1C11 was found safe and well tolerated.[75]

Anti-integrins. Angiogenesis depends on the adhesive interactions of endothelial cells with the surrounding extracellular matrix. The adhesion receptor integrin $\alpha V\beta 3$ is expressed on blood vessels during wound healing and in the tumor stroma, but not, or to a much lower level, in normal tissues.[76] Inhibition of $\alpha V\beta 3$ function with the monoclonal antibody LM609 blocked endothelial cell adhesion, migration and sprouting *in vitro* and angiogenesis *in vivo* in the chicken chorioallantoic membrane model and in a human skin and tumor transplantation model.[76–78] Quiescent and preexisting vessels were not perturbed. LM609 was subsequently humanized and affinity maturated. For affinity maturation peptide libraries for all CDRs were expressed in phages and screened to identify variants with improved binding to $\alpha V\beta 3$. Each peptide in the library contained a single mutation, and all 20 amino acids were introduced at each CDR residue, resulting in the expression of 2336 unique clones. Subsequent expression and screening of these clones and of all combinatorial variants of the optimal mutations identified, resulted in the generation of a reconstituted antibody (MEDI 522 or Vitaxin®) with greater than 90-fold improved affinity.[79] Vitaxin® was tested in clinical trials shortly thereafter. Patients received escalating doses up to 4.0 mg/kg/week. Treatment was well tolerated with little or no toxicity. The most common side effect was infusion-related fever, which could be controlled with antipyretics. Vitaxin® has a half-life of over five days with no tendency of accumulation over 6 weeks of therapy. In this study, three patients demonstrated disease stabilization.[80] Subsequent phase II studies in patients with metastatic solid cancers confirmed good tolerability, but failed to demonstrate objective antitumor responses. No immune response to Vitaxin® was observed in any of these patients.[81]

## 4.2. *Recombinant antibodies*

Genetic engineering techniques have allowed the generation of recombinant antibodies with the purpose to improve their therapeutic properties *in vivo*. High-affinity recombinant antibodies can efficiently target a tumor tissue and rapidly clear from normal tissues and from the circulation, thereby resulting in greatly improved tumor selectivity. Recombinant antibodies were first produced from existing monoclonal antibodies by cloning the variable heavy (VH) and light (VL) chain domains from the corresponding hybridoma line, joining via a synthetic linker and expressing as single-chain variable fragments (scFv) of 25–27 kDa molecular weight. Subsequently, molecular biology techniques have allowed the generation of large repertoires of antibody fragments from VH and VL germline genes, thus bypassing immunization and hybridoma technology. Libraries of scFv can be expressed at the surface of filamentous phages and screened for binding to the desired antigen. ScFv recovered from phage expression libraries generally show moderate binding strength ($Kd = 10^{-5}$–$10^{-8}$ M).[82] Due to their ease of isolation, the possibility of obtaining fully human sequences, and their ability to recognize virtually any epitope and highly conserved antigens, scFv molecules are rapidly becoming commonplace in biomedical research. The monovalent nature of the scFv molecule and their moderate to low affinity, however, often results in transient interactions with the target antigen. scFv fragments can be improved by mutating crucial residues of CDR (to increase affinity) or by increasing the number of binding sites (to increase avidity).[83] Dimeric scFv (i.e. diabodies, 55 kDa) with improved tissue targeting and pharmacokinetic properties have been engineered. Further gains were obtained by generating larger recombinant fragments, such as the minibody, that self-assembles into a bivalent dimer of 80 kDa. Recombinant antibodies are being developed and tested for diagnostic (i.e. targeting imaging contrast agents into tumor tissues), and for therapeutic purposes (i.e. targeting drugs or radioisotopes to tumor sites).[84,85]

A further step in the modification of monoclonal antibodies was achieved with the generation of bispecific and bifunctional antibodies, produced by conventional biochemical methods or by genetic engineering. Furthermore, through genetic engineering, it has been possible to add additional protein domains (e.g. toxins, enzymes) in order to create "designer antibody molecules" with totally novel functions. Recombinant antibodies produced by genetic engineering techniques have proved useful in the immunodiagnostic and immunotherapy of cancer and other diseases and some of them entered clinical testing.[86] To illustrate the possible applications of recombinant antibody techniques in cancer research and clinical oncology we will describe here a scFV specific for $EDB^{+}$-fibronectin.

Anti-$EDB^{+}$-fibronectin. The scFv antibody L19 specific for the ED-B domain of fibronectin, a marker of angiogenesis highly conserved in rodents, primates, and human,[87] was generated by screening a human scFv phage display library against the antigen eluted from a two-dimensional gel spot. The original antibody was affinity maturated by combinatorial mutation of six residues in the VH chain, giving rise to a scFv with very high affinity ($Kd = 54\,\text{pM}$).[88] Radiolabeled L19 efficiently localized in experimental tumors in mice (8.2% of the injected dose localized 3 hr after injection). Because of the rapid clearance of the antibody fragment from the circulation a high tumor-to-blood ratios was observed at later time points.[89] *Ex vivo* micro-autoradiographic analysis, revealed that L19 localizes around tumor blood vessels, but not around normal vessels, consistent with immunohistological data.[90] $^{123}$I-labelled dimeric L19 was subsequently tested in humans for its ability to target primary tumors and metastases. Dimeric L19 antibody selectively localized in tumor lesions in lung and colorectal cancers in the absence of detectable side effects.[91] L19 was fused to bioactive molecules, including IL-2, IL-12, TNF or IFN$\gamma$ and tested in mice for antitumor activity. The fusion proteins efficiently targeted the tumor vessels resulting in the accumulation of high levels of the cytokine within the tumors. As a consequence there was a dramatic enhancement of the therapeutic properties of the cytokine compared to free cytokine or to the cytokine fused to a control protein.[25,92–94] Since the EDB domain targeted by the L19 scFv is a marker of tumor angiogenesis abundantly expressed in a variety of aggressive solid tumors, but undetectable in most normal tissues, these observations may open the possibility of administering L19-cytokine conjugates for therapeutic purposed in human cancer.

Additional angiogenesis-relevant molecules targeted by recombinant antibodies include VEGF-A,[95] angiopoietin-2,[96] Flk-1/KDR[97] laminin,[98] and activated vitronectin.[99]

## 5. Soluble VEGF-Rs (Traps)

Natural and recombinant soluble VEGF-R1 and -R2 retain their ability to bind VEGF-A resulting in the neutralization of its biological activity. This observation suggested the possibility of using soluble VEGF-R1 or -R2 as VEGFs antagonists. Since VEGF-R1 has a 10 times higher affinity to VEGF-A compared to VEGF-R2, a soluble decoy receptor was initially created by fusing the first three Ig domains of VEGF-R1 to an Ig Fc region. This fusion protein retained the binding affinity to VEGF-A of the native VEGF-R1 (around 10 pM) and inhibited VEGF-A across different species, but required administration at high doses to achieve maximal inhibitory effects (up to 25 mg/kg compared to 0.1–0.5 mg/kg dose or bevacizumab), mostly due to poor *in vivo* pharmacokinetic properties.[100] Attempts to extend *in vivo* half-life while maintaining high affinity resulted in the generation of a Fc fusion construct consisting of the second Ig domains of VEGF-R1 and of the third Ig domain of VEGF-R2.[101] This hybrid molecule has enhanced pharmacokinetic properties and binds VEGF-A with a Kd = 1–10 pM and PlGF-2 with a Kd = 45 pM.[101] VEGF Trap has demonstrated marked efficacy in suppressing angiogenesis and shrinking tumors in preclinical animal models and is currently being tested in phase I clinical trials in patients with advanced solid cancers.[102,103] More recently a VEGF-R3-immunoglobulin (VEGF R3-Ig) fusion protein, which binds VEGF-C and -D and inhibits VEGF-R3 signaling, was reported to suppress tumor lymphangiogenesis and inhibit metastasis to regional lymph nodes.[104]

## 6. Small Peptides

### 6.1. *Vascular targeting peptides*

Compared to the normal, quiescent vasculature, the tumor vasculature expresses a different pattern of molecules at its surface. Molecularly, these differences are due to changes in expression of growth factor receptors, cell adhesion receptors, proteases, or morphogenic molecules. *In vivo* screening of phage libraries that display random peptide sequences on their surfaces has led to the identification of short sequences with homing capacities for the tumor vasculature.[105,106] Tumor-targeting sequences identified with this strategy include cRGD-4C, a RGD-based integrin targeting motif; NGR, a motif binding to aminopeptidase N (see below); CGKRK and CDTRL, two motifs binding to unknown receptors in skin carcinomas; CRGRRST, a motif binding to a platelet-derived growth factor-receptor $\beta$ (PDGFR$\beta$) associated molecule in vessels of experimental tumors; and cCGNKRTRC, a motif binding to tumor lymphatics.[107] Some of these peptides have been evaluated as tool to target contrast agents, drugs and other therapeutic materials to solid tumors. Such a peptide-based targeting strategy is currently being explored in the clinic to potentially improve the efficacy of existing anticancer drugs and to reduce their side effects.[105,106]

NGR. The tripeptidic sequence NGR has been extensively studied for its tumor-targeting properties and is described here in more details. This motif was originally identified by *in vivo* selection of phage display libraries homing to tumor blood vessels.[108] Subsequently it was shown that the receptor for the NGR peptides in the tumor vasculature is aminopeptidase N (also called CD13), a cell-surface proteinase expressed on a variety of cells and up-regulated in endothelial cells within mouse and human tumors.[109] Tissue immunostaining with different antibodies suggested that different isoforms of aminopeptidase N are expressed in tumor-associated blood vessels, leukocytes, and epithelia. Direct binding assays and competitive inhibition experiments with antiaminopeptidase N antibodies showed that only the aminopeptidase N isoform expressed in tumor blood vessels functions as a receptor for the NGR motif.[110] Aminopeptidase N antagonists specifically inhibited angiogenesis in chorioallantoic membranes and in the retina and suppressed tumor angiogenesis and tumor growth in mice.[109] When coupled to the anticancer drug doxorubicin, the NGR motif enhanced the efficacy of the drug against human breast cancer xenografts in nude mice while reduced its toxicity.[108] Administration of TNF coupled to a cyclic NGR peptide, resulted in efficient TNF targeting to tumor vessels and to enhanced vascular permeability at TNF doses $10^5$-fold lower than the doses required for nontargeted TNF to exert the same effect. Doxorubicin and melphalan coupling to NGR resulted in enhanced drug accumulation in the tumor tissue and enhanced therapeutic efficacy, with no evidence of increased toxicity.[111] Coupling of linear GNGRG peptide to doxorubicin-containing liposomes resulted in targeted delivery of the liposomes to the tumors, and markedly improved therapeutic index compared to free doxorubicin. Thus NGR peptides appear to be useful tumor-endothelium targeting moieties to improve the efficacy of chemotherapeutic drugs.[111]

## 6.2. *Function-blocking peptides*

A number of peptides have been developed to block the function of receptors expressed on angiogenic endothelial cells. The most successful peptides of this category are peptides inhibiting adhesion molecules, in particular integrins.

Cilengitide ($\alpha V\beta 3$ integrin blocking peptide). In the mid 80s, it was reported that the tripeptidic sequence RGD is a ligand recognition sequence for several integrins, including $\alpha 5\beta 1$, $\alpha V\beta 3$, and $\alpha iib\beta 3$. Subsequently it was shown that soluble, linear RGD peptides were able to inhibit integrin-mediated cell adhesion to extracellular matrix ligands, such as fibronectin or fibrinogen, *in vitro* and *in vivo*.[112] These discoveries opened the possibility of using RGD-based peptides for therapeutic purposes in pathologies involving integrin-dependent adhesion events, including cancer invasion, spreading, and tumor angiogenesis.[113] Linear RGD peptides, however, have low affinity for integrins, they nonselectively block multiple integrins, and are sensitive to degradation by peptidases, thereby preventing their use *in vivo* for therapeutic purposes. Subsequently, attempts to improve peptide stability and integrin binding specificity and affinity lead to the development of low molecular weight cyclic RGD peptides.[114–116] One of the peptides, EMD 66203, cyclo- (RGDfV), selectively inhibited integrin $\alpha V\beta 3$, and blocked cytokine- and tumor-induced angiogenesis by promoting apoptosis of proliferating endothelial cells.[76,77] Chemical modification and optimization of the initial structure yielded the compound cyclo-(RGDfNMeV) with improved potency and solubility (EMD 121974, Cilengitide).[117,118] The $IC_{50}$ of Cilengitide to inhibit $\alpha V\beta 3$-dependent cell adhesion to vitronectin is around 1 $\mu$M. Cilengitide also efficiently blocks $\alpha V\beta 5$-dependent adhesion while it does not interfere with $\alpha IIb\beta 3$ binding to fibrinogen. Other RGD-binding integrins, such as $\alpha 5\beta 1$, $\alpha 1\beta 1$, $\alpha 2\beta 1$, $\alpha 3\beta 1$ are not significantly affected by cilengitide. Preclinical testing revealed that Cilengitide inhibited the growth of tumors, including melanoma,[119] and brain tumors[120] in a dose- and time-dependent manner. The antitumor effect of Cilengitide is due in part to the direct inhibition of $\alpha V\beta 3$ integrin on tumor cells and in part to the suppression or tumor angiogenesis.[5,76,77] Preclinical studies also demonstrated that Cilengitide significantly enhanced the effect of radiotherapy in a human breast cancer tumor model lacking functional p53, compared with single-modality therapy with either agent. The enhanced therapeutic synergy was achieved without additional toxicity.[121] Cilengitide was also coupled to cytotoxic drugs, such as doxorubicin, and successfully use for tumor drug targeting.[122] Cilengitide was tested in cancer patients for its toxicity and efficacy. Phase I trials revealed that cilengitide has no acute dose-limiting toxicity at doses up to 1.6 $g/m^2$ after administration as short infusion, and no cumulative toxicity after repeated administration. At 120 $mg/m^2$/infusion, peak plasma concentrations were attained that optimally inhibited tumor growth in preclinical models. Half-life was short, ranging from 3 to 5 hr.[123] Phase II trials in combination with chemotherapy are ongoing.

$\alpha5\beta1$ integrin blocking peptide. Inhibitory peptides targeting $\alpha5\beta1$, another integrin expressed on angiogenic endothelial cells,[5] have been identified by screening phage display libraries engineered to express cyclic-RGD-based[124] and non-RGD-based peptides.[125] One of these peptides, ATN-161 (Ac-PHSCN), is derived from the PHSCN synergy region of fibronectin, a region that potentiates the binding of fibronectin to $\alpha5\beta1$.[126] ATN-161 acts by interfering with this interaction.[126] ATN-161 significantly reduced the *in vivo* growth of human xenografts (HT29) that do not express integrin $\alpha5\beta1$ suggesting that ATN-161 may act by suppressing tumor angiogenesis.[127] Subsequently it was shown that combined administration of ATN-161 with 5-fluorouracil (5-FU) chemotherapy significantly reduced the number the volume and number of liver metastases, decreased microvascular density and improved overall survival in the murine CT26 colon cancer model.[128]

Anginex. A combinatorial approach was used to design $\beta$-sheet-forming 33-meric peptides containing short sequences from the $\beta$-sheet domains of antiangiogenic proteins. One of these designed peptides, Anginex, was found to potently inhibit proliferation, adhesion, migration sprouting and to induce apoptosis of vascular endothelial cells *in vitro*, and to suppress angiogenesis *in vivo*.[129] Five hydrophobic residues and two lysines within Anginex appear to be crucial to its activity. In the MA148 human ovarian carcinoma mouse model, anginex alone inhibited tumor growth by 70%, while when combined with a suboptimal dose of carboplatin, tumors regressed to an impalpable state. Assessment of microvessel density suggested that the antitumor activity of Anginex is mediated by inhibition of tumor angiogenesis. No sign of toxicity was observed in any of the studies.[130] Recently it was reported that Anginex enhanced the antitumor effect of radiotherapy causing significant tumor regression or delay in tumor growth compared to the effects of either treatment alone.[131] Anginex is the first peptide designed *de novo* to inhibit angiogenesis. Anginex entered clinical testing for therapeutic use against cancer and other angiogenesis-dependent diseases.

## 7. Conclusions and Perspectives

Proteins and peptides have greatly contributed to the understanding and the modulation of the molecular mechanisms of tumor angiogenesis for experimental purposes. Several proteins (antibodies), polypeptides and peptides entered clinical testing as antiangiogenic drugs, and a few of them are now approved for clinical use (e.g. IFN$\alpha$, TNF, bevacizumab). There is no question that several antiangiogenic drugs that will be approved in the future will be again proteins or peptides.

## Acknowledgments

The author is grateful to Prof. FJ Lejeune for continuous support and discussion and Robert Driscoll for critical reading of the manuscript. Work in our laboratory is supported by funds from the Molecular Oncology Program of the National Center for Competence in Research (NCCR), a research instrument of the Swiss National Science Foundation, the Swiss Cancer League/Oncosuisse, the Swiss National Science Foundation, *Fondazione San Salvatore*, the Leenards Foundation, the *Fondation de la Banque Cantonale Vaudoise*, the *Gertraud Hagmann Stiftung für Malignomforschung* and by the Medic Foundation. We apologize to those colleagues whose work could not be cited due to space limitations.

## References

1. Folkman J (2002) Role of angiogenesis in tumor growth and metastasis. *Semin Oncol* **29**:15–18.
2. Carmeliet P and Jain RK (2000) Angiogenesis in cancer and other diseases. *Nature* **407**:249–257.
3. Carmeliet P (2003) Angiogenesis in health and disease. *Nat Med* **9**:653–660.

4. Cavallaro U and Christofori G (2000) Molecular mechanisms of tumor angiogenesis and tumor progression. *J Neuro Oncol* **50**:63–70.
5. Ruegg C and Mariotti A (2003) Vascular integrins: Pleiotropic adhesion and signaling molecules in vascular homeostasis and angiogenesis. *Cell Mol Life Sci* **60**:1135–1157.
6. Bicknell R and Harris AL (2004) Novel angiogenic signaling pathways and vascular targets. *Annu Rev Pharmacol Toxicol* **44**:219–238.
7. Kerbel R and Folkman J (2002) Clinical translation of angiogenesis inhibitors. *Nat Rev Cancer* **2**:727–739.
8. Rüegg C, Driscol R, P. W-G and Stupp R (2003) Translational research in tumor angiogenesis: From bench to bedside and back to the bench. *Bulletin Suisse du Cancer* **23**:3–7.
9. Bouma-ter Steege JC, Mayo KH and Griffioen AW (2001) Angiostatic proteins and peptides. *Crit Rev Eukaryot Gene Expr* **11**:319–334.
10. Carswell EA, Old LJ, Kassel RL, Green S, Fiore N and Williamson B (1975) An endotoxin-induced serum factor that causes necrosis of tumors. *Proc Natl Acad Sci USA* **72**:3666–3670.
11. Aggarwal BB and Natarajan K (1996) Tumor necrosis factors: Developments during the last decade. *Eur Cytokine Netw* **7**:93–124.
12. Moss ML, Jin SL, Milla ME, Bickett DM, Burkhart W, Carter HL, Chen WJ, Clay WC, Didsbury JR, Hassler D, Hoffman CR, Kost TA, Lambert MH, Leesnitzer MA, McCauley P, McGeehan G, Mitchell J, Moyer M, Pahel G, Rocque W, Overton LK, Schoenen F, Seaton T, Su JL, Becherer JD *et al.* (1997) Cloning of a disintegrin metalloproteinase that processes precursor tumour-necrosis factor-alpha. *Nature* **385**:733–736.
13. Black RA, Rauch CT, Kozlosky CJ, Peschon JJ, Slack JL, Wolfson MF, Castner BJ, Stocking KL, Reddy P, Srinivasan S, Nelson N, Boiani N, Schooley KA, Gerhart M, Davis R, Fitzner JN, Johnson RS, Paxton RJ, March CJ and Cerretti DP (1997) A metalloproteinase disintegrin that releases tumour-necrosis factor- alpha from cells. *Nature* **385**:729–733.
14. Wallach D, Varfolomeev EE, Malinin NL, Goltsev YV, Kovalenko AV and Boldin MP (1999) Tumor necrosis factor receptor and Fas signaling mechanisms. *Annu Rev Immunol* **17**:331–367.
15. Lienard D, Ewalenko P, Delmotte JJ, Renard N and Lejeune FJ (1992) High-dose recombinant tumor necrosis factor alpha in combination with interferon gamma and melphalan in isolation perfusion of the limbs for melanoma and sarcoma. *J Clin Oncol* **10**:52–60.
16. Lejeune F, Ruegg C and Liénard D (1998) Clinical applications of TNF-alpha in cancer. *Curr Opin Immunol* **10**:573–580.

17. Rüegg C, Dormond O, Oguey D and Lejeune FJ (2000) Tumor necrosis factor: Clinical use and mechanisms of action. *Drug Resist Update* **3**:271–276.
18. Folli S, Pelegrin A, Chalandon Y, Yao X, Buchegger F, Lienard D, Lejeune F and Mach JP (1993) Tumor-necrosis factor can enhance radio-antibody uptake in human colon carcinoma xenografts by increasing vascular permeability. *Int J Cancer* **53**:829–836.
19. van der Veen AH, de Wilt JH, Eggermont AM, van Tiel ST, Seynhaeve AL and ten Hagen TL (2000) TNF-alpha augments intratumoural concentrations of doxorubicin in TNF-alpha-based isolated limb perfusion in rat sarcoma models and enhances anti-tumour effects. *Br J Cancer* **82**:973–980.
20. Blum MS, Toninelli E, Anderson JM, Balda MS, Zhou J, O'Donnell L, Pardi R and Bender JR (1997) Cytoskeletal rearrangement mediates human microvascular endothelial tight junction modulation by cytokines. *Am J Physiol* **273**:H286–294.
21. Petrache I, Birukova A, Ramirez SI, Garcia JG and Verin AD (2003) The role of the microtubules in tumor necrosis factor-alpha-induced endothelial cell permeability. *Am J Respir Cell Mol Biol* **28**:574–581.
22. Romer LH, McLean NV, Yan HC, Daise M, Sun J and DeLisser HM (1995) IFN-gamma and TNF-alpha induce redistribution of PECAM-1 (CD31) on human endothelial cells. *J Immunol* **154**:6582–6592.
23. Lampugnani MG, Resnati M, Raiteri M, Pigott R, Pisacane A, Houen G, Ruco LP and Dejana E (1992) A novel endothelial-specific membrane protein is a marker of cell-cell contacts. *J Cell Biol* **118**:1511–1522.
24. Ruegg C, Yilmaz A, Bieler G, Bamat J, Chaubert P and Lejeune FJ (1998) Evidence for the involvement of endothelial cell integrin alpha-v-beta-3 in the disruption of the tumor vasculature induced by tnf and ifn-gamma. *Nat Med* **4**:408–414.
25. Halin C, Gafner V, Villani ME, Borsi L, Berndt A, Kosmehl H, Zardi L and Neri D (2003) Synergistic therapeutic effects of a tumor targeting antibody fragment, fused to interleukin 12 and to tumor necrosis factor alpha. *Cancer Res* **63**:3202–3210.
26. Curnis F, Sacchi A and Corti A (2002) Improving chemotherapeutic drug penetration in tumors by vascular targeting and barrier alteration. *J Clin Invest* **110**:475–482.
27. Lejeune FJ (2002) Clinical use of TNF revisited: Improving penetration of anti-cancer agents by increasing vascular permeability. *J Clin Invest* **110**:433–435.
28. Ezekowitz RA, Mulliken JB and Folkman J (1992) Interferon alfa-2a therapy for life-threatening hemangiomas of infancy. *N Engl J Med* **326**:1456–1463.

29. Hong YK, Chung DS, Joe YA, Yang YJ, Kim KM, Park YS, Yung WK and Kang JK (2000) Efficient inhibition of in vivo human malignant glioma growth and angiogenesis by interferon-beta treatment at early stage of tumor development. *Clin Cancer Res* **6**:3354–3360.
30. Dinney CP, Bielenberg DR, Perrotte P, Reich R, Eve BY, Bucana CD and Fidler IJ (1998) Inhibition of basic fibroblast growth factor expression, angiogenesis, and growth of human bladder carcinoma in mice by systemic interferon-alpha administration. *Cancer Res* **58**:808–814.
31. Watanabe N, Niitsu Y, Yamauchi N, Umeno H, Sone H, Neda H and Urushizaki I (1988) Antitumor synergism between recombinant human tumor necrosis factor and recombinant human interferon-$\gamma$. *J Biol Response Mod* **7**:24–31.
32. Kohno T, Mizukami H, Suzuki M, Saga Y, Takei Y, Shimpo M, Matsushita T, Okada T, Hanazono Y, Kume A, Sato I and Ozawa K (2003) Interleukin-10-mediated inhibition of angiogenesis and tumor growth in mice bearing VEGF-producing ovarian cancer. *Cancer Res* **63**:5091–5094.
33. Voest EE, Kenyon BM, O'Reilly MS, Truitt G, D'Amato RJ and Folkman J (1995) Inhibition of angiogenesis *in vivo* by interleukin 12. *J Natl Cancer Inst* **87**:581–586.
34. Zhai Y, Ni J, Jiang GW, Lu J, Xing L, Lincoln C, Carter KC, Janat F, Kozak D, Xu S, Rojas L, Aggarwal BB, Ruben S, Li LY, Gentz R and Yu GL (1999) VEGI, a novel cytokine of the tumor necrosis factor family, is an angiogenesis inhibitor that suppresses the growth of colon carcinomas *in vivo*. *Faseb J* **13**:181–189.
35. Gengrinovitch S, Greenberg SM, Cohen T, Gitay-Goren H, Rockwell P, Maione TE, Levi BZ and Neufeld G (1995) Platelet factor-4 inhibits the mitogenic activity of VEGF121 and VEGF165 using several concurrent mechanisms. *J Biol Chem* **270**:15059–15065.
36. Regulier E, Paul S, Marigliano M, Kintz J, Poitevin Y, Ledoux C, Roecklin D, Cauet G, Calenda V and Homann HE (2001) Adenovirus-mediated delivery of antiangiogenic genes as an antitumor approach. *Cancer Gene Ther* **8**:45–54.
37. Shellenberger TD, Wang M, Gujrati M, Jayakumar A, Strieter RM, Burdick MD, Ioannides CG, Efferson CL, El-Naggar AK, Roberts D, Clayman GL and Frederick MJ (2004) BRAK/CXCL14 is a potent inhibitor of angiogenesis and a chemotactic factor for immature dendritic cells. *Cancer Res* **64**:8262–8270.
38. O'Reilly MS, Holmgren L, Shing Y, Chen C, Rosenthal RA, Moses M, Lane WS, Cao Y, Sage EH and Folkman J (1994) Angiostatin: A novel angiogenesis inhibitor that mediates the suppression of metastases by a Lewis lung carcinoma. *Cell* **79**:315–328.
39. Folkman J (1995) Angiogenesis inhibitors generated by tumors. *Mol Med* **1**:120–122.

40. O'Reilly MS, Holmgren L, Chen C and Folkman J (1996) Angiostatin induces and sustains dormancy of human primary tumors in mice. *Nat Med* **2**:689–692.
41. Soff GA (2000) Angiostatin and angiostatin-related proteins. *Cancer Metastasis Rev* **19**:97–107.
42. Beerepoot LV, Witteveen EO, Groenewegen G, Fogler WE, Sim BK, Sidor C, Zonnenberg BA, Schramel F, Gebbink MF and Voest EE (2003) Recombinant human angiostatin by twice-daily subcutaneous injection in advanced cancer: A pharmacokinetic and long-term safety study. *Clin Cancer Res* **9**:4025–4033.
43. O'Reilly MS, Boehm T, Shing Y, Fukai N, Vasios G, Lane WS, Flynn E, Birkhead JR, Olsen BR and Folkman J (1997) Endostatin: An endogenous inhibitor of angiogenesis and tumor growth. *Cell* **88**:277–285.
44. Sorensen DR and Read TA (2002) Delivery of endostatin in experimental cancer therapy. *Int J Exp Pathol* **83**:265–274.
45. Herbst RS, Hess KR, Tran HT, Tseng JE, Mullani NA, Charnsangavej C, Madden T, Davis DW, McConkey DJ, O'Reilly MS, Ellis LM, Pluda J, Hong WK and Abbruzzese JL (2002) Phase I study of recombinant human endostatin in patients with advanced solid tumors. *J Clin Oncol* **20**:3792–3803.
46. Maeshima Y, Colorado PC, Torre A, Holthaus KA, Grunkemeyer JA, Ericksen MB, Hopfer H, Xiao Y, Stillman IE and Kalluri R (2000) Distinct antitumor properties of a type IV collagen domain derived from basement membrane. *J Biol Chem* **275**:21340–21348.
47. Maeshima Y, Manfredi M, Reimer C, Holthaus KA, Hopfer H, Chandamuri BR, Kharbanda S and Kalluri R (2001) Identification of the anti-angiogenic site within vascular basement membrane-derived tumstatin. *J Biol Chem* **276**:15240–15248.
48. Kaur B, Brat DJ, Devi NS and Van Meir EG (2005) Vasculostatin, a proteolytic fragment of Brain Angiogenesis Inhibitor 1, is an antiangiogenic and antitumorigenic factor. *Oncogene*.
49. Frazier WA (1991) Thrombospondins. *Curr Opin Cell Biol* **3**:792–799.
50. Bornstein P (1992) Thrombospondins: Structure and regulation of expression. *Faseb J* **6**:3290–3299.
51. Vailhe B and Feige JJ (2003) Thrombospondins as anti-angiogenic therapeutic agents. *Curr Pharm Des* **9**:583–588.
52. Iruela-Arispe ML, Luque A and Lee N (2004) Thrombospondin modules and angiogenesis. *Int J Biochem Cell Biol* **36**:1070–1078.
53. Noh YH, Matsuda K, Hong YK, Kunstfeld R, Riccardi L, Koch M, Oura H, Dadras SS, Streit M and Detmar M (2003) An N-terminal 80 kDa recombinant fragment of human thrombospondin-2 inhibits vascular endothelial growth factor induced endothelial cell migration *in vitro* and tumor growth and angiogenesis *in vivo*. *J Invest Dermatol* **121**:1536–1543.

54. Haviv F, Bradley MF, Kalvin DM, Schneider AJ, Davidson DJ, Majest SM, McKay LM, Haskell CJ, Bell RL, Nguyen B, Marsh KC, Surber BW, Uchic JT, Ferrero J, Wang YC, Leal J, Record RD, Hodde J, Badylak SF, Lesniewski RR and Henkin J (2005) Thrombospondin-1 mimetic peptide inhibitors of angiogenesis and tumor growth: Design, synthesis, and optimization of pharmacokinetics and biological activities. *J Med Chem* **48**:2838–2846.
55. Rehn M, Veikkola T, Kukk-Valdre E, Nakamura H, Ilmonen M, Lombardo C, Pihlajaniemi T, Alitalo K and Vuori K (2001) Interaction of endostatin with integrins implicated in angiogenesis. *Proc Natl Acad Sci USA* **98**:1024–1029.
56. Sudhakar A, Sugimoto H, Yang C, Lively J, Zeisberg M and Kalluri R (2003) Human tumstatin and human endostatin exhibit distinct antiangiogenic activities mediated by alpha v beta 3 and alpha 5 beta 1 integrins. *Proc Natl Acad Sci USA* **100**:4766–4771.
57. Bix G and Iozzo RV (2005) Matrix revolutions: "Tails" of basement-membrane components with angiostatic functions. *Trends Cell Biol* **15**:52–60.
58. Maeshima Y, Sudhakar A, Lively JC, Ueki K, Kharbanda S, Kahn CR, Sonenberg N, Hynes RO and Kalluri R (2002) Tumstatin, an endothelial cell-specific inhibitor of protein synthesis. *Science* **295**:140–143.
59. Eder JP, Jr, Supko JG, Clark JW, Puchalski TA, Garcia-Carbonero R, Ryan DP, Shulman LN, Proper J, Kirvan M, Rattner B, Connors S, Keogan MT, Janicek MJ, Fogler WE, Schnipper L, Kinchla N, Sidor C, Phillips E, Folkman J and Kufe DW (2002) Phase I clinical trial of recombinant human endostatin administered as a short intravenous infusion repeated daily. *J Clin Oncol* **20**:3772–3784.
60. Dell'Eva R, Pfeffer U, Indraccolo S, Albini A and Noonan D (2002) Inhibition of tumor angiogenesis by angiostatin: From recombinant protein to gene therapy. *Endothelium* **9**:3–10.
61. Ross JS, Schenkein DP, Pietrusko R, Rolfe M, Linette GP, Stec J, Stagliano NE, Ginsburg GS, Symmans WF, Pusztai L and Hortobagyi GN (2004) Targeted therapies for cancer 2004. *Am J Clin Pathol* **122**:598–609.
62. Presta LG, Chen H, O'Connor SJ, Chisholm V, Meng YG, Krummen L, Winkler M and Ferrara N (1997) Humanization of an anti-vascular endothelial growth factor monoclonal antibody for the therapy of solid tumors and other disorders. *Cancer Res* **57**:4593–4599.
63. Ferrara N, Hillan KJ, Gerber HP and Novotny W (2004) Discovery and development of bevacizumab, an anti-VEGF antibody for treating cancer. *Nat Rev Drug Discov* **3**:391–400.
64. Yang J, Haworth L, Steinberg S, Rosenberg S and Novotny W (2002) A randomized double-blind placebo-controlled trial of bevacizumab (anti-VEGF antibody) demonstrating a prolongation in time to progression in patients

with metastatic renal cancer. *Annual Meeting of the Am. Soc. Clin. Oncol. 2000, 20–23 May; New Orleans, LA, USA. URL: http://www.asco.org*

65. Kabbinavar F, Hurwitz HI, Fehrenbacher L, Meropol NJ, Novotny WF, Lieberman G, Griffing S and Bergsland E (2003) Phase II, randomized trial comparing bevacizumab plus fluorouracil (FU)/leucovorin (LV) with FU/LV alone in patients with metastatic colorectal cancer. *J Clin Oncol* **21**:60–65
66. Hurwitz H, Fehrenbacher L, Novotny W, Cartwright T, Hainsworth J, Heim W, Berlin J, Baron A, Griffing S, Holmgren E, Ferrara N, Fyfe G, Rogers B, Ross R and Kabbinavar F (2004) Bevacizumab plus irinotecan, fluorouracil, and leucovorin for metastatic colorectal cancer. *N Engl J Med* **350**: 2335–2342.
67. Olszewski AJ, Grossbard ML and Kozuch PS (2005) The horizon of antiangiogenic therapy for colorectal cancer. *Oncology (Huntingt)* **19**:297–306; discussion 306, 308, 317–233.
68. Jayson GC, Mulatero C, Ranson M, Zweit J, Jackson A, Broughton L, Wagstaff J, Hakansson L, Groenewegen G, Lawrance J, Tang M, Wauk L, Levitt D, Marreaud S, Lehmann FF, Herold M and Zwierzina H (2005) Phase I investigation of recombinant anti-human vascular endothelial growth factor antibody in patients with advanced cancer. *Eur J Cancer* **41**:555–563.
69. Prewett M, Huber J, Li Y, Santiago A, O'Connor W, King K, Overholser J, Hooper A, Pytowski B, Witte L, Bohlen P and Hicklin DJ (1999) Antivascular endothelial growth factor receptor (fetal liver kinase 1) monoclonal antibody inhibits tumor angiogenesis and growth of several mouse and human tumors. *Cancer Res* **59**:5209–5218.
70. Shaheen RM, Tseng WW, Vellagas R, Liu W, Ahmad SA, Jung YD, Reinmuth N, Drazan KE, Bucana CD, Hicklin DJ and Ellis LM (2001) Effects of an antibody to vascular endothelial growth factor receptor-2 on survival, tumor vascularity, and apoptosis in a murine model of colon carcinomatosis. *Int J Oncol* **18**:221–226.
71. Zhang L, Yu D, Hicklin DJ, Hannay JA, Ellis LM and Pollock RE (2002) Combined anti-fetal liver kinase 1 monoclonal antibody and continuous low-dose doxorubicin inhibits angiogenesis and growth of human soft tissue sarcoma xenografts by induction of endothelial cell apoptosis. *Cancer Res* **62**:2034–2042.
72. Bruns CJ, Shrader M, Harbison MT, Portera C, Solorzano CC, Jauch KW, Hicklin DJ, Radinsky R and Ellis LM (2002) Effect of the vascular endothelial growth factor receptor-2 antibody DC101 plus gemcitabine on growth, metastasis and angiogenesis of human pancreatic cancer growing orthotopically in nude mice. *Int J Cancer* **102**:101–108.

73. Witte L, Hicklin DJ, Zhu Z, Pytowski B, Kotanides H, Rockwell P and Bohlen P (1998) Monoclonal antibodies targeting the VEGF receptor-2 (Flk1/KDR) as an anti-angiogenic therapeutic strategy. *Cancer Metastasis Rev* **17**:155–161.
74. Hunt S (2001) Technology evaluation: IMC-1C11, ImClone Systems. *Curr Opin Mol Ther* **3**:418–424.
75. Posey JA, Ng TC, Yang B, Khazaeli MB, Carpenter MD, Fox F, Needle M, Waksal H and LoBuglio AF (2003) A phase I study of anti-kinase insert domain-containing receptor antibody, IMC-1C11, in patients with liver metastases from colorectal carcinoma. *Clin Cancer Res* **9**:1323–1332.
76. Brooks PC, Clark RA and Cheresh DA (1994) Requirement of vascular integrin alpha v beta 3 for angiogenesis. *Science* **264**:569–571.
77. Brooks PC, Montgomery AM, Rosenfeld M, Reisfeld RA, Hu T, Klier G and Cheresh DA (1994) Integrin alpha v beta 3 antagonists promote tumor regression by inducing apoptosis of angiogenic blood vessels. *Cell* **79**:1157–1164.
78. Brooks PC, Stromblad S, Klemke R, Visscher D, Sarkar FH and Cheresh DA (1995) Antiintegrin alpha v beta 3 blocks human breast cancer growth and angiogenesis in human skin. *J Clin Invest* **96**:1815–1822.
79. Wu H, Beuerlein G, Nie Y, Smith H, Lee BA, Hensler M, Huse WD and Watkins JD (1998) Stepwise *in vitro* affinity maturation of Vitaxin, an alpha v beta3-specific humanized mAb. *Proc Natl Acad Sci USA* **95**:6037–6042.
80. Gutheil JC, Campbell TN, Pierce PR, Watkins JD, Huse WD, Bodkin DJ and Cheresh DA (2000) Targeted antiangiogenic therapy for cancer using Vitaxin: A humanized monoclonal antibody to the integrin alpha v beta3. *Clin Cancer Res* **6**:3056–3061.
81. Posey JA, Khazaeli MB, DelGrosso A, Saleh MN, Lin CY, Huse W and LoBuglio AF (2001) A pilot trial of Vitaxin, a humanized anti-vitronectin receptor (anti alpha v beta 3) antibody in patients with metastatic cancer. *Cancer Biother Radiopharm* **16**:125–132.
82. McCafferty J, Griffiths AD, Winter G and Chiswell DJ (1990) Phage antibodies: Filamentous phage displaying antibody variable domains. *Nat* **348**:552–554.
83. Pini A and Bracci L (2000) Phage display of antibody fragments. *Curr Protein Pept Sci* **1**:155–169.
84. Wu AM and Yazaki PJ (2000) Designer genes: Recombinant antibody fragments for biological imaging. *Q J Nucl Med* **44**:268–283.
85. Adams GP and Schier R (1999) Generating improved single-chain Fv molecules for tumor targeting. *J Immunol Methods* **231**:249–260.
86. Kriangkum J, Xu B, Nagata LP, Fulton RE and Suresh MR (2001) Bispecific and bifunctional single chain recombinant antibodies. *Biomol Eng* **18**:31–40.

87. Castellani P, Viale G, Dorcaratto A, Nicolo G, Kaczmarek J, Querze G and Zardi L (1994) The fibronectin isoform containing the ED-B oncofetal domain: A marker of angiogenesis. *Int J Cancer* **59**:612–618.
88. Pini A, Viti F, Santucci A, Carnemolla B, Zardi L, Neri P and Neri D (1998) Design and use of a phage display library. Human antibodies with subnanomolar affinity against a marker of angiogenesis eluted from a two-dimensional gel. *J Biol Chem* **273**:21769–21776.
89. Tarli L, Balza E, Viti F, Borsi L, Castellani P, Berndorff D, Dinkelborg L, Neri D and Zardi L (1999) A high-affinity human antibody that targets tumoral blood vessels. *Blood* **94**:192–198.
90. Demartis S, Tarli L, Borsi L, Zardi L and Neri D (2001) Selective targeting of tumour neovasculature by a radiohalogenated human antibody fragment specific for the ED-B domain of fibronectin. *Eur J Nucl Med* **28**:534–539.
91. Santimaria M, Moscatelli G, Viale GL, Giovannoni L, Neri G, Viti F, Leprini A, Borsi L, Castellani P, Zardi L, Neri D and Riva P (2003) Immunoscintigraphic detection of the ED-B domain of fibronectin, a marker of angiogenesis, in patients with cancer. *Clin Cancer Res* **9**:571–579.
92. Carnemolla B, Borsi L, Balza E, Castellani P, Meazza R, Berndt A, Ferrini S, Kosmehl H, Neri D and Zardi L (2002) Enhancement of the antitumor properties of interleukin-2 by its targeted delivery to the tumor blood vessel extracellular matrix. *Blood* **99**:1659–1665.
93. Borsi L, Balza E, Carnemolla B, Sassi F, Castellani P, Berndt A, Kosmehl H, Biro A, Siri A, Orecchia P, Grassi J, Neri D and Zardi L (2003) Selective targeted delivery of TNFalpha to tumor blood vessels. *Blood* **102**:4384–4392.
94. Ebbinghaus C, Ronca R, Kaspar M, Grabulovski D, Berndt A, Kosmehl H, Zardi L and Neri D (2005) Engineered vascular-targeting antibody-interferon-gamma fusion protein for cancer therapy. *Int J Cancer* **116**:304–313.
95. Vitaliti A, Wittmer M, Steiner R, Wyder L, Neri D and Klemenz R (2000) Inhibition of tumor angiogenesis by a single-chain antibody directed against vascular endothelial growth factor. *Cancer Res* **60**:4311–4314.
96. Cai M, Zhang H and Hui R (2003) Single chain Fv antibody against angiopoietin-2 inhibits VEGF-induced endothelial cell proliferation and migration *in vitro*. *Biochem Biophys Res Commun* **309**:946–951.
97. Zhu Z, Lu D, Kotanides H, Santiago A, Jimenez X, Simcox T, Hicklin DJ, Bohlen P and Witte L (1999) Inhibition of vascular endothelial growth factor induced mitogenesis of human endothelial cells by a chimeric anti-kinase insert domain-containing receptor antibody. *Cancer Lett* **136**:203–213.
98. Sanz L, Kristensen P, Blanco B, Facteau S, Russell SJ, Winter G and Alvarez-Vallina L (2002) Single-chain antibody-based gene therapy: Inhibition of

tumor growth by in situ production of phage-derived human antibody fragments blocking functionally active sites of cell-associated matrices. *Gene Ther* **9**:1049–1053.

99. Bloemendal HJ, de Boer HC, Koop EA, van Dongen AJ, Goldschmeding R, Landman WJ, Logtenberg T, Gebbink MF and Voest EE (2004) Activated vitronectin as a target for anticancer therapy with human antibodies. *Cancer Immunol Immunother* **53**:799–808.
100. Gerber HP, Kowalski J, Sherman D, Eberhard DA and Ferrara N (2000) Complete inhibition of rhabdomyosarcoma xenograft growth and neovascularization requires blockade of both tumor and host vascular endothelial growth factor. *Cancer Res* **60**:6253–6258.
101. Holash J, Davis S, Papadopoulos N, Croll SD, Ho L, Russell M, Boland P, Leidich R, Hylton D, Burova E, Ioffe E, Huang T, Radziejewski C, Bailey K, Fandl JP, Daly T, Wiegand SJ, Yancopoulos GD and Rudge JS (2002) VEGF-Trap: A VEGF blocker with potent antitumor effects. *Proc Natl Acad Sci USA* **99**:11393–11398.
102. Huang J, Frischer JS, Serur A, Kadenhe A, Yokoi A, McCrudden KW, New T, O'Toole K, Zabski S, Rudge JS, Holash J, Yancopoulos GD, Yamashiro DJ and Kandel JJ (2003) Regression of established tumors and metastases by potent vascular endothelial growth factor blockade. *Proc Natl Acad Sci USA* **100**: 7785–7790.
103. Konner J and Dupont J (2004) Use of soluble recombinant decoy receptor vascular endothelial growth factor trap (VEGF Trap) to inhibit vascular endothelial growth factor activity. *Clin Colorectal Cancer* **4** (Suppl 2):S81–85.
104. He Y, Kozaki K, Karpanen T, Koshikawa K, Yla-Herttuala S, Takahashi T and Alitalo K (2002) Suppression of tumor lymphangiogenesis and lymph node metastasis by blocking vascular endothelial growth factor receptor 3 signaling. *J Natl Cancer Inst* **94**:819–825.
105. Ruoslahti E (2000) Targeting tumor vasculature with homing peptides from phage display. *Semin Cancer Biol* **10**:435–442.
106. Pasqualini R (1999) Vascular targeting with phage peptide libraries. *Q J Nucl Med* **43**:159–162.
107. Ruoslahti E (2004) Vascular zip codes in angiogenesis and metastasis. *Biochem Soc Trans* **32**:397–402.
108. Arap W, Pasqualini R and Ruoslahti E (1998) Cancer treatment by targeted drug delivery to tumor vasculature in a mouse model. *Science* **279**:377–380.
109. Pasqualini R, Koivunen E, Kain R, Lahdenranta J, Sakamoto M, Stryhn A, Ashmun RA, Shapiro LH, Arap W and Ruoslahti E (2000) Aminopeptidase N is a receptor for tumor-homing peptides and a target for inhibiting angiogenesis. *Cancer Res* **60**:722–727.

110. Curnis F, Arrigoni G, Sacchi A, Fischetti L, Arap W, Pasqualini R and Corti A (2002) Differential binding of drugs containing the NGR motif to CD13 isoforms in tumor vessels, epithelia, and myeloid cells. *Cancer Res* **62**:867–874.
111. Corti A and Ponzoni M (2004) Tumor vascular targeting with tumor necrosis factor alpha and chemotherapeutic drugs. *Ann NY Acad Sci* **1028**:104–112.
112. Ruoslahti E and Pierschbacher MD (1987) New perspectives in cell adhesion: RGD and integrins. *Science* **238**:491–497.
113. Ruoslahti E (1996) RGD and other recognition sequences for integrins. *Annu Rev Cell Dev Biol* **12**:697–715.
114. Gurrath M, Muller G, Kessler H, Aumailley M and Timpl R (1992) Conformation/activity studies of rationally designed potent anti-adhesive RGD peptides. *Eur J Biochem* **210**:911–921.
115. Aumailley M, Gurrath M, Muller G, Calvete J, Timpl R and Kessler H (1991) Arg-Gly-Asp constrained within cyclic pentapeptides. Strong and selective inhibitors of cell adhesion to vitronectin and laminin fragment P1. *FEBS Lett* **291**:50–54.
116. Pfaff M, Tangemann K, Muller B, Gurrath M, Muller G, Kessler H, Timpl R and Engel J (1994) Selective recognition of cyclic RGD peptides of NMR defined conformation by alpha IIb beta 3, alpha V beta 3, and alpha 5 beta 1 integrins. *J Biol Chem* **269**:20233–20238.
117. Dechantsreiter MA, Planker E, Matha B, Lohof E, Holzemann G, Jonczyk A, Goodman SL and Kessler H (1999) N-Methylated cyclic RGD peptides as highly active and selective alpha(V)beta(3) integrin antagonists. *J Med Chem* **42**:3033–3040.
118. Smith JW (2003) Cilengitide Merck. *Curr Opin Investig Drugs* **4**:741–745.
119. Mitjans F, Meyer T, Fittschen C, Goodman S, Jonczyk A, Marshall JF, Reyes G and Piulats J (2000) *In vivo* therapy of malignant melanoma by means of antagonists of alphav integrins. *Int J Cancer* **87**:716–723.
120. Taga T, Suzuki A, Gonzalez-Gomez I, Gilles FH, Stins M, Shimada H, Barsky L, Weinberg KI and Laug WE (2002) alpha v-Integrin antagonist EMD 121974 induces apoptosis in brain tumor cells growing on vitronectin and tenascin. *Int J Cancer* **98**:690–697.
121. Burke PA, DeNardo SJ, Miers LA, Lamborn KR, Matzku S and DeNardo GL (2002) Cilengitide targeting of alpha(v)beta(3) integrin receptor synergizes with radioimmunotherapy to increase efficacy and apoptosis in breast cancer xenografts. *Cancer Res* **62**:4263–4272.
122. Burkhart DJ, Kalet BT, Coleman MP, Post GC and Koch TH (2004) Doxorubicin-formaldehyde conjugates targeting alphavbeta3 integrin. *Mol Cancer Ther* **3**:1593–1604.

123. Eskens FA, Dumez H, Hoekstra R, Perschl A, Brindley C, Bottcher S, Wynendaele W, Drevs J, Verweij J and van Oosterom AT (2003) Phase I and pharmacokinetic study of continuous twice weekly intravenous administration of Cilengitide (EMD 121974), a novel inhibitor of the integrins alphavbeta3 and alphavbeta5 in patients with advanced solid tumours. *Eur J Cancer* **39**:917–926.
124. Koivunen E, Gay DA and Ruoslahti E (1993) Selection of peptides binding to the alpha 5 beta 1 integrin from phage display library. *J Biol Chem* **268**:20205–20210.
125. Pasqualini R, Koivunen E and Ruoslahti E (1995) A peptide isolated from phage display libraries is a structural and functional mimic of an RGD-binding site on integrins. *J Cell Biol* **130**:1189–1196.
126. Livant DL, Brabec RK, Pienta KJ, Allen DL, Kurachi K, Markwart S and Upadhyaya A (2000) Anti-invasive, antitumorigenic, and antimetastatic activities of the PHSCN sequence in prostate carcinoma. *Cancer Res* **60**:309–320.
127. Schreiner C, Bauer J, Margolis M and Juliano RL (1991) Expression and role of integrins in adhesion of human colonic carcinoma cells to extracellular matrix components. *Clin Exp Metastasis* **9**:163–178.
128. Stoeltzing O, Liu W, Reinmuth N, Fan F, Parry GC, Parikh AA, McCarty MF, Bucana CD, Mazar AP and Ellis LM (2003) Inhibition of integrin alpha5beta1 function with a small peptide (ATN-161) plus continuous 5-FU infusion reduces colorectal liver metastases and improves survival in mice. *Int J Cancer* **104**:496–503.
129. Griffioen AW, van der Schaft DW, Barendsz-Janson AF, Cox A, Struijker Boudier HA and Hillen HF and Mayo KH (2001) Anginex, a designed peptide that inhibits angiogenesis. *Biochem J* **354**:233–242.
130. Dings RP, Yokoyama Y, Ramakrishnan S, Griffioen AW and Mayo KH (2003) The designed angiostatic peptide anginex synergistically improves chemotherapy and antiangiogenesis therapy with angiostatin. *Cancer Res* **63**:382–385.
131. Dings RP, Williams BW, Song CW, Griffioen AW, Mayo KH and Griffin RJ (2005) Anginex synergizes with radiation therapy to inhibit tumor growth by radiosensitizing endothelial cells. *Int J Cancer* **115**:312–319.
132. Corada M, Zanetta L, Orsenigo F, Breviario F, Lampugnani MG, Bernasconi S, Liao F, Hicklin DJ, Bohlen P and Dejana E (2002) A monoclonal antibody to vascular endothelial-cadherin inhibits tumor angiogenesis without side effects on endothelial permeability. *Blood* **100**:905–911.

# 13

# Utilizing Lymphatic Transport in Enhancing the Delivery of Drugs, Including Proteins, and Peptides, to Metastatic Tumors

*Ellen K. Wasan and Kishor M. Wasan*

## Abstract

The gastrointestinal lymphatic system is a specific transport pathway through which dietary lipids, fat-soluble vitamins, and water-insoluble peptide type molecules can gain access to the systemic circulation. Drugs transported by way of the gastrointestinal lymphatic system bypass the liver and avoid potential hepatic first-pass metabolism. Lymphatic delivery of immunomodulatory and protein and peptide drugs used in the treatment of cancer cell metastases presents an opportunity to maximize therapeutic benefit while minimizing general systemic drug exposure. Furthermore, lymphatic drug transport may promote drug incorporation into the body's lipid-handling system, thus offering the potential to manipulate drug distribution and residence time within the body and reduce the invasiveness of treatment for the patient. This chapter will discuss the potential utilization of lymphatic transport in enhancing the delivery of drugs, including proteins and peptides, to metastatic tumors.

## 1. Introduction

For many cancers, particularly those affecting older adults, cancer mortality has changed little in the past 40 years, despite of so much effort focused on discovering the underlying cellular pathways that lead to cancer and the development of new drugs.[1] Part of the problem lies in our lack of control of metastatic disease. Second, many new drugs in development are difficult to formulate due to poor water solubility. This dual problem represents an opportunity to explore the role of lipid-based drug delivery to achieve lymphatic targeting of hydrophobic anticancer compounds to sites of lymph node metastasis. The subject of metastasis in lymph nodes has received intense scrutiny and the mechanisms involved are finally revealing therapeutic targets.[2] The lymphatic transport pathway is accessible via the oral route of drug delivery, for example, with the use of certain lipid-based excipients, which have the added advantage of reducing the hepatic first-pass effect for susceptible drugs.[3] In this way, maximal drug exposure of lymph node metastases is possible. Examples of hydrophobic drugs useful in metastatic breast cancer, for example, include capecitabine, tamoxifen, vinorelbine, and docetaxel. These agents have low water solubility and capecitabine and tamoxifen are presently used orally. Capecitabine bioavailability is thought to be increased by food, which may suggest that lipid-based formulations may further improve its absorption. This compound is a prodrug metabolized in the liver and in tumor cells to the active compounds fluorouracil, 5′-DFCR, 5′-DFUR, FdUMP, and FUTP. Reducing hepatic metabolism of capecitabine to its active metabolites while improving delivery to lymph node metastases may spare some of the toxic effects of the active metabolites on normal tissues. Tamoxifen is an antiestrogen compound that is well absorbed orally and in widespread use in the adjuvant treatment of breast cancer. Docetaxel, a microtubule stabilizer, is not presently used orally due to poor absorption. This is possibly due to extensive metabolism by hepatic CYP3A4 enzyme, which results significant interpatient pharmacokinetic variability. An oral lipid-based formulation that reduces first-pass effect and improves intestinal absorption may circumvent this problem, while promoting delivery to the lymphatics.

The gastrointestinal lymphatic system is a specific transport pathway through which dietary lipids,[4–10] fat-soluble vitamins and water-insoluble peptide type molecules (i.e. cyclosporine)[11,12] can gain access to the systemic circulation.[13] Drugs transported by way of the gastrointestinal lymphatic system bypass the liver and avoid potential hepatic first-pass metabolism. Lymphatic delivery of immunomodulatory agents and low therapeutic index drugs used in the treatment of cancer cell metastases presents an opportunity to maximize therapeutic benefit while minimizing general systemic drug exposure.[14,15] Furthermore, lymphatic drug transport may promote drug incorporation into the body's lipid-handling system, thus offering the potential to manipulate drug distribution and residence time within the body.

## 2. Lipid Absorption from the Small Intestine and the Lymphatic System

### 2.1. *Lipid absorption from the small intestine*

Most dietary lipids are absorbed from the jejunum region of the small intestine, with the exception of bile salts, which remain in the small intestine lumen in order to facilitate further digestion (Fig. 1).[16–18] The bile salts are finally absorbed in the distal ileum[17] and are transported back to the liver by the portal blood in a cycle that constitutes the enterohepatic recirculation.[18–20] Lipid absorption occurs when the micellar solution of lipids and bile salts comes into contact with the microvillus membrane of the enterocytes.[21] The lipids are transported across the enterocyte membrane by primarily an energy independent process, which relies on the maintenance of an inward diffusion gradient. This gradient can partly be achieved by the attachment of the fatty acids to specific intracellular binding proteins. However, the ultimate driving force for absorption probably comes from the rapid re-esterification of the lipids, which is an ATP-dependent process depending upon activation of fatty acids to acyl-CoA esters.[21]

The major digestive products of triglycerides (TG) are monoglyceride and fatty acid while the major digestive product of biliary and dietary phosphatidylcholine is lysophosphatidylcholine. These digestive products are absorbed primarily by the enterocytes through simple diffusion. However, the absorption of cholesterol by the enterocytes is specific, since $\beta$-sitosterol (a plant sterol), a molecule that bears considerable resemblance to cholesterol, is poorly absorbed. This specificity requires energy as the deprivation

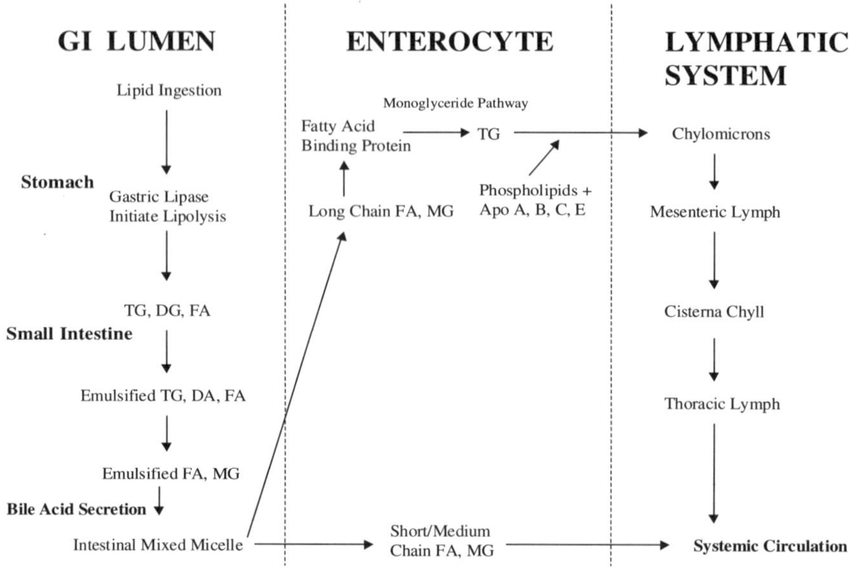


**Fig. 1** Schematic diagram describing the sequential steps in digestion of triglycerides (TG) and subsequent absorption through the portal blood and intestinal lymphatics. MG, monoglyceride; DG, diglyceride; FA, fatty acids. Adopted from Refs. 4 and 5.

of blood supply results in free permeability of different sterols.[21] Recent studies by Repa *et al.* have suggested that oxysterol receptors and the bile acid receptor are retinoid X receptor (RXR) heterodimeric partners that mediate cholesterol uptake by regulating expression of the reverse cholesterol transporter, ABC1, and the rate-limiting enzyme of bile acid synthesis, CYP7A1, respectively.[22] These heterodimers serve as key regulators of cholesterol homeostasis by governing reverse cholesterol transport from peripheral tissues, bile acid synthesis in liver, and cholesterol absorption within the intestine.[22]

After entry into the enterocytes, monoglycerides, fatty acids, and cholesterol are transported within the cell to the endoplasmic reticulum by fatty acid-binding protein and sterol carrier protein, respectively. Through the monoglyceride pathway, the digestive byproducts of triglycerides, monoglycerides, and fatty acids are resythesized to form triglyceride in the endoplasmic reticulum. This triglyceride is then transported to the Golgi

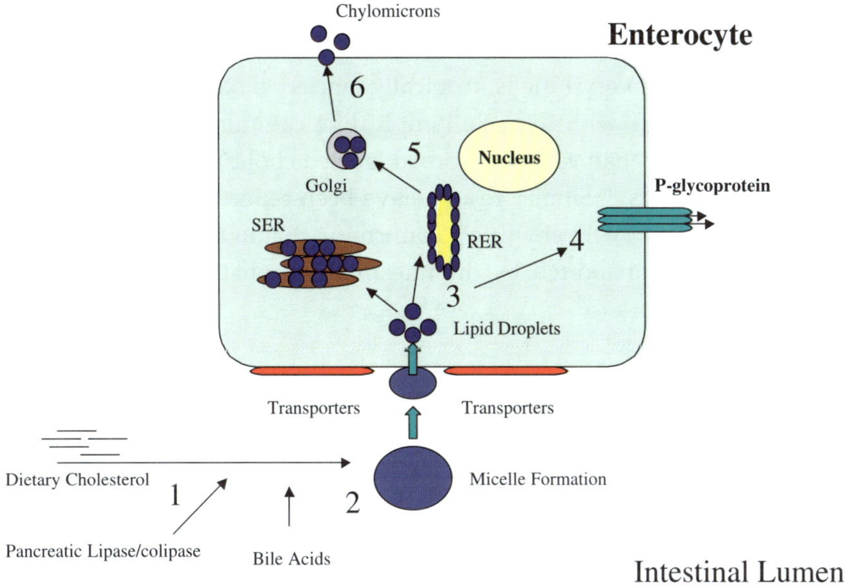


**Fig. 2** Transport of lipids through the enterocytes and the formation of chylomicrons.

apparatus where it is packaged into chylomicrons and released into the lymphatics (Fig. 2).[19–21]

The transport and metabolism of the absorbed cholesterol is much lower than that of triacylglycerols. The estimated half-life for absorbed cholesterol in the enterocyte is about 12 hr. During absorption the cholesterol becomes incorporated into the membranes of the enterocytes and diluted with endogenous cholesterol. A large proportion of the cholesterol that is transported from the enterocyte is esterified, mainly with oleic acid. The rate of esterification of cholesterol may regulate the rate of lymphatic transport of cholesterol. Two enzymes have been proposed to be involved in the esterification, cholesterol esterase, and acyl-CoA cholesterol acyltransferase (ACAT).

In studies investigating intestinal absorption of cholesterol conducted in the early 20th century it was reported that adding cholesterol to the diet increased the concentration of cholesterol in intestinal lymph.[23] However, not until the 1950s was the quantitative significance of this lymphatic pathway known.[24] Biggs *et al.* demonstrated that following an intragastric dose of [$^3$H]cholesterol very little isotropically labeled cholesterol appeared in the plasma of rats with thoracic lymph duct cannulas.[25] Chaikoff *et al.* recovered greater than 94% of absorbed labeled cholesterol in the thoracic duct lymph of rats.[26] Similar results have been reported in rabbits[27] and in a human subject with chyluria[28] confirming that in mammals absorbed cholesterol are transported by the intestinal lymphatics and not by the portal system.

### 2.2. *Intestinal lymphatic system*

The lymphatic system is an elaborate network of specialized vessels distributed throughout the vascular regions of the body. The primary and well recognized function of the lymphatics is to drain the capillary beds and return extracellular fluid to the systemic circulation, thus maintaining the body's water balance. However, the structure and function of the lymphatics throughout the body are not uniform and in specific areas the lymphatics perform a specialized role.[29–31] For example, the intestinal lymphatic system is responsible for the transport of dietary fat[32] and lipid soluble vitamins to the systemic circulation.[29–31]

## 3. Lymphatic Vessels Functioning as Transport Routes for Malignant Cells

When blood circulates throughout the vasculature, fluid and components of the blood (i.e. proteins) leak out. A network of lymphatic vessels collects the extravasated fluid (which does not contain red blood cells) from the tissues and transfers it, as lymph, *via* the collecting lymphatic vessels and thoracic duct back into the venous circulation. The lymphatics serve as a route of transport for a variety of cells including white blood cells, antigen-presenting, and lymphoid cells.[33] However, these lymph vessels may also serve as an escape route for malignant cancerous cells from their resident tumor. Metastatic spread of tumor cells is though to involve a number of complicated and interrelated events, from the detachment of the cancerous cells from the original tumor site, microinvasion into stromal tissues, intravasation into both lymphatic and blood vessels, and the extravasation growth in secondary sites. The exact mechanism of metastatic spread within the lymphatic system (and possibly further dissemination from there) has not yet been fully elucidated, however, regional lymph nodes are involved in many advanced cancers in a manner corresponding to the lymphatic drainage patterns of those tissues. As Alitalo and Carmelie clearly point out in their excellent review of lymphangiogensis in cancer,[33] *"Metastatic spread of tumors to distant organs-not the local growth of the primary tumor-is what usually kills the patient."*

# 4. Biological and Pharmaceutical Factors Affecting Lymphatic Drug Delivery

## 4.1. *Contribution of lymphatic transport to the increased absorption of water-insoluble drugs into the systemic circulation*

The majority of orally administered drugs gain access to the systemic circulation by direct absorption into the portal blood (Fig. 3).[4] However, for some water-insoluble compounds, transport by way of the intestinal lymphatic system may provide an additional route of access to the systemic circulation.[6] Exogenous compounds absorbed through the intestinal lymph appear to be generally transported in association with the lipid core of intestinal lipoproteins (predominantly triglyceride-rich chylomicrons) thereby requiring co-administered lipid to stimulate lipoprotein formation. Delivery into the bloodstream by way of the intestinal lymphatics has been suspected to contribute to the overall absorption of a number of highly lipophilic compounds[34–40] including cyclosporine,[11,12] naftifine,[41] probucol,[42] mepitiostane,[43–47] halofantrine,[48–50] testosterone undecanoate,[34] and polychlorinated biphenyls.[51]

Lymph from the intestinal lymphatic system (as well as hepatic and lumbar lymph) drains through the thoracic lymph duct into the left internal jugular vein and then to the systemic circulation.[20,29,30] Thus, the transport of drug by way of the intestinal lymphatic system may increase the percentage of drug that can gain access to the systemic circulation. In addition, the process of intestinal lymphatic drug transport often continues over

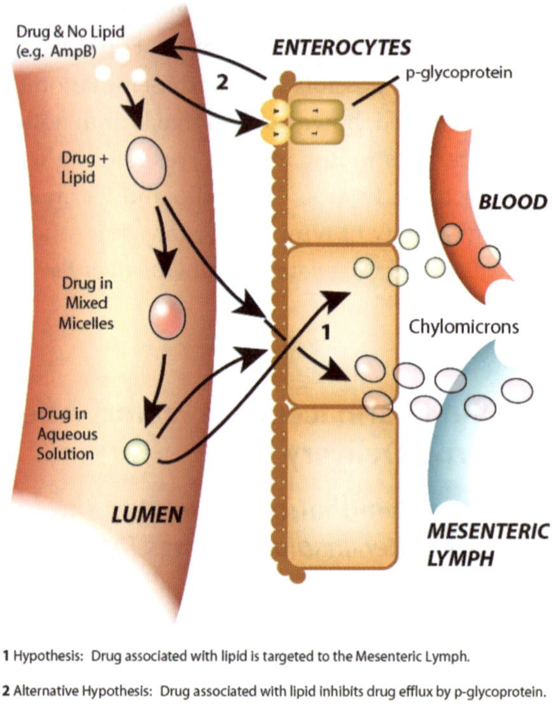


**Fig. 3** Transport of drug in lipid from the lumen into the lymph *via* the enterocyte.

time periods longer than typically observed for drug absorption through the portal vein into the systemic circulation. Consequently, drug transport through the lymph may be utilized to prolong the time course of drug delivery to the systemic circulation. For cancer treatment, a longer exposure of the tumor to the drug may produce an additional benefit, even if at the expense of a lower systemic concentration. Preliminary findings published by Hauss *et al.*[52] suggest that the incorporation of a water-insoluble agent, ontazolast (a potent inhibitor of leukotriene $B_4$), into lipid-based formulations composed of a mixture of mono-, di-, and triglycerides increased the amount of drug that reached the systemic circulation (Table 1), and was transported through the lymph (Table 2). Charman and colleagues in working with another hydrophobic compound, halofantrine, have made similar observations about the effects of triglycerides in drug absorption.[48,53]

**Table 1** Selected plasma noncompartmental pharmacokinetics parameters after oral gavage administration of ontazolast (100 mg/kg) to rats in various formulations.

| Formulation into blood | **$AUC_{(0\text{–}8\,hr)}$ (ng hr/ml) | $T_{max}$ (hr) | $C_{max}$ (ng/ml) | Absorption (% of original dose) |
|---|---|---|---|---|
| Suspension ($n = 3$) | 65 ± 15 | 3.5 ± 1.3 | 16 ± 2.3 | 0.5 |
| Peceol ($n = 6$) | 528 ± 68* | 4.6 ± 0.6 | 137 ± 34* | 5.3 |
| SEDDS 50/50 ($n = 6$) | 752 ± 236* | 2.0 ± 0.24 | 164 ± 35* | 7.0 |
| SEDDS 20/80 ($n = 6$) | 877 ± 104* | 1.8 ± 0.5 | 345 ± 83* | 7.8 |

*$p < 0.05$ vs. suspension formulation using normal scores ranks. **$AUC_{(0\text{–}8\,hr)}$ for 2 mg/kg intravenous dose is 240 ± 40 ng hr/ml. Peceol is lymphotropic solubilizing agent comprised of a mixture of mon-, and diglycerides of oleic acid, which closely resembles the end-products of intestinal lipid disgestion,[31] SEDDS is a self-emulsifying drug delivery system, 50/50 represents equal proportions of Peceol and Gelucire 44/14 and 20/80 represents two part of Peceol for every eight parts of Gelucire 44/14. Gelucire is a liquid excipient, which contains 19% mono-, di-, and triglycerides and 81% of PEG 1500 mono- and diesters and free PEG 1500.[30] Data reported as mean ± standard error of the mean.
*Source:* Ref. [52].

**Table 2** Effect of formulation on the lymphatic transport of ontazolast and triglyceride in rats following a 100 mg/kg oral gavage dose of ontazolast.

| Formulation (ml) | Amount (μg) of ontazolast transported lymphatically by time interval | | | Total lymph output | |
|---|---|---|---|---|---|
| | 0–5 hr | 5–8 hr | 8–24 hr | 0–8 hr | 8–24 hr |
| Suspension ($n = 3$) | 4.8 ± 2 | 1.1 ± 0.5 | 1.7 ± 0.1 | 24.5 ± 5.6 | 67.1 ± 27.8 |
| Peceol ($n = 6$) | 217 ± 45* | 45 ± 4* | 12 ± 4* | 34.3 ± 7.2 | 75.6 ± 15.2 |
| SEDDS 50/50 ($n = 6$) | 192 ± 8* | 16 ± 8* | 6 ± 4 | 23.9 ± 4.0 | 52.5 ± 8.7 |
| SEDDS 20/80 ($n = 6$) | 157 ± 29* | 13 ± 4* | 11 ± 4* | 17.3 ± 4.1 | 65.7 ± 15.2 |

*$p < 0.05$ vs. suspension formulation using normal scores ranks. Peceol and SEDDS are described above in Table 1. Data reported as mean ± standard error of the mean.
*Source:* Ref. [52].

## 4.2. *Evaluation and assessment of intestinal lymphatic transport*

A number of animal models have been described for the assessment of intestinal lymphatic drug transport.[4,5,14,52,54,55] Lymphatic transport studies are commonly first conducted in the laboratory rat, with larger more complicated models (i.e. dog or pig) subsequently investigated.[4,5,14,55] However, the utility of lymph duct fistulation in large animals is limited by considerable logistical and economic constraints. Ideally, sampling strategies for lymphatic transport studies should provide the capacity to estimate both the extent of lymphatic transport, as well as the extent of portal blood absorption in order to estimate the overall bioavailability of the drug/formulation. This strategy enables the unambiguous determination of the extent of lymphatic transport, relative to absorption via the portal blood, and the total bioavailability of the drug/formulation. As lymphatic transport can be affected by experimental factors such as the site of lymphatic cannulation or the period of fasting prior to dosing,[4–6,13,14,55–57] it is important to standardize procedures when comparing studies.

The triple-cannulated anaesthetized rat model (where the mesenteric lymph duct, jugular vein, and duodenum are accessed) has been used for the assessment of lymphatic transport.[52,54] General anesthesia precludes oral dosing in the anaesthetized model and consequently drug and lipid formulations are administered intraduodenally.[4,5,14,55] This limitation thus circumvents the inherent emulsifying action of the stomach and the potential effects of lipids on gastric emptying. Thus, the conscious rat model best represents the *in vivo* situation in terms of both lack of anaesthetic effects and the ability to administer drug formulations orally. Previously reported methods for collecting lymph from the rat required total restraint of the animal and fluid replacement, by intravenous or intraduodenal infusion, to maintain lymph output.[16,58,59] Hauss *et al.*[52,54] have developed a rat model to allow collection of mesenteric lymph for 5 days from conscious, minimally restrained animals with a fully patent cannulae and no signs of physical distress. This model obviates the need for total restraint or general anesthesia, both of which are known to influence intestinal lymphatic transport of test compounds in unpredictable ways.[54] Animals are provided free access to an electrolyte solution, which they consume in sufficient quantity to maintain adequate lymph output without the need for the previously required infusions for fluid replacement. The rat is the appropriate experimental animal to investigate oral absorption and lymphatic transport because thee intestinal characteristics (i.e. anatomical, metabolic, and biochemical characteristics) of these animals are similar to those found in humans.[60–62] Specifically, the intestinal processing and absorption of dietary lipids are similar in rats and humans, although gastric emptying is quite different.[63]

## 4.3. *Proposed mechanisms that govern the lymphatic transport of water-insoluble drugs*

Although the mechanisms by which drugs gain access into the intestinal lymphatic system through the enterocyte are not fully elucidated, there is growing evidence that supports our hypothesis that the majority of drugs transported by the lymphatics are associated with the triglyceride core of chylomicrons.[21,57,64–67] In addition Charman and Stella proposed two important factors, which appear to be prerequisites for the lymphatic transport of water-insoluble drugs, which are the drug's diffusion/partition behavior and lipid solubility.[68]

### 4.3.1. *Diffusion and partition behavior of water-insoluble drugs*

The extent of a drug's partitioning between the portal blood and intestinal lymph may be estimated from a comparison of the relative rates of drug mass transfer by each route. In this regard, the rate of fluid flow in the intestinal lymphatic system is approximately 500-fold less than that in the portal blood,[30,69,70] and during peak lipid transport, the lipid content of the lymph is only of the order of 1–2% (w/v). Thus, the effective mass ratio between lymph lipid and the portal blood is of the order of 1:50,000. Consequently, the selective lymphatic transport of small molecular weight, water-soluble drugs is unlikely if the route of absorption (portal blood *vs* lymph) is governed by the relative rates of fluid flow. However, this ratio suggests that for similar extents of absorption and transport by the portal blood and intestinal lymph (not taking into account metabolic conversion, chemical stability, and/or bioavailability considerations) a candidate molecule should have a log octanol/water partition coefficient (log $p$) in the region of five (highly water-insoluble). Hauss *et al.*[52] reported that when Ontazolast, which has an octanol/water log $p = 4.0$ was incorporated into lipid-based formulations composed of a mixture of mono-, di-, and triglycerides, a significantly greater amount of drug was transported by the lymph than suspension control. Caliph *et al.* studied the effects of short-, medium-, and long-chain fatty acid-based vehicles on the absolute oral bioavailability and intestinal lymphatic transport of halofantrine.[53] They reported that increases in lymphatic drug transport appeared to correlate with increases in lymphatic lipid transport.[53] These initial studies provide evidence that lymphatic transport contributes to the overall oral absorption of water-insoluble compounds incorporated into lipid-based formulations. However, a more comprehensive investigation of these initial findings needs to be done, such as whether or not the drugs are infact incorporated into chylomicrons, how they are processed within enterocytes, and the mechanism by which the drug is transported into the lymph.

### 4.3.2. *Lipid solubility of water-insoluble drugs*

In addition to a high-octanol/water partition coefficient being a prerequisite for lymphatic transport, lipid solubility is a further important parameter to consider. Charman and Stella reported the relationship between lipid solubility and lymphatic transport of two highly water-insoluble compounds (DDT, and hexachlorobenzene, HCB) which have similar octanol/water partition coefficients yet different solubilities.[68,71] Both compounds would be regarded as highly lipophilic as evidenced by their high-octanol/water partition coefficients (DDT, 6.2 vs. HCB 6.5), however, the 13-fold higher triglyceride solubility of DDT compared with HCB (DDT, $9.75 \pm 0.15$ vs. HCB $0.75 \pm 0.05$ solubility in peanut oil (g solute/100 ml) is reflected in the 14.6-fold increase in the extent of intestinal lymph transport reported in an anesthetized rat model.[68,71] Hauss *et al.* previously observed that when ontazolast concentration in the lymph was correlated to chylomicron triglyceride in the lymph, self-emulsifying drug delivery systems (SEDDs formulations, consisting of mainly mixed triglycerides with surfactants as a minor component), which promote more rapid absorption of ontazolast, also favored lymphatic drug transport (Table 3).[52] These findings suggest that solubility in chylomicron triglycerides may be a determining factor for promoting lymphatic transport. However, the way in which these drugs may be associated with chylomicron triglycerides at the molecular level is presently unknown. Taken together, these studies provide preliminary evidence that triglyceride solubility may play a major role in promoting the lymphatic transport and increased oral absorption of lipophilic compounds.

**Table 3** Effect of formulation on the lymphatic transport of triglyceride (TG) and ontazolast in rats following a 100 mg/kg oral gavage dose of ontazolast.

| Formulation | Cumulative amount of TG (mg) transported in mesenteric lymph (exogenous and endogenous lipid) | | Total TG transported from exogenous lipid (% of dose, 0–8 hr) | Total ontazolast transport (μg, 0–8 hr) |
|---|---|---|---|---|
| | 0–5 hr | 5–8 hr | | |
| Suspension ($n = 4$) | 21.9 ± 4.9 | 52.8 ± 17.9 | ND | 6.0 ± 2.0 |
| Peceol ($n = 6$) | 128.9 ± 26.9* | 210.7 ± 39.9* | 51.8 ± 10.8 | 260 ± 40* |
| SEDDS 50/50 ($n = 6$) | 101.0 ± 6.7* | 158.1 ± 10.8* | 38.3 ± 1.4 | 220 ± 10* |
| SEDDS 20/80 ($n = 6$) | 67.2 ± 7.0* | 131.0 ± 11.3* | 31.1 ± 2.5 | 170 ± 30* |

*$p < 0.05$ vs. suspension formulation using normal scores ranks. Peceol and SEDDS are described above in Table 1; ND, nondetectable. Data reported as mean ± standard error of the mean.
*Source:* Ref. [52].

## 5. Utilization of Lymphatic System in the Delivery of Drugs, Including Proteins and Peptides to Metastatic Tumors

The most direct means of targeting anticancer drugs to the lymphatic system for the treatment of metastatic disease in lymph nodes is by infusing the drugs directly into the affected region rather than by intravenous injection. For example, hepatic artery infusion in the case of hepatic metastases of colorectal cancer[72] or retroperitoneal infusion of 5-FU for gynecological cancer metastatic to regional lymph nodes.[73] Subcutaneous or locoregional administration of micron-size particulates also targets selective uptake into the lymphatics. An example is the use of styrene maleic acid neocarzinostatin (SMANC) in Lipodiol, an oily lymphatic contrast agent used for lymphangiography,[74,75] which may also reach metastatic lymph nodes by the arterial route.[74] The further advantage of this method is that the extent of uptake into the lymphatic system can be evaluated by x-ray due to the presence of the contrast agent. Considerable work has been done using various types of activated carbon particles on which the anticancer drug (e.g. mitomycin C) has been adsorbed and the particles are injected subcutaneously near the tumor site or placed in the area of tumor resection during surgery.[76,77] Lymph nodes that have taken up the particles are visibly stained black and so can easily be identified.[78] A comparison of drug levels in axillary lymph nodes of breast cancer patients to whom carboplatin was adminstered either intravenously in saline or subcutaneously adsorbed onto carbon particles was performed by Chen *et al.*[79] Platinum concentrations were measured over time in 154 lymph nodes from 30 patients receiving the carboplatin-carbon particles and 121 lymph nodes from 30 patients receiving intravenous carboplatin in solution. Approximately half of the lymph nodes from both groups contained histopathologically identified metastases (in 23 patients) following radical mastectomy with lymph node resection at various timepoints after drug administration. At all timepoints (1–36 hr), platinum levels were significantly higher in lymph nodes where patients received the carboplatin-carbon particles rather than intravenous carboplatin solution. Interestingly, the presence of metastases in the lymph nodes did not significantly influence drug levels in the lymph nodes in this study.

Hagiwara *et al.* also targeted aclarubicin to the peri-tumoral region and regional lymphatics by local injection of the drug adsorbed on activated carbon particles in patients with breast cancer.[80] They found that aclarubicin concentrations were significantly higher in the peritumoral regions and regional lymph nodes and were significantly lower in the plasma in patients when administered as drug adsorbed onto activated carbon particles compared to patients administered drug in aqueous solution.[80] DiResta *et al.* enhanced doxorubincin and cyclophosphamide uptake into solid tumors using an "artificial lymphatic system" in neuroblastoma and sarcoma dual-tumor models.[81] Videira *et al.* demonstrated that solid lipid nanoparticles may be effective in delivering a greater concentration of drug to the lymphatics upon pulmonary administration.[82]

Regional administration of liposomes, drug-containing lipid bilayer vesicles in the micron or submicron range, has also been used for "lymphatic chemotherapy" with doxorubicin for gastric carcinoma injected into the gastric submucosa[83] and for various chemotherapy drugs by the subcutaneous route.[84]

The oral route of administration provides another route of lymphatic access, and one that is certainly less invasive and more the focus of this chapter. Deak and Csaky originally designed a set of experiments to study the factors that regulate the absorption of nutrients and drugs into the intestinal lymphatic system in normal and cirrhotic rats.[85] In these studies test compounds were administered by *in situ* jejunal luminal perfusion or systemic intravenous infusion and concentrations within intestinal lymph, venous plasma, and intestinal perfusate were determined. They found that the absorption of compounds from the interstitial space into the intestinal lymphatic system is determined by the compound permeability through the capillaries of the portal circulation and by their lipid solubility. Compounds too large to be absorbed into the portal circulation are primarily absorbed into the intestinal lymphatics.

The study of protein absorption *via* the lymphatics goes back nearly 30 years to the early work of Gallo-Torres[35] and recent work of Ikeda *et al.*[86] In these early studies the investigators found that up to 30% of the original doses of d,1-3,4-3H2-a-tocopheryl nicotinate,[35] alpha-, gamma-, and delta-tocotrienols and alpha-tocopherol[86] are absorbed from the gastrointestinal tract through the lymphatic system. In recent years, a number of protein-like and peptide-like compounds have been developed for oral administration. However, oral administration of these compounds is often limited by their instability in the gastrointestinal environment and poor absorption from the gut.[87]

To promote the absorption of protein and peptide like drugs, a number of groups have reported the use of unsaturated fatty acids with absorption-enhancing activities and less harmful properties to the GI membranes in hydrolysates of natural oil.[87] These unsaturated fatty acids affect the permeability of these compounds in enterocytes by inducing disorder in the cell membrane's interior due to the interaction of these fatty acids with the polar head group of phospholipids.[87]

A second approach is the development of a drug delivery system that specifically targets the intestinal lymphatics.[87] Once such delivery system was developed using the anticancer agent, bleomycin. Bleomycin, by the combined effects of an ion-pair complex with dextran sulfate and an absorption enhancer (i.e. unsaturated fatty acids), was found in high concentrations within the lymph following oral administration. Its mechanism may be due to a molecular sieving effect in the blood-lymph barrier in the intestinal tissues.[87]

A third approach in improving the intestinal absorption of protein/peptide-like compounds is by synthesizing novel lipophilic derivatives of peptides by a chemical modification with fatty acids, while maintaining their pharmacological activity.[87] This has been attempted with peptides such as thyrotropin-releasing hormone, tetragastrin, enkephalin, calcitonin, and insulin. In each of these cases the stability and permeability of these peptides were improved by acylation with fatty acids having an appropriate carbon chain length. This resulted in greater absorption of these compounds from the GI tract via the lymphatic system.

Mimicking the absorption process of dietary fats, lipid conjugates, and lipid microspheres have also been used to target the lymphatic route. Yanagawa *et al.*[88] demonstrated when the polypeptide, cyclosporine (CSA, an immunosuppresive agent widely used in organ transplantation), was incorporated into lipid microspheres consisting of olive and soybean oil and egg lecithin and administered orally to rats, CSA concentrations were 46 times greater in the lymph ducts 2 hr postadministration compared to conventional CSA in cremophore therapy. Sato *et al.*[89] further reported that when a CSA derivative, dihydrocyclosporine D, was emulsified with a lipophilic emulsifier, namely milk fat globule membrane, this preparation significantly enhanced the blood and lymphatic fluid concentrations of the CSA derivative after intraduodenal dosing in rats.

Taken together these studies report a number of strategies for improving the delivery and GI absorption of several drugs, including protein and peptide-like drugs, via the intestinal lymphatic system. Special attention may be needed regarding potential adverse effects on lymph nodes and lymphatic vessels themselves, which to date has not been well studied. With further investigation and research the potential exists where the use of these approaches will help improve the therapeutic effectiveness of these types of drugs.

## 6. Clinical Significance

By improving the bioavailability of poorly soluble drugs and directing them to the lymphatics, tumor cells in the lymph nodes will be maximally exposed and should result in an improvement in the control of lymphatic metastases. Improved absorption of these anticancer compounds into the lymphatics will also provide a sustained systemic exposure of nonlymphatic cancerous disease sites to the drug as well as the drug leaves the lymphatics to enter the bloodstream.

## References


1. National Cancer Institute (2004) Cancer Statistics, http://www.nci.nih.gov/statistics/
2. Nathanson S (2003) Insights into the mechanisms of lymph node metastasis. *Cancer* **98**(2):413–423.
3. Risovic V, Sachs-Barrable K, Boyd M and Wasan KM (2004) Potential mechanisms by which Peceol increases the gastrointestinal absorption of amphotericin B. *Drug Dev Ind Pharm* **30**(7):767–774.
4. Pouton CW and Charman WN (1997) The potential of oily formulations for drug delivery to the gastrointestinal tract. *Adv Drug Deliv Rev* **25**:1–2.
5. Porter CJH and Charman WN (1997) Uptake of drugs into the intestinal lymphatics after oral administration. *Adv Drug Delivery Rev* **25**:71–89.
6. Hofmann AF and Borgstrom B (1986) Phyisco-chemical state of lipids in intestinal content during their digestion and absorption. *Feb Proc* **21**:43–50.
7. Serajuddin ATM, Sheen PC, Mufson D, Berstein DF and Augustine MA (1988) Effect of vehicle amphiphilicity on the dissolution and bioavailability of a poorly water-soluble drug from solid dispersions. *J Pharm Sci* **77**:414–417.
8. Sheen PC, Kim EI, Petillo JJ and Serajuddin ATM (1991) Bioavailability of a poorly water-soluble drug from tablet and solid dispersion in humans. *J Pharm Sci* **80**:712–714.
9. Shiau Y-F (1981) Mechanism of intestinal fat absorption. *Am J Physiol* **240**:G1–G9.
10. Borgstrom B, Dahlqvist A, Lundh G and Sjovall J (1957) Studies of intestinal digestion and absorption in the human. *J Clin Invest* **36**:1121–1136.
11. Ueda CT, Lemaire M, Gsell G and Nussbaumer K (1983) Intestinal lymphatic absorption of cyclosporin a following oral administration in an olive oil solution to rats. *Biopharm Drug Dispos* **4**:113–124.
12. Reymond JP, Sucker H and Vonderscher J (1988) *In vivo* model for ciclosporin intestinal absorption in lipid vehicles. *Pharm Res* **5**:677–679.
13. Thomson ABR, Keelan M, Garg ML and Clandinin MT (1989) Intestinal aspects of lipid absorption: In review. *Can J Physiol Pharmacol* **67**:179–191.
14. Charman WN and Porter CJH (1996) Lipophilic prodrugs designed for intestinal lymphatic transport. *Adv Drug Delivery Rev* **19**:149–169.
15. Tucker G (1993) Drug delivery: Lymphing along? *Lancet* **341**:1314–1315.
16. Carey MC, Small DM and Bliss CM (1983) Lipid digestion and absorption. *Ann Rev Physiol* **45**:651–677.
17. Thomson AB, Schoeller C, Keelan M, Smith L and Clandinin MT (1993) Lipid absorption: Passing through the unstirred layers, brush border membrane, and beyond. *Can J Physiol Pharmacol* **71**:531–555.

18. Tso P and Simmonds WJ (1984) The absorption of lipid and lipoprotein synthesis. *Lipid Research Methodol*, 191–216.
19. Tso P and Balint JA (1986) Formation and transport of chylomicrons by enterocytes to the lymphatics. *Am J Physiol* **13**:G715–G726.
20. Tso P (1994) Intestinal lipid absorption in *Physiology of the Gastrointestinal Tract* Johnson LR (ed.). Raven Press: New York, 1867–1908.
21. Davis RA and Vance JE (1996) Structure, assembly and secretion of lipoproteins in *Biochemistry of Lipids, Lipoproteins and Membranes* Vance DE and Vance JE (eds.). Elsevier: New York, pp. 473–493.
22. Repa JJ, Turley SD, Lobaccaro JMA, Medina J, Li L, Lustig K, Shan B, Heyman RA, Dietschy M and Mangelsdorf DJ (2000) Regulation of absorption and ABC1-mediated efflux of cholesterol by RXR heterodimers. *Science* **289**:1524–1529.
23. Mueller JH (1916) The mechanism of cholesterol absorption. *J Biol Chem* **27**: 463–480.
24. Wilson MD and Rudel LL (1994) Review of cholesterol absorption with emphasis on dietary and biliary cholesterol. *J Lipid Res* **35**:943–955.
25. Biggs MW, Friedman M and Byers SO (1951) Intestinal lymphatic transport of absorbed cholesterol. *Proc Soc Exp Biol Med* **78**:641–643.
26. Chaikoff IL, Bloom B, Siperstein MD, Kiyasu JY, Reinhardt WO, Dauben WG and Eastham JF (1952) $C^{14}$-Cholesterol. I. Lymphatic transport of absorbed cholesterol-4-$C^{14}$. *J Biol Chem* **194**:407–412.
27. Rudel LL, Morris MD and Felts JM (1972) The transport of exogenous cholesterol in the rabbit. I. Role of cholesterol ester of lymph chylomicra and lymph very low density lipoproteins in absorption. *J Clin Invest* **51**:2686–2692.
28. Hellman L, Frazell EL and Rosenfeld RS (1960) Direct measurement of cholesterol absorption via the thoracic duct in man. *J Clin Invest* **39**:1288–1294.
29. Davidson NO (1994) Cellular and molecular mechanisms of small intestinal lipid transport in *Physiology of the GastroIntestinal Tract*. Johnson LR (ed.) Raven Press: New York, pp. 1909–1934.
30. Johnson JM (1976) Triglyceride biosynthesis in the intestinal mucosa in *Lipid Absorption: Biochemical and Clinical Aspects*. Rommel K, Goebell H and Bohmer R (eds.) Lancaster: MTP Press, pp. 85–94.
31. Zilversmit DB (1965) The composition and structure of lymph chylomicrons in dog, rat and man. *J Clin Invest* **44**:1610–1622.
32. Olszewski WL (2003) The lymphatic system in body homeostasis: Physiological conditions. *Lymphat Res Biol* **1**(1):11–24.
33. Alitalo K and Carmelie P (2002) Molecular mechanisms of lymphangiogenesis in health and disease. *Cancer Cell* **1**:219–227.

34. Horst HJ, Holte WJ, Dennis M, Coert A, Geelen J and Voigt KD (1976) Lymphatic absorption and metabolism of orally administered testosterone undecanoate in man. *Klin Wschr* **54**:875–879.
35. Gallo-Torres HE (1970) Intestinal absorption and lymphatic transport of d-1,3,4-$^3H_2$-$\alpha$-tocopheryl nicotinate in the rat. *Int J Vitam Nutr Res* **40**: 505–514.
36. Sugihara J, Furuuchi S, Ando H, Takashima K and Harigaya S (1988) Studies on the intestinal lymphatic absorption of drugs. II. Glyceride prodrugs for improving the lymphatic absorption of naproxen and nicotinic acid. *J Pharmacobio Dyn* **11**:555–562.
37. Sugihara J and Furuuchi S (1988) Lymphatic absorption of hypolipidemic compound 1-*O*-[p-(myristyloxy)-$\alpha$-methylcinnamoyl]glycerol (LK-903). *J Pharmacobio Dyn* **11**:121–130.
38. Adachi I, Liu H, Horkoshi I and Ueno M (1993) Possiblity of lymphatic absorption of epidermal growth factor from intestine. *Yakugaku Zasshi* **113**: 256–263.
39. Hauss DJ, Mehta S and Radebaugh GW (1994) Targeted lymphatic transport and modified systemic distribution of CI-976, a lipophilic lipid-regulator drug, via a formulation approach. *Int J Pharm* **108**:85–93.
40. Barnwell SG, Laudanski T, Story MJ, Mallinson CB, Harris RJ, Cole SK, Keating M and Attwood D (1992) Improved oral bioavailability of propranolol in healthy human volunteers using a liver bypass drug delivery system containing oleic acid. *Int J Pharm* **88**:423–432.
41. Grimus RC and Schuster I (1984) The role of the lymphatic transport in the enteral absorption of naftifine by the rat. *Xenobiotica* **14**:287–294.
42. Palin KJ and Wilson CJ (1984) The effect of different oils on the absorption of probucol in the rat. *J Pharm Pharmacol* **36**:641–643.
43. Ichihashi T, Kinoshita H, Shimamura K and Yamada H (1991) Absorption and disposition of epithiosteroids in rats (1): Route of administration and plasma levels of epitiostanol. *Xenobiotica* **21**:865–872.
44. Ichihashi T, Kinoshita H and Yamada H (1991) Absorption and disposition of epithiosteroids in rats (2): Avoidance of first-pass metabolism of mepitiostane by lymphatic absorption. *Xenobiotica* **21**:873–880.
45. Ichihashi T, Kinoshita H, Takagishi Y and Yamada H (1992) Effect of oily vehicles on absorption of mepitiostane by the lymphatic system in rats. *J Pharm Pharmacol* **44**:560–564.
46. Ichihashi T, Kinoshita H, Takagishi Y and Yamada H (1992) Effect of bile on absorption of mepitiostane by the lymphatic system in rats. *J Pharm Pharmacol* **44**:566–569.

47. Ichihashi T, Kinoshita H, Takagishi Y and Yamada H (1991) Intrinsic lymphatic partition rate of mepitiostane, epitiostanol and oleic acid absorbed from rat intestine. *Pharm Res* **8**:1302–1306.
48. Porter CJH, Charman SA and Charman WN (1996) Lymphatic transport of halofantrine in the triple-cannulated anaesthetized rat model: Effect of lipid vehicle dispersion. *J Pharm Sci* **85**:351–356.
49. Humberstone AJ, Porter CJH and Charman WN (1996) A physicochemical basis for the effect of food on the absolute bioavailability of halofantrine. *J Pharm Sci* **85**:525–529.
50. Wasan KM, Ramaswamy M, MacIntosh M, Porter CJH and Charman WN (1999) Differences in the lipoprotein distribution of halofantrine are regulated by lipoprotein apolar lipid and protein concentration and lipid transfer protein I activity: *in vitro* studies in normolipidemic and dyslipidemic human plasmas. *J Pharm Sci* **88**:185–190.
51. Busbee DL, Yoo J-SH, Norman JO and Joe CO (1985) Polychlorinated biphenyl uptake and transport by lymph and plasma components. *Proc Soc Exp Biol Med* **179**:116–122.
52. Hauss DJ, Fogal SE, Ficorilli JV, Price CA, Roy T, Jayaraj AA and Keirns JJ (1998) Lipid-based delivery systems for improving the bioavailability and lymphatic transport of a poorly water-soluble $LTB_4$ inhibitor. *J Pharm Sci* **87**: 165–169.
53. Caliph SM, Charman WN and Porter CJ (2000) Effect of short-, medium-, and long-chain fatty acid-based vehicles on the absolute oral bioavailability and intestinal lymphatic transport of halofantrine and assessment of mass balance in lymph-cannulated and non-cannulated rats. *J Pharm Sci* **89**:1073–1084.
54. Hauss DJ, Fogal SE and Ficorilli JV (1998) Chronic collection of mesenteric lymph from conscious, tethered rats. *Contemp Top Lab Anim Sci* **37**:56–58.
55. White DG, Story MJ and Barnwell SG (1991) An experimental animal model for studying the effects of a novel lymphatic drug delivery system for propranolol. *Int J Pharm* **69**:169–174.
56. Levet-Trafit B, Gruyer MS, Marjanovic M and Chou RC (1996) Estimation of oral drug absorption in man based on intestine permeability in rats. *Life Sci* **58**:PL359–363.
57. Liu H-X, Adachi I, Horkoshi I and Ueno M (1995) Mechanism of promotion of lymphatic drug absorption by milk fat globule membrane. *Int J Pharm* **118**:55–64.
58. Chang CM and Bodmeier R (1997) Binding of drugs to monoglyceride-based drug delivery systems. *Int J Pharm* **147**:135–142.
59. Shah NH, Carvajal MT, Patel CI, Infeld MH and Malick AW (1994) Self-emulsifying drug delivery systems (SEDDS) with polyglycolyzed glycerides

for improving *in vitro* dissolution and oral absorption of lipophilic drugs. *Int J Pharm* **106**:15–23.
60. Fagerholm U, Johansson M and Lennernas H (1996) Comparison between permeability coefficients in rat and human jejunum. *Pharm Res* **13**:1336–1342.
61. Kararli TT (1995) Comparison of the gastrointestinal anatomy, physiology, and biochemistry of humans and commonly used laboratory animals. *Biopharm Drug Dispos* **16**:351–380.
62. Soria I and Zimmerman CL (1996) The validation of the intestinal permeability approach to predict oral fraction of dose absorbed in humans and rats. *Biopharm Drug Dispos* **17**:817–818.
63. Pahl MV, Oveisi F, Khamiseh G and Vaziri ND (1998) Intestinal absorption and biliary secretion of cholesterol in rats with nephrotic syndrome. *Nephrol Dial Transplant* **13**:1446–1451.
64. McDonald GB, Saunders DR, Weidman M and Fisher L (1980) Portal venous transport of long-chain fatty acids absorbed from rat intestine. *Am J Physiol* **239**:G141–G150.
65. Bugaut M and Carlier H (1987) Role of intestinal hydrolases, endogenous substrates and chyloportal partition in fat absorption in *Fat Absorption*, Vol. 1 Kuksis A (ed.). CRC Press: Boca Raton, pp. 197–232.
66. Simmonds WJ (1972) The role of micellar solubilisation in lipid absorption. *Aust J Exp Biol Med Sci* **50**:403–421.
67. Muranishi S (1991) Drug targeting towards the lymphatics. *Adv Drug Deliv Res* **21**:1–38.
68. Charman WN and Stella VJ (1986) Estimating the maximal potential for intestinal lymphatic transport of lipophilic drug molecules. *Int J Pharm* **34**:175–178.
69. Bollman JL, Cava JC and Grindley JH (1948) Techniques for the collection of lymph from the liver, small intestine and thoracic duct of the rat. *J Lab Clin Med* **33**:1349–1352.
70. Reininger EJ and Saperstein LA (1957) Effect of digestion on the distribution of blood flow in the rat. *Science* **126**:1176.
71. Sieber SM (1976) The lymphatic absorption of p,p-DDT and some structurally-related compounds in the rat. *Pharmacology* **14**:443–454.
72. Patt YZ and Mavligit GM (1991) Arterial chemotherapy in the management of colorectal cancer: An overview. *Semin Oncol* **18**(5):478–490.
73. Zeyi C, Zongbing Y, Shiying Z, Zhilan P, Jinghua W and Peiyi L (1996) Chemotherapy for lymphatic metastatic gynecological cancer via pelvic retroperitoneal cannulation: A preliminary report. *Gynecol Oncol* **63**(3):358–363.
74. Ohtsuka N, Konno T, Miyauchi Y and Maeda H (1987) Anticancer effects of arterial adminstration of the anticancer agent SMANCS with lipiodol on mesenteric lymph nodes. *Cancer* **59**(9):1560–1565.

75. Kuruppu D, Christophi C, Maeda H and O'Brien PE (2002) Changes in the microvascular architecture of colorectal liver metastases following the administration of SMANCS/lipiodol. *J Surg Res* **103**(1):47–54.
76. Kin S, Nakase Y, Fukuda K, Ito T, Shimomura K, Takagi T, Toma A, Fujiyama J, Okamoto K, Sakakura C, Otsuiji E, Kitamura K, Itoi H, Hagiwara A and Yamagishi H (2002) Endoscopic local injection of anticancer dugs bound to carbon particles for treatment of advanced rectal cancer, when surgical treatment was contraindicated. *Gan Kagaku Ryoho* **29**:2413–2415.
77. Takahashi T, Hagiwara A, Shomotsuma M, Sawai K and Yamaguchi T (1995) Prophylaxis and treatment of perionteal carcinomatosis: Intraperitoneal chemotherapy with mitomycin C bound to activated carbon particles. *World J Surg* **19**:565–569.
78. Yokoto T, Saito T, Narushima Y, Iwamoto K, Iizuka M, Hagiwara A, Sawai K, Kikuchi S, Kunii Y and Yamauchi H (2000) Lymph node staining with activated carbon CH40: A new method for axillary lymph node dissection in breast cancer. *Can J Surg* **43**:191–196.
79. Chen J, Wang L, Yao Q, Ling R, Li K and Wang H (2004) Drug concentrations in axillary lymph nodes after lymphatic chemotherapy on patients with breast cancer. *Breast Cancer Res* **6**:R474–R477.
80. Hagiwara A, Takahashi T, Sawai K, Sakakura C, Shirasu M, Ohgaki M, Imanishi T, Yamasaki J, Takemoto Y and Kageyama N (1997) Selective drug delivery to peri-tumoral region and regional lymphatics by local injection of aclarubicin adsorbed on activated carbon particles in patients with breast cancer-a pilot study. *Anticancer Drugs* **8**(7):666–670.
81. DiResta GR, Lee J, Healey JH, Larson SM and Arbit E (2000) Enhancing the uptake of chemotherapeutic drugs into tumors using an artificial lymphatic system. *Ann Biomed Eng* **28**(5):556–564.
82. Videira MA, Botelho MF, Santos AC, Gouveia LF, de Lima JJ and Almeida AJ (2002) Lymphatic uptake of pulmonary delivered radiolabelled solid lipid nanoparticles. *J Drug Target* **10**(8):607–613.
83. Akomo Y, Mizuno I, Yotsuyanagi T, Ichino T, Tanimoto N, Yamamoto T, Hagata M, Takeyama H, Shinagawa N and Yura J (1994) Chmotherapy targeting regional lymph nodes by gastric submucosal injection of liposomal adriamycin in patient with gastric carcinoma. *Jpn J Cancer Res* **85**:652–658.
84. Oussoren C and Storm G (2001) Liposomes to target the lymphatics by subcutaneous administration. *Adv Drug Deliv Rev* **50**:143–156.
85. Deak ST and Csaky TZ (1984) Factors regulating the exchange of nutrients and drugs between lymph and blood in the small intestine. *Microcirc Endothelium Lymphatics* **1**(5):569–588.

86. Ikeda I, Imasato Y, Sasaki E and Sugano M (1996) Lymphatic transport of alpha-, gamma- and delta-tocotrienols and alpha-tocopherol in rats. *Int J Vitam Nutr Res* **66**:217–221.
87. Muranishi S (1997) Delivery system design for improvement of intestinal absorption of peptide drugs. *Yakugaku Zasshi* **117**:394–414.
88. Yanagawa A, Iwayama T, Saotome T, Shoji Y, Takano K, Oka H, Nakagawa T and Mizushima Y (1989) Selective transfer of cyclosporin to thoracic lymphatic systems by the application of lipid microspheres. *J Microencapsul* **6**:161–164.
89. Sato H, Liu HX, Adachi I, Ueno M, Lemaire M and Horikoshi I (1994) Enhancement of the intestinal absorption of a cyclosporine derivative by milk fat globule membrane. *Biol Pharm Bull* **17**:1526–1528.

# 14

# Delivery of Protein and Peptide Drugs to Brain Tumors

*Herbert B. Newton*

## Abstract

Primary and metastatic brain tumors remain a significant health problem and affect between 130 000 and 185 000 people each year in the USA. Conventional treatment with surgical resection, irradiation, and cytotoxic chemotherapy continues to be inadequate, with a median survival of approximately 1 year for most patients. Advances in molecular pharmacology and the biology of the blood–brain barrier (BBB) are improving the ability of researchers to target therapeutic peptides, proteins, and small molecules to the brain and brain tumors. Simple methods to improve BBB penetration and brain delivery include chemical modification, glycosylation, and pegylation. Drug delivery vehicles such as nanoparticles and liposomes are also under study. These vehicles can be modified by pegylation or conjugation with a targeting vector to improve delivery across the BBB. Targeting vectors include natural ligands (e.g. epidermal growth factor) or monoclonal antibodies to receptors (e.g. transferrin, insulin, interleukin). Other vector-mediated delivery approaches involve the conjugation of a therapeutic peptide or protein with a targeting molecule that can induce transcytosis across BBB endothelial cells. The most commonly used vectors are peptidomimetic antibodies to endothelial receptors, such as the transferrin and insulin receptors.

# 1. Introduction

Brain tumors remain a significant health problem in the USA and worldwide. Overall, they comprise some of the most malignant tumors known to affect humans and are generally refractory to all modalities of treatment. It is estimated that between 30 000 and 35 000 new cases of primary brain tumors (PBT) will be diagnosed in the upcoming year in the USA (1–2% of newly diagnosed cancers overall).[1–3] The majority of these tumors will be malignant gliomas of neuroectodermal origin (see Tables 1 and 2).[4] Metastatic brain tumors (MBT) are even more common and affect between 100 000 and 150 000 new patients each year in this country.[5] Most studies suggest that approximately 14 per 100 000 people in the USA will be diagnosed with a PBT each year. Among this cohort with newly diagnosed tumors, 6–8 per 100 000 will have a high-grade neoplasm. Recent

**Table 1** WHO classification: Tumors of the central nervous system.

| |
|---|
| Tumors of neuroepithelial tissue |
| Tumors of cranial nerves and spinal nerves |
| Tumors of the meninges |
| Lymphomas and hemopoietic neoplasms |
| Germ cell tumors |
| Tumors of the sellar region |
| Cysts and tumor-like lesions |
| Metastatic tumors |

Data derived from Ref. [4].

**Table 2** WHO classification: Tumors of neuroepithelial tissue.

| |
|---|
| Astrocytic tumors |
| Oligodendroglial tumors |
| Ependymal tumors |
| Mixed gliomas |
| Choroid plexus tumors |
| Neuronal and mixed neuronal-glial tumors |
| Pineal parenchymal tumors |
| Neuroepithelial tumors of uncertain origin |
| Embryonal tumors |

Data derived from Ref. [4].

epidemiological studies suggest that incidence rates of PBT may be increasing for children and patients 70 years or older.[6] The exact cause of this trend remains unclear.

The prognosis and survival of patients with brain tumors remains poor.[1–3,5] Although an uncommon neoplasm, PBT are among the top 10 causes of cancer-related deaths in the USA and account for 2.4% of all yearly cancer-related deaths.[7] The median survival for a patient with glioblastoma multiforme (GBM) is approximately 12–14 months, and hasn't improved substantially over the past 30 years. For patients with a low-grade astrocytoma or oligodendroglioma, the median survival is still significantly curtailed and is about 6–10 years. It is obvious from these survival data that the current treatment of brain tumor patients (i.e. surgical resection, irradiation, chemotherapy) is inadequate and unable to overcome the malignant biology of these neoplasms. Although chemotherapy has been demonstrated to provide a survival benefit to high-grade glioma patients in several recent meta-analyses, the results have been modest, at best.[8,9] Explanations for the poor results of chemotherapy include the nonspecific, nontargeted nature of most of the currently used drugs (e.g. alkylating agents, DNA topoisomerase inhibitors, spindle poisons), inadequate delivery of many drugs across the blood–brain and blood–tumor barriers (BBB, BTB), and the high degree of intrinsic chemotherapy resistance prevalent among high-grade brain tumors. It is obvious, therefore, that new treatment modalities must be developed that have a more molecular, "targeted" mechanism of action, with the ability to overcome the malignant phenotype of brain tumors and further improve survival.[10–14] Included in this new arsenal of therapeutics will be protein- and peptide-based drugs with variable cellular targets of intervention. The rest of this chapter will discuss some of the inherent difficulties in the delivery of protein and peptide-based drugs to brain tumors and then outline the many different ways these drugs are being developed to exploit new information regarding signal transduction pathways and the malignant phenotype.

## 2. Biology of the BBB, BBB-Transport Systems, and BTB

The BBB was initially thought to reside at the level of the astrocytic foot processes that ensheath the capillary endothelial cells within the brain. Recent pharmacological and genomic studies have shown that the endothelial cells of the central nervous system (CNS) are more specialized than vascular cells of systemic organs, and that these specialized features form the anatomical and physiological basis for the functional properties of the BBB.[15–17] The BBB segregates the circulating blood from the cells and interstitial fluid of the CNS, functioning as a protective filtering mechanism. Anatomically, the basis of the BBB resides in the microvascular wall of CNS endothelial cells, which is composed of two plasma membranes in series (i.e. luminal and abluminal membranes, separated by approximately 3 $\mu$m of endothelial cytosol), forming a continuous endothelium with the presence of epithelial-like, high-resistance tight junctions that eliminate pores typically noted in the walls of systemic capillaries. The endothelial tight junctions are comprised of six to eight junctional strands, which form a "zipper-like" structure that adjoins adjacent CNS capillary endothelial cells. Electrical resistance is very high across BBB tight junctions, on the order of 2000 $\Omega$ cm$^2$, which is similar to other resistant epithelial barriers, such as urinary bladder epithelium.[15] The presence of the tight junctions removes a para-cellular route for free diffusion of substances across the capillary. CNS capillaries also have few endocytic vesicles, which removes a transcellular route for free diffusion of substances into the interstitium. Astrocytic foot processes invest approximately 99% of CNS capillaries, but do not contribute to the physiological activity of the BBB. However, the presence of the astrocytic processes is still important, since they secrete trophic factors necessary for the differentiation, health, and function of the BBB endothelial cells.[15] Vascular pericytes may also play a role in the differentiation and maintenance of BBB endothelial cells, through the release of various soluble factors.

The BBB has several specialized mechanisms for transport of substances across endothelial cells and into the brain interstitium.[15,17–19] These include carrier systems for monosaccharides (D-glucose), monocarboxylic acids (L-lactate), neutral amino acids (L-system [phenylalanine, leucine] and A-system [glycine, GABA]), basic amino acids (arginine), acidic amino acids (aspartate, glutamate), amines (choline), purine bases (adenine, guanine), nucleosides (adenosine, guanosine), vitamins (thiamine), and hormones (triiodothyronine). In addition, there are also transport systems designed for the efflux of substances from endothelial cells, including the p-glycoprotein and active organic acid efflux systems.[19,20] P-glycoprotein is present in high concentrations in the luminal membrane of brain endothelial cells and is responsible for the efflux of cytotoxic drugs such as vincristine and vinblastine. Although these drugs easily penetrate the luminal membrane, they are pumped out by the transporter and not allowed to concentrate within the brain. This mechanism explains why these drugs have a lower brain penetration than would be predicted from their lipid partition coefficients. P-glycoprotein appears to have a broad specificity, without a clear quantitative structure-activity relationship for targeted substances. The organic acid transporter is present in the choroid plexus and may also be active in brain endothelial cells. In the choroid plexus, it is known to transport compounds such as penicillin and methotrexate out of the cerebrospinal fluid (CSF). The other mechanisms for transport of compounds across the BBB are receptor-mediated transcytosis (e.g. transferrin, insulin, and insulin-like growth factors) and adsorptive endocytosis.[19] Receptor-mediated transcytosis is a slow process (i.e. minutes to hours) by which endothelial cells endocytose macromolecules that are bound to a receptor on the luminal surface, followed by diffusion of the compound across the cell and exocytosis on the abluminal side. Adsorptive endocytosis is a slow, nonselective process in which endothelial cells endocytose macromolecules nonspecifically bound to the luminal surface, with subsequent discharge on the abluminal side. The macromolecules are usually positively charged proteins (e.g. cationized albumin or histone) that accumulate on the membrane surface by coulombic interactions with negatively charged moieties on the endothelial membrane.

Within brain tumors, the integrity of the BTB is variable and may be relatively intact or quite permeable.[21,22] Recent investigations in animal models would suggest that there may be three distinct microvascular populations in brain tumors. The first group consists of continuous, nonfenestrated capillaries similar to those of normal brain. These are the type of vessels present in low-grade gliomas and typically do not enhance after the administration of contrast agents during neuro-imaging studies. The second microvessel population consists of continuous, fenestrated capillaries, which have increased permeability to small but not to large molecules. The third capillary population contains interendothelial gaps that may be quite large (up to 1 $\mu$m in rat RG-2 glioma and human D-54 glioma), allowing the passage of relatively large compounds in comparison to normal brain.

In addition to the anatomic barriers mentioned above, the BBB and BTB also have a metabolic barrier in the form of proteolytic enzymes within the endothelial cells.[15–18] Several proteolytic enzymes have been identified, including aminopeptidase A, aminopeptidase M, glutamyl aminopeptidase, neutral endopeptidase 24.11, endopeptidase 24.15, and $\gamma$–glutamylpeptidase. The expression of these enzymes in brain vasculature appears to be elevated in comparison to systemic vessels.

## 3. Overview of Drug Delivery and Distribution

For a chemotherapy agent to be effective, it must be exposed to brain tumor cells at concentrations that can overcome intrinsic resistance mechanisms.[23] This is often difficult for many drugs, due to the privileged location of central nervous system tumors, which are protected by the BBB and BTB. For example, it is estimated for most drugs, regardless of lipid solubility status, that less than 1% of the circulating compound will be able to reach the vasculature of a CNS tumor after systemic administration.[22] In general, drugs with the best penetration into the nervous system and brain tumors tend to be nonionized, small molecules that are very lipid soluble.[24] Lipid solubility (i.e. lipophilicity; equivalent to lipid/aqueous partition coefficient) is an essential factor for allowing drugs to have a high brain capillary permeability ($P_C$) and be able to transfer across the intact BBB or to regions of relatively intact BTB.[25–27] The lipophilicity of a given drug or compound is inversely correlated with the degree of hydrogen bond formation that occurs with surrounding water molecules.[28] The presence of certain chemical moieties within the drug potentiates the likelihood of hydrogen bond formation, such as terminal amides (four bonds), primary amines or amides (three bonds), and hydroxyl groups (two bonds). Once the cumulative number of hydrogen bonds exceeds eight to ten, the ability of any compound to transport through the BBB or BTB is significantly impaired. This is true even for small peptides (e.g. di- or tripeptides), due to the high hydrogen bond-forming potential of the amide groups. For small drugs, increasing lipophilicity has a positive impact on $P_C$ and drug transfer into the brain. However, for large drugs above 400 Da (e.g. many proteins and peptides), or for those with strong polarity, the $P_C$ will remain low regardless of the degree of lipophilicity. The two main approaches for improving the lipid solubility of drugs are blocking functional groups with hydrogen bond-forming potential (e.g. hydroxyl groups) and increasing the number of methylene groups in a given compound (e.g. adding fatty acyl side chains).[28]

Other factors of importance in the final distribution pattern of a given drug are the regional blood flow, tumor capillary permeability, tumor vascular surface area, drug stability, and intracerebral drug metabolism.[25–27] Once the drug has transferred across the brain or tumor capillaries, further distribution will be determined by convection (i.e. via bulk flow) and diffusion (i.e. down concentration gradients). Convection is defined as the movement of solute in the direction of solvent flow and is related to the concentration of solute, the hydraulic conductivity of the solvent, and the regional hydrostatic pressure gradient.[29] For most tumors, there is a convective flux from the interstitium to surrounding normal tissues that is 10–1000 times greater than in normal tissues. Diffusion is the movement of a substance along a concentration gradient and is related to the diffusion coefficient of the fluid medium and the concentration gradient of the substance.[29] *In vitro* experiments suggest that tumor blood vessels have higher interstitial diffusion coefficients than normal vessels, but the differential can be quite variable. In general, larger molecules (i.e. molecular weight greater than 800 Da) move more efficiently with convection, while smaller molecules move more easily by diffusion. The interstitial fluid pressure (IFP) of a given tumor is related to intra-tumoral convection currents and contributes to the ease or difficulty of ingress of drugs from the interstitial spaces.[30] Larger tumors tend to have elevated IFP, which generate a pressure gradient and an outward convection of interstitial fluid towards the periphery of the mass. This outward flow of interstitial fluid works against the gradient of drug diffusion and impedes the delivery of drugs to neoplastic cells. Taking all of these different variables into consideration, it is apparent that the most efficient way to deliver protein and peptide drugs to brain tumors will be via convective flow approaches or through the use of transport vectors that allow passage through the BBB and BTB.

## 4. Chemical Modification, Glycosylation, Pegylation, and Other Delivery Approaches

Numerous methods have been attempted to overcome the limitations of the BBB and improve the delivery of proteins, peptides, and other compounds to the brain.[22,24,31–40] Initial research suggested that peptides and protein drugs could be delivered to the brain *via* the transnasal and intraventricular (or subarachnoid) routes.[24,33–35] More recent studies of transnasal drug administration have shown that this route does not bypass the BBB and that only small, highly lipid soluble compounds will be able to enter the CSF surrounding the olfactory bulb. In addition, once drugs have reached the CSF *via* the transnasal route, they have poor penetration into the brain parenchyma (including brain tumors). Similarly, peptide or protein drugs administered by intraventricular or subarachnoid methods (ommaya reservoir or lumbar puncture) will also have poor distribution within the CSF and only affect brain parenchyma within 1–2 mm of the CSF/brain interface. This distribution pattern arises from the rapid rates of CSF and interstitial fluid bulk flow out of the brain relative to the slow rates of drug or peptide diffusion into brain parenchyma from the surface.

As mentioned above, one of the most simple and direct methods to improve BBB penetration is to chemically modify compounds to reduce hydrogen bond potential and increase lipophilicity.[22,24,31–40] The addition of hydrophobic moieties such as methyl groups are capable of improving lipophilicity and brain penetration, and have been applied to drugs such as barbiturates and small peptides.[41] For example, methylation studies of DPDPE, a cyclized peptide with analgesic properties, have shown that dimethylation of the N-terminal tyrosine can lead to increased analgesia, while trimethylation of the phenylalanine group increased transport across the BBB, without any alteration of analgesic activity.[38] These studies illustrate the potential for variability between different isomers after chemical modification to augment lipophilicity. The site and type of modification are vital for both the improvement of delivery into the brain and for the maintenance of intrinsic biological activity. In addition, when considering the modification of larger peptides and proteins, it is critical to understand how intramolecular hydrogen bonding affects the tertiary molecular structure of the compound. The inadvertent alteration of internal hydrogen bonds could potentially affect the tertiary structure of a given peptide or protein and reduce its ability to interact with therapeutic targets.

Another chemical method to improve delivery of peptide drugs to the brain and brain tumors is glycosylation.[38,42] Glycosylation involves the addition of a carbohydrate moiety to a protein (glycoprotein) or peptide (glycopeptide). Studies of small glycopeptides (e.g. analgesics) have demonstrated improvements in the stability, hydrophobicity, and bioavailability of the compounds. For example, glycosylation of opioid peptides enhances BBB penetration and improves analgesic activity.[43] Other studies of glycopeptide BBB permeability demonstrate up to a threefold increase in the rate of brain delivery in comparison to the unglycosylated parent compounds. The type of glycosylation and the placement of the moiety are both important factors and can dramatically affect the pharmacological properties of the new compound. The properties of mono-, di-, and triglycosylated compounds are typically different and can vary dramatically in their tissue distribution patterns, BBB permeability, and peptide/receptor interactions.[38] In addition, the placement of the carbohydrate moiety may reduce or alter drug activity if placed within or close to the active site (e.g. phenol rings of tyrosine or phenylalanine).

Pegylation involves the conjugation of polyethylene glycol (PEG) moieties to surface amino acids of proteins, peptides, nanoparticles, and other compounds (e.g. liposomes; see below), in an attempt to improve the biopharmaceutical properties of these drugs (for a more detailed description of pegylation technology and applications, see Chaps. 4 and 5).[44–46] Most compounds are pegylated on the $\varepsilon$-amino groups of lysine residues, which usually allows for the biological activity to be maintained. However, in some cases, the active site can be affected by steric hindrance from attached PEG moieties, with diminution of biological activity. In these cases, an alternative to amino-directed pegylation, which may improve retention of biological activity, is carboxyl-directed pegylation. The majority of protein, peptide, and similar drugs have a very short circulating half-life due to rapid degradation by proteolytic enzymes, clearance by the kidneys and other organs, and activity of neutralizing antibodies. Pegylation is able to modify the structural characteristics of a given compound to increase its solubility, overall size and hydrodynamic volume, mask some of the protein and glycosylated surfaces, alter the charge pattern, and shield key epitopes. Enlarging the compound slows down kidney ultrafiltration, thereby allowing better accumulation into solid tumors and other permeable tissues by the passive enhanced permeation and retention mechanism (for more details, see Chap. 3). Masking of charged regions and glycosylation sites reduces the phagocytosis of compounds by the reticuloendothelial system (RES) and liver cells. Protein shielding also reduces proteolysis within the serum and tissues, and hinders immune surveillance of surface epitopes.

Numerous drugs are under development using pegylation technology, which have implications for oncology, neurological therapeutics, and brain tumor treatment.[44–46] An example of peptides being developed for treatment of neurological disorders are the neurotropins, including nerve growth factor (NGF), brain-derived neurotrophic factor (BDNF), and neurotropin-3.[24,25,33,34,47] Similar to other drugs, neurotropins are rapidly cleared from the plasma compartment by the liver and kidneys, reducing the plasma AUC and subsequent brain uptake. Carboxyl-directed pegylation of BDNF results in improved pharmacokinetics; most importantly, a reduction of hepatic and plasma clearance by 67 ($PEG^{2000}$)–91% ($PEG^{5000}$).[47] Biological activity of BDNF was partially reduced with the 5000 Da PEG moiety, but equivalent to native BDNF after pegylation with the smaller 2000 Da moiety. In the field of oncology, pegylation has been applied to supportive medications and therapeutics.[48] Pegfilgastrim is a pegylated version of the granulocyte colony-stimulating factor, filgastrim, with enhanced bioavailability, a less frequent dosing schedule, and full biological activity. Mono-pegylation can also be used as a linker molecule between peptide growth factors (e.g. epidermal growth factor, EGF) and BBB drug delivery vectors (see Vector-Mediated Section, below).[49] The BBB delivery vector allows passage of EGF across the CNS endothelial cells, while the extended PEG linker minimizes steric hindrance of EGF biological activity.

## 5. Nanoparticle Drug Delivery Approaches

Nanoparticles are solid colloidal matrix-like particles made of polymers, which have simple procedures for preparation, high physical stability, and the ability to provide sustained release of peptides, proteins, genes, and other drugs to the brain and systemic organs (also see Chap. 7).[50–52] Nanoparticles used for drug delivery range in size from 10 to 1000 nm, with most in the range of 50–300 nm. Because of their size, most nanoparticles are not able to diffuse through the BBB and enter the brain parenchyma without some form of structural modification. The most commonly used polymers for production of nanoparticles are polybutylcyanoacrylate (PBCA), polylactide homopolymers (PLA), and poly(lactide-co-glycolide) heteropolymers (PLGA). PBCA nanoparticles coated with the nonionic surfactant polysorbate 80 have shown the capacity to deliver drugs to the brain (e.g. doxorubicin, loperamide).[50,52] The mechanism of transport into the brain remains unclear, but may be related to endocytosis across BBB endothelial cells or via nanoparticle-induced nonspecific BBB permeabilization. When PLA and PLGA nanoparticles are administered intravenously, they are rapidly cleared by the liver and RES, with a plasma half-life of 2–3 min. Opsonization occurs quickly within the plasma, with targeting from complement proteins, apolipoproteins, fibronectins, and immunoglobulins. Pegylation of PLA and PLGA improves the pharmacokinetic profile of these compounds by reducing opsonization, phagocytosis, and clearance by the liver and RES.[52,53] Conventional cytotoxic drugs (e.g. doxorubicin, irinotecan, paclitaxel) and peptides have been loaded without difficulty into pegylated PLA and PLGA nanoparticles. The drug entrapment efficiency depends on the lipophilicity of a given drug, since the nanoparticles have a hydrophobic core. Proteins are more complicated to entrap, since numerous steps in the process (i.e. storage, temperature, shear forces, exposure to surfactants) are capable of damaging the tertiary structure of the protein and integrity of the active site. Because of the delicacy of the process, each nanoparticle formulation of protein is unique and requires specific parameters. Once PLA and PLGA nanoparticles have been delivered to their site of activity, they are biodegradable via autocatalytic cleavage of the ester bonds through spontaneous hydrolysis into oligomers and D,L-lactic acid and glycolic acid monomers.

Strategies for brain targeting of nanoparticles include the coating of PBCA with polysorbate 80 and the covalent conjugation of protein ligands to pegylated PLA and PLGA.[50–52] The linkages occur at the free end of 1–2% of the PEG strands that surround the nanoparticle, using biotinylation, amine-reactive, or thiol-reactive chemical methods. The protein ligands usually consist of monoclonal antibodies capable of targeting receptors that are highly expressed in CNS tissues and transported across BBB endothelial cells, such as the transferrin receptor (see Vector-Mediated Delivery Section below).[54]

Application of nanoparticle technology to brain tumors has reached the stage of preclinical testing *in vitro* and in animal models. Several authors have reported the use of polysorbate 80-coated PBCA in animals.[55,56] Gelperina *et al.* initially treated normal and GBM containing rats with intravenous doxorubicin-loaded PBCA nanoparticles (100–400 mg/kg).[55] The drug was well tolerated in the healthy and diseased animals across all drug dosages, and appeared to be less toxic than free doxorubicin. In addition, the doxorubicin was shown to have excellent penetration through the BBB and into brain parenchyma. When tested in a rat model against other forms of chemotherapy, animals treated with doxorubicin-loaded nanoparticles had significantly longer survival times, with over 20% demonstrating a long-term remission.[56] Paclitaxel is a drug with known activity against primary and metastatic brain tumors. However, its efficacy is usually limited by poor penetration through the BBB and tumor cell efflux by p-glycoprotein. In a study of cetyl alcohol/polysorbate nanoparticles loaded with paclitaxel, Koziara *et al.* found the drug to improve the brain uptake of paclitaxel.[57] In addition, paclitaxel-loaded nanoparticles were toxic to p-glycoprotein-expressing tumor cell lines (U-118 and HCT-15). In a more recent study of doxorubicin-loaded PBCA nanoparticles, the drug was tested in rats with 9L-gliosarcoma in comparison to free doxorubicin.[58] Although the maximum tolerated dose of nanoparticulate doxorubicin was 1.5 times higher than that of free doxorubicin, the survival times were equivalent between the two groups.

## 6. Liposomal Drug Delivery Approaches

Liposomes are vesicles in which an aqueous inner volume is entirely enclosed by a phospholipid membrane bilayer that ranges in size from 50 to 100 nm in diameter.[59–61] Similar to nanoparticles, liposomes are used as vehicles for the delivery of therapeutic agents, including small molecules, peptides, proteins, cancer chemotherapeutics, and nucleic acid-based drugs (also see Chap. 8). This technology allows for the entrapment of up to 10 000 small molecules in a single 100 nm liposome. The pharmacokinetics of conventional liposomes after an intravenous injection is very poor, and is characterized by rapid systemic plasma clearance. After injection, the liposomes are opsonized by plasma proteins (e.g. complement, immunoglobulins) and targeted for phagocytosis and removal by the liver, spleen, and other reticuloendothelial organs. The plasma half-life of liposomes can be extended by the incorporation of gangliosides (e.g. $GM_1$) or via pegylation. Pegylated liposomes, also known as "sterically stabilized" or "stealth" liposomes, have been studied in detail and have a much improved pharmacokinetic profile.[62] The PEG coating prevents binding of opsonins and significantly reduces recognition by phagocytic cells of the liver and RES. Pegylation results in a 50-fold decrease in the plasma volume of distribution, a 200-fold decrease in the systemic plasma clearance, and a nearly 100-fold increase in the area under the concentration-time curve. The extended circulation half-life of

PEG-liposomes, in combination with the increased permeability of tumor vasculature, results in passive accumulation of liposomes and associated drugs in systemic solid tumors. However, this effect does not occur to any substantial degree with brain tumors, since even small liposomes (i.e. 50–80 nm) are too large to cross the BBB and BTB.

The most common method to improve delivery of sterically stabilized liposomes to the brain and brain tumors is via conjugation of an appropriate targeting vector, such as modified immunoglobulins and proteins that are able to undergo absorptive-mediated or receptor-mediated transcytosis through the endothelial cells of the BBB (see Vector-Mediated Delivery section below).[33–36,59–61] Several brain-targeting vectors are available and include cationized albumin, the OX26 monoclonal antibody to the rat transferrin receptor, and monoclonal antibodies to the insulin receptor. The vectors can be attached to the liposome by several methods.[61] Direct methods involve covalent attachment to surface phospholipids through amino-reactive or disulfide bond linkages. However, this approach is often compromised by steric hindrance of the PEG chains, so that the attached vector cannot interact appropriately with its receptor. Indirect methods involve the attachment of vectors to the terminus of the PEG chains. With this technique, the PEG moiety is used as a spacer that minimizes steric hindrance and improves target binding efficiency by a factor of two to three. For example, conjugation of the OX26 monoclonal antibody to sterically stabilized immunoliposomes improves the degree of brain delivery, such that 0.03% of an injected dose per gram tissue can be observed in brain parenchyma at 60 min after an intravenous injection.[63] Control experiments using unspecified $IgG_{2a}$ isotype antibodies eliminates the brain targeting effect. Other modifications of immunoliposomes include the attachment of two or more different peptidomimetic antibodies or targeting fragments to increase liposomal delivery to both extracellular epitopes as well as intracellular compartments. For example, the OX26 antibody could be used to target transfer across the BBB vasculature via transferrin receptors, while another antibody to the human insulin receptor could be used to target transfer across the plasma membrane and nuclear membrane of brain tumor cells.[64]

Liposomes meet most, if not all, of the criteria suggested for an ideal drug delivery system for cancer chemotherapeutic drugs, including localized delivery to tumors and decreased toxicity to normal tissues. Thus far, liposomal technology has progressed the farthest with preparations of anthracycline drugs, especially doxorubicin.[59,65–67] Pegylated liposomal doxorubicin has demonstrated activity in numerous animal model systems, including tumors of the breast, lung, ovaries, prostate, colon, bladder, and pancreas, as well as lymphoma, sarcoma, and myeloma. In the clinical setting, liposomal doxorubicin has been approved in the USA for treatment of Kaposi's sarcoma and advanced ovarian carcinoma, as well as for metastatic breast cancer in Canada and Europe. Clinical studies suggest a reduction in gastrointestinal toxicity, cardiotoxicity, and alopecia for most patients in comparison to conventional doxorubicin. Preclinical and clinical studies of liposomal doxorubicin have also been applied to brain tumors. Several authors have reported its use in rat glioma models.[68,69] Intravenous liposomal doxorubicin resulted in significantly higher levels of drug in the serum and CSF in comparison to conventional drug. In addition, the survival of animals treated with the liposomal preparation was significantly longer than that of animals treated with conventional doxorubicin. The distribution of liposomal doxorubicin within brain tumors and adjacent infiltrated normal brain can be improved over conventional intravenous dosing by the use of convection-enhanced delivery (CED; see CED section below).[70]

Clinical applications were initially investigated by Koukourakis *et al.* in a small series of patients with primary and metastatic brain tumors.[71] Five patients with GBM and ten patients with MBT were treated with $^{99m}$Tc-DTPA radiolabelled liposomal doxorubicin (25 mg/m$^2$) concurrently during external beam irradiation. Accumulation of radiolabelled liposomal doxorubicin was 13–19 times higher in GBM and 7–13 times higher in MBT, as compared to surrounding normal brain. Responses were reported for the MBT cohort, including four complete responses (CR; two breast, two lung) and four partial responses (PR; one breast, three lung). In a review of 13 evaluable patients with recurrent malignant glioma (six GBM, six AA, one gliosarcoma), treatment consisted of intravenous liposomal doxorubicin (20 mg/m$^2$) every 2 weeks.[72] There were no CR or PR; seven of the patients had stabilization of disease after the initiation of doxorubicin, with a median time to progression (TTP) of 11 weeks and a median overall survival of 40 weeks. In a series of phase II trials with liposomal doxorubicin, alone or in combination with tamoxifen, Hau *et al.* treated a total of 40 patients with recurrent high-grade gliomas.[73] The activity of liposomal doxorubicin was equivalent between the two groups, with a

12-week progression-free survival of 15% and a median overall survival of 26 weeks after treatment onset. For the combined cohort, there was one patient with a CR, one with a PR, and 14 with stabilized disease. In another combination chemotherapy approach, liposomal doxorubicin (40 mg/mg$^2$, day 1) was added to temozolomide (200 mg/m$^2$, days 1–5) in a phase II trial of 22 patients with recurrent GBM.[74] Objective responses included one patient with a CR and three with PR, while another 11 patients had stabilized disease. The median TTP and overall survival were 3.2 and 8.2 months, respectively. The 6-month progression-free survival rate was 32%, which compares favorably to studies of GBM patients using temozolomide alone (20%).

## 7. Convection-Enhanced Delivery Approaches

Convection-enhanced delivery is a relatively new method of drug delivery to the brain, that may be able to circumvent the BBB and BTB, in which drugs are continuously infused directly into the brain tumor using an external pump system and distributed throughout the brain by bulk flow.[22] An important aspect of CED is that distribution of the therapeutic molecule is determined by the infusion parameters and the type of tissue being infused (i.e. brain tumor and surrounding parenchyma), and less by the size or concentration of the drug in the infusate. Large therapeutic agents, such as monoclonal antibodies, immunotoxins, liposomes, antisense oligonucleotides, and viral vectors, can be infused using CED and attain significant concentrations in tumors and surrounding neural tissues.[22,75–77] Clinical trials are being developed to apply CED techniques to the problem of drug delivery of large molecular therapeutic agents (e.g. liposomes, peptides, immunotoxins, proteins) into brain tumors.

## 8. Immunoglobulin-Based Delivery Approaches

In this approach, immunoglobulins are used to target specific antigens on the surface of brain tumor cells to allow delivery of therapeutics, such as radiopharmaceuticals and immunotoxins. The field of radioimmunotherapy has expanded in the past few decades with improvements in monoclonal antibody technology.[78] Initial attempts with this treatment modality utilized $^{123}$I-labeled antibodies against the EGF receptor, which is often overexpressed on brain tumor cells.[79] In a cohort of ten patients with recurrent AA and GBM, antibody therapy resulted in clinical improvement in six patients that lasted from 6 months to 3 years, including two patients with tumor shrinkage on radiologic studies. Recent work by several investigators has focused on the use of antibodies to tenascin, a tumor-associated extracellular matrix glycoprotein that is ubiquitously expressed in gliomas and other solid tumors.[80–83] Riva *et al.* have reported their experience using the $^{131}$I-labeled antitenascin antibodies BC-2 and BC-4, in a series of 62 patients with newly diagnosed and recurrent malignant gliomas.[80] The radiolabeled antibodies were instilled into the resection cavity using an indwelling Rickham catheter, placed after maximal surgical removal of the tumor. Responses included three patients with CR, nine with PR, and 11 with stabilized disease. The overall median survival of the cohort was 23 months. The median TTP was 12 months for newly diagnosed GBM patients and 5 months for patients with recurrent disease. Investigators at Duke have performed phase I trials using $^{131}$I-labeled antitenascin antibody 81C6 in patients with recurrent and newly diagnosed malignant gliomas.[81,82] The treatment was well tolerated with maximum tolerated doses ranging from 100 to 120 mCi; higher doses resulted in delayed neurological toxicity. Median survival after treatment for patients with recurrent and newly diagnosed GBM was 52 and 69 weeks, respectively. In a phase II trial by the same group, 120 mCi of $^{131}$I-labeled antitenascin antibody 81C6 was administered to 33 newly diagnosed patients with malignant glioma.[83] The median survival for the entire cohort and the GBM sub-group was 86.7 and 79.4 weeks, respectively. These survival data exceeded that of historical controls treated with conventional irradiation and chemotherapy.

The other immunoglobulin-based treatment approach under intense investigation is immunotoxin therapy, in which a targeting ligand or carrier is combined with a bacterial cytotoxin.[84,85] With this method, monoclonal antibodies are developed that target receptors and other tumor-associated antigens on the surface of brain tumor cells that are overexpressed in comparison to normal cells. Numerous target receptors have been evaluated for suitability, including the transferrin receptor, EGF receptor, fibroblast growth factor receptor, interleukin-4 receptor, and interleukin-13 receptor. The monoclonal antibody must be able to bind to the receptor with appropriate specificity and then induce antigen/receptor internalization, thereby allowing access of the cytotoxin to the cytoplasmic organelles. Several bacterial cytotoxins have been evaluated for use as antibrain immunotoxins and include the diptheria toxin (DT), pseudomonas exotoxin (PE), and the ricin A chain. All of the cytotoxins are cellular poisons that disrupt de novo protein synthesis.[87] Following internalization, the cytotoxins are activated after proteolytic cleavage within endosomes, then translocate to the cystosol where they reach the endoplasmic reticulum and either inactivate the ribosomes (ricin A chain) or neutralize the activity of elongation factor 2 through an irreversible ADP-ribosylation reaction (DP and PE). Cytotoxins are quite potent, with very low $IC_{50}$'s (i.e. femtomolar range in some cases) and the ability to kill brain tumor cells during low-oxygen conditions. Preclinical studies of immunotoxins in cell cultures and animal models demonstrate the consistent ability to kill brain tumor cells and prolong survival of test animals.[84–86] Clinical studies have begun to be reported, using a variety of targeting ligands.[84,85,88–93] One of the first studies was described by Laske *et al.*, using a monoclonal antibody to the transferrin receptor linked to a genetically modified DT (Tf-CRM 107).[88] The Tf-CRM 107 was administered by regional high-flow interstitial microinfusion to 18 patients with malignant brain tumors (GBM 10, AA 5). Treatment was well tolerated and resulted in objective responses in nine patients (CR 2, PR 7), with post-infusion survival ranging from 11 to 142 weeks. The median time to tumor progression was 38 weeks. Similar results were noted in a phase II study and a phase III trial is under consideration.[89] In a different variation, Weber *et al.* have developed a chimeric recombinant fusion protein composed of a circularly permuted interleukin-4 and a truncated form of PE.[91] In a phase I study, 31 patients with GBM and AA were treated with escalating doses of IL4-PE ($6\,\mu g/ml \times 40\,ml$ to $9\,\mu g/ml \times 100\,ml$), administered by intratumoral CED. No systemic drug-related toxicity was noted. The overall median survival was 8.2 months, with a median survival of 5.8 months in the GBM cohort. Several patients had MRI changes after treatment (i.e. decreased signal intensity) suggestive of tumor necrosis. In a similar phase I trial, a different chimeric recombinant immunotoxin was tested, composed of transforming growth factor-$\alpha$ (TGF-$\alpha$) and a mutated form of the PE (TP-38).[90] Twenty patients with recurrent malignant brain tumors were treated with TP-38, administered by CED into the resection cavity margins at doses ranging from 25 to 100 ng/ml. The treatment was fairly well tolerated, without reaching a maximum tolerated dose. For the entire cohort and a sub-group with minimal residual disease at the onset of therapy, the median survival times were 23 and 31.9 weeks, respectively. Several groups have been investigating the activity in phase I/II trials of a chimeric recombinant immunotoxin composed of interleukin-13 and a truncated form of PE, in which the binding domain has been deleted (IL13-PE38QQR).[92,93] The IL13-PE38QQR was delivered by CED in both studies, either into the resection cavity (intratumoral) or the peritumoral brain parenchyma. In the study by Prados *et al.*, 74 patients received treatment, with a dose-limiting toxicity of nonspecific necrosis.[93] For the GBM cohort, the overall survival was longer in patients receiving peritumoral infusions as compared to those receiving intratumoral infusions (45.9 vs. 37.1 weeks). Patients also had prolonged survival when the CED infusions were through two or more catheters, with optimal placement. The study by Weingart *et al.* was similar, except that patients did not undergo surgical resection prior to catheter placement and infusion of IL13-PE38QQR.[92] Despite the lack of resection, the infusions were well tolerated in most patients with the use of corticosteroids.

## 9. Vector-Mediated Delivery Approaches

Vector-mediated delivery involves the conjugation of a peptide, protein, or other large neuropharmaceutical to a targeting molecule, which has affinity for receptors specific to the target tissue, which in this case are the CNS vascular endothelial cells that form the BBB and BTB.[24,31,34,35,38] Vectors that have been used for this purpose include cationized albumin, liposomes, nanoparticles, endogenous ligands (e.g. transferrin, EGF), and monoclonal antibodies to receptors (e.g. transferrin, insulin, EGF). The vector is attached to the therapeutic compound by a disulphide bond or, in some cases, a chimeric peptide or fusion protein is generated to combine the molecules. Receptor-specific peptidomimetic monoclonal antibodies are endocytosing antibodies that bind to exofacial epitopes on the receptor that are distinct from the binding site of the endogenous ligand. This type of epitope allows the antibody to cross the BBB without interfering with the transport of the endogenous ligand. In addition, because the transport is mediated solely by the vector portion of the chimeric peptide, the attached drug is allowed to cross the BBB without recognition by endothelial receptors. A more recent method of attachment involves biotinylation of the neurotherapeutic and fusion protein preparation of either avidin or streptavidin with the transport vector.[34,94] Each reaction is prepared in a separate vial and then mixed just before administration, resulting in the instantaneous formation of the drug-vector complex. After attachment of the vector to the specific receptor on the luminal side of the endothelial cell membrane, the vector-drug complex is endocytosed by receptor-mediated or adsorptive transcytosis, diffuses across the endothelial cytosol, and then undergoes exocytosis at the abluminal side of the membrane for entry into the brain. If disulfide bonds are present, they are cleaved by disulfide reductases in the brain, thereby releasing the pharmacologically active drug.

The transferrin receptor has been the most frequent target for vector-mediated transport approaches.[24,31,34,35,38] These receptors are constitutively expressed in BBB vessels in comparison to systemic capillaries and are involved in the transport of iron into the brain. As mentioned above, the OX26 antibody targets transferrin receptors and has been used successfully to increase brain uptake of peptides, proteins, and small molecule chemotherapeutics (e.g. methotrexate) in a receptor-mediated fashion.[54,59–61,95,96] Other studies have shown that the use of OX26 leads to an 18-fold higher accumulation in the brain when compared to nonspecific IgG antibodies. Several peptides have been linked to OX26 and investigated in animal models, including vasoactive intestinal peptide, BDNF, NGF, and EGF. In each case, linkage to OX26 improved brain uptake of the compound. In addition, OX26 has been used to improve brain targeting for immunoliposomes and delivery of therapeutic genes.

In an attempt to develop a vector-mediated system that would allow drug targeting to the brain in humans, investigators have begun to study the human insulin receptor (HIR), which is ubiquitous on the abluminal membrane of BBB vessels.[24,35] The 83-14 murine monoclonal antibody is an insulin peptidomimetic monoclonal antibody that binds to an exofacial epitope on the $\alpha$–subunit of the HIR.[97] The 83-14 antibody binds to the HIR on isolated human brain capillaries with high affinity and undergoes receptor-mediated transcytosis through the BBB *in vivo* in Rhesus monkeys. In comparison to antibodies to the transferrin receptor, the 83-14 antibody has a 9-fold greater ability to penetrate the BBB in the primate brain. Intavenous administration of 111-indium labeled 83-14 antibody to the living primate brain demonstrates robust uptake and suggests that this vector would have a similar ability to target the human brain. Studies are ongoing to evaluate the ability of the 83-14 antibody to deliver therapeutic drugs to the brain and brain tumors.

## Acknowledgments

The author would like to thank Ryan Smith for research assistance. Dr. Newton was supported in part by National Cancer Institute grant, CA 16058 and the Dardinger Neuro-Oncology Center Endowment Fund.

## References

1. Newton HB (1994) Primary brain tumors: Review of etiology, diagnosis, and treatment. *Am Fam Phys* **49**:787–797.
2. Davis FG and McCarthy BJ (2001) Current epidemiological trends and surveillance issues in brain tumors. *Exp Rev Anticancer Ther* **1**:395–401.
3. Wrensch M, Minn Y, Chew T, Bondy M and Berger MS (2002) Epidemiology of primary brain tumors: Current concepts and review of the literature. *Neuro Oncol* **4**:278–299.
4. Kleihues P, Louis DN, Scheithauer BW and *et al.* (2002) The WHO classification of tumors of the nervous system. *J Neuropathol Exp Neurol* **61**:215–225.
5. Newton HB (1999) Neurological complications of systemic cancer. *Am Fam Phys* **59**:878–886.
6. Legler JM, Ries LA, Smith MA and *et al.* (1999) Cancer surveillance series [corrected]: Brain and other central nervous system cancers: Recent trends in incidence and mortality. *J Natl Cancer Inst* **91**:1382–1390.
7. ACS (American Cancer Society): Cancer facts and figures 2002. American Cancer Society: Atlanta. 2002.
8. Fine HA, Dear KBG, Loeffler JS, Black PM, Canellos GP (1993) Meta-analysis of radiation therapy with and without adjuvant chemotherapy for malignant gliomas in adults. *Cancer* **71**:2585–2597.
9. Stewart LA, Burdett S, Parmar MKB and *et al.* (2002) Chemotherapy in adult high-grade glioma: A systematic review and meta-analysis of individual patient data from 12 randomized trials. *Lancet* **359**:1011–1018.
10. Newton HB (2003) Molecular neuro-oncology and the development of "targeted" therapeutic strategies for brain tumors. Part 1 — growth factor and ras signaling pathways. *Exp Rev Anticancer Ther* **3**:595–614.
11. Newton HB (2004) Molecular neuro-oncology and the development of "targeted" therapeutic strategies for brain tumors. Part 2 — PI3K/Akt/PTEN, mTOR, SHH/PTCH, and angiogenesis. *Exp Rev Anticancer Ther* **4**:105–128.

12. Newton HB (2004) Molecular neuro-oncology and the development of "targeted" therapeutic strategies for brain tumors. Part 3 — brain tumor invasiveness. *Exp Rev Anticancer Ther* **4**:803–821.
13. Newton HB (2005) Molecular neuro-oncology and the development of "targeted" therapeutic strategies for brain tumors. Part 4 — p53 signaling pathways. *Exp Rev Anticancer Ther* **5**:177–191.
14. Newton HB (2005) Molecular neuro-oncology and the development of "targeted" therapeutic strategies for brain tumors. Part 5 — apoptosis and cell cycle. *Exp Rev Anticancer Ther* **5**:355–378.
15. Pardridge WM (1999) Blood-brain barrier biology and methodology. *J Neurovirol* **5**:556–569.
16. Pardridge WM (2005) Molecular biology of the blood-brain barrier. *Mol Biotechnol* **30**:57–70.
17. Shusta EV (2005) Blood-brain barrier genomics, proteomics, and new transporter discovery. *NeuroRx* **2**:151–161.
18. Tsuji A and Tamai II (1999) Carrier-mediated or specialized transport of drugs across the blood-brain barrier. *Adv Drug Deliv Rev* **36**:277–290.
19. de Boer AG, van der Sandt ICJ and Gaillard PJ (2003) The role of drug transporters at the blood-brain barrier. *Ann Rev Pharmacol Toxicol* **43**:629–656.
20. Borst P, Evers R, Kool M and Wijnholds J (2000) A family of drug transporters: The multidrug resistance-associated proteins. *J Natl Cancer Inst* **92**:1295–1302.
21. Schlageterr KE, Molnar P, Lapin GD and Groothuis DR (1999) Microvesesel organization and structure in experimental brain tumors: Microvessel populations with distinctive structural and functional properties. *Microvasc Res* **58**:213–328.
22. Groothuis DR (2000) The blood-brain and blood-tumor barriers: A review of strategies for increasing drug delivery. *Neuro Oncol* **2**:45–59.
23. Parney IF and Prados MD (2005) Chemotherapy principles in *Textbook of Neuro-Oncology*, Vol. 11 Berger MS and Prados MD (eds.), Elsevier Saunders: Philadelphia, 75–79.
24. Pardridge WM (1998) CNS drug design based on principles of blood-brain barrier transport. *J Neurochem* **70**:1781–1792.
25. Rapoport SI, Ohno K and Pettigrew KD (1979) Drug entry into the brain. *Brain Res* **172**:354–359.
26. Levin VA (1980) Relationship of octanol/water partition coefficient and molecular weight to rat brain capillary permeability. *J Med Chem* **23**:682–684.
27. Blasberg RG and Groothuis DR (1986) Chemotherapy of brain tumors: Physiological and pharmacokinetic considerations. *Semin Oncol* **13**:70–82.
28. Diamond JM and Wright EM (1969) Molecular forces governing non-electrolyte permeation through cell membranes. *Proc Royal Soc Lond [Biol]* **172**:273–316.

29. Jain RK (1987) Transport of molecules in the interstitium: A review. *Cancer Res* **47**:3039–3051.
30. Stohrer M, Boucher Y, Stangassinger M and Jain RK (2000) Oncotic pressure in solid tumors is elevated. *Cancer Res* **60**:4251–4255.
31. Abbott NJ and Romero IA (1996) Transporting therapeutics across the blood-brain barrier. *Mol Med Today March* 106–113.
32. Kroll RA and Neuwelt EA (1998) Outwitting the blood-brain barrier for therapeutic purposes: Osmotic opening and other means. *Neurosurg* **42**:1083–1100.
33. Pardridge WM (2001) Brain drug targeting and gene technologies. *Jpn J Pharmacol* **87**:97–103.
34. Pardridge WM (2002) Drug and gene targeting to the brain with molecular Trojan horses. *Nat Rev Drug Discov* **1**:131–139.
35. Pardridge WM (2002) Targeting Neurotherapeutic agents through the blood-brain barrier. *Arch Neurol* **59**:35–40.
36. Demeule M, Régina A, Annabi B, Bertrand Y, Bojanowski MW and Béleveau R (2004) Brain endothelial cells as pharmacological targets in brain tumors. *Mol Neurobiol* **30**:157–183.
37. McCarty JH (2005) Cell biology of the neurovascular unit: implications for drug delivery across the blood-brain barrier. *Assay Drugs Dev Technol* **3**:89–95.
38. Egleton RD and Davis TP (2005) Development of neuropeptide drugs that cross the blood-brain barrier. *NeuroRx* **2**:44–53.
39. Tsuji A (2005) Small molecular drug transfer across the blood-brain barrier via carrier-mediated transport systems. *NeuroRx* **2**:54–62.
40. Cucullo L, Aumayr B, Rapp E and Janigro D (2005) Drug delivery and *in vitro* models of the blood-brain barrier. *Curr Opin Drug Discov Devel* **8**:89–99.
41. Chikhale EG, Ng KY, Burton PS and Borchardt RT (1994) Hydrogen bonding potential as a determinant of the *in vitro* and in situ blood-brain permeability of peptides. *Pharm Res* **11**:412–419.
42. Poduslo JF and Curran GL (1994) Glycation increases the permeability of proteins across the blood-nerve and blood-brain barriers. *Brain Res Mol Brain Res* **23**:157–162.
43. Egleton RD, Mitchell SA, Huber JD and *et al.* (2000) Improved bioavailability to the brain of glycosylated Met-enkephalin analogs. *Brain Res* **881**:37–46.
44. Caliceti P and Veronese FM (2003) Pharmacokinetic and biodistribution properties of poly(ethyleneglycol)-protein conjugates. *Adv Drug Deliv Rev* **26**:1261–1277.
45. Harris JM and Chess RB (2003) Effect of pegylation on pharmaceuticals. *Nat Rev Drug Discov* **2**:214–221.
46. Morpurgo M and Veronese FM (2004) Conjugates of peptides and proteins to polyethylene glycols. *Methods Mol Biol* **283**:45–70.

47. Pardridge WM, Wu D and Sakane T (1998) Combined use of carboxyl-directed protein pegylation and vector-mediated blood-brain barrier drug delivery system optimizes brain uptake of brain-derived neurotropic factor following intravenous administration. *Pharm Res* **15**:576–582.
48. Molineux G (2003) Pegylation: Engineering improved biopharmaceuticals for oncology. *Pharmacother* **23**:3S–8S.
49. Deguche Y, Kurihara A and Pardridge WM (1999) Retention of biologic activity of human epidermal growth factor following conjugation to a blood-brain barrier drug delivery vector via an extended poly(ethylene glycol) linker. *Bioconjugate Chem* **10**:32–37.
50. Kreuter J (2001) Nanoparticulate systems for brain delivery of drugs. *Adv Drug Del Rev* **47**:65–81.
51. Lockman PR, Mumper RJ, Khan MA and Allen DD (2002) Nanoparticle technology for drug delivery across the blood-brain barier. *Drug Dev Ind Pharm* **28**:1–13.
52. Olivier JC (2005) Drug transport to brain with targeted nanoparticles. *NeuroRx* **2**:108–119.
53. Gref R, Domb A, Quellec P and *et al*. (1995) The controlled intravenous delivery of drugs using PEG-coated sterically stabilized nanospheres. *Adv Drug Deliv Rev* **16**:215–233.
54. Olivier JC, Huertas R, Lee HJ, Calon F and Pardridge WM (2002) Synthesis of pegylated immunonanoparticles. *Pharm Res* **19**:1137–1143.
55. Gelperina SE, Khalansky AS, Skidan IN and *et al*. (2002) Toxicological studies of doxorubicin bound to polysorbate 80-coated poly(butyl cyanoacrylate) nanoparticles in healthy rats and rats with intracranial glioblastoma. *Toxicol Lett* **126**:131–141.
56. Steininger SC, Kreuter J, Khalanskyi AS and *et al*. (2004) Chemotherapy of glioblastoma in rats using doxorubicin-loaded nanoparticles. *Int J Cancer* **109**:759–767.
57. Koziara JM, Lockman PR, Allen DD and Mumper RJ (2004) Paclitaxel nanoparticles for the potential treatment of brain tumors. *J Control Rel* **99**:259–269.
58. Brigger I, Morizet J, Laudani L and *et al*. (2004) Negative preclinical results with stealth nanospheres-encapsulated doxorubicin in an orthotopic murine brain tumor model. *J Control Rel* **100**:29–40.
59. Di Paolo A (2004) Liposomal anticancer therapy: Pharmacokinetic and clinical aspects. *J Chemother* **16**:90–93.
60. Fenske DB and Cullis PR (2005) Entrapment of small molecules and nucleic acid-based drugs in liposomes. *Methods Enzymol* **391**:7–40.
61. Schnyder A and Huwyler J (2005) Drug transport to brain with targeted liposomes. *NeuroRx* **2**:99–107.

62. Allen TM (1994) Long-circulating (sterically stabilized) liposomes for targeted drug delivery. *Trends Pharmacol Sci* **15**:215–220.
63. Huwyler J, Wu D and Pardridge WM (1996) Brain drug delivery of small molecules using immunoliposomes. *Proc Natl Acad Sci USA* **93**:14164–14169.
64. Zhang Y, Zhu C and Pardridge WM (2002) Antisense gene therapy of brain cancer with an artificial virus gene delivery system. *Mol Ther* **6**:67–72.
65. Abraham SA, Waterhouse DN, Mayer LD, Cullis PR, Madden TD and Bally MB (2005) The liposomal formulation of doxorubicin. *Meth Enzymol* **391**:71–97.
66. Allen TM and Martin FJ (2004) Advantages of liposomal delivery systems for anthracyclines. *Sem Oncol* **31**:5–15.
67. Vail DM, Amantea MA, Colbern GT, Martin FJ, Hilger RA and Working PK (2004) Pegylated liposomal doxorubicin: proof of principle using preclinical animal models and pharmacokinetic studies. *Sem Oncol* **31**:16–35.
68. Siegal T, Horowitz A and Gabizon A (1995) Doxorubicin encapsulated in sterically stabilized liposomes for the treatment of a brain tumor model: Biodistribution and therapeutic efficacy. *J Neurosurg* **83**:1029–1037.
69. Sharma US, Sharma A, Chau RI and Straubinger RM (1997) Liposome-mediated therapy of intracranial brain tumors in a rat model. *Pharm Res* **14**:992–998.
70. Saito R, Bringas JR, McKnight TR and *et al.* (2004) Distribution of liposomes into brain and rat brain tumor models by convection-enhanced delivery monitored with magnetic resonance imaging. *Cancer Res* **64**:2572–2579.
71. Koukourakis MI, Koukouraki S, Fezoulidis I and *et al.* (2000) High intratumoral accumulation of stealth liposomal doxorubicin (Caelyx) in glioblastomas and in metastatic brain tumors. *Br J Cancer* **83**:1281–1286.
72. Fabel K, Dietrich J, Hau P and *et al.* (2001) Long-term stabilization in patients with malignant glioma after treatment with liposomal doxorubicin. *Cancer* **92**:1936–1942.
73. Hau P, Fabel K, Baumgart U and *et al.* (2004) Pegylated liposomal doxorubicin-efficacy in patients with recurrent high-grade glioma. *Cancer* **100**:1199–1207.
74. Chua SL, Rosenthal RA, Wong SS and *et al.* (2004) Phase 2 study of temozolomide and caelyx in patients with recurrent glioblastoma multiforme. *Neuro-Oncol* **6**:38–43.
75. Bobo RH, Laske DW, Akbasak A, Morrison PF, Dedrick RL and Oldfield EH (1994) Convection-enhanced delivery of macromolecules in the brain. *Proc Natl Acad Sci* **91**:2-76–2080.
76. Lieberman DM, Laske DW, Morrison PF, Bandiewicz KS and Oldfield EH (1995) Convection-enhanced distribution of large molecules in gray matter during interstitial drug infusion. *J Neurosurg* **82**:1021–1029.

77. Mamot C, Nguyen JB, Pourdehnad M and *et al.* (2004) Extensive distribution of liposomes in rodent brains and brain tumors following convection-enhanced delivery. *J Neuro Oncol* **68**:1–9.
78. Meredith RF and LoBuglio AF (1997) Recent progress in radioimmunotherapy of cancer. *Oncol* **11**:979–987.
79. Kalofonos HP, Pawlikowska TR, Hemingway A and *et al.* (1989) Antibody guided diagnosis and therapy of brain gliomas using radiolabeled monoclonal antibodies against epidermal growth factor receptor and placental alkaline phosphatase. *J Nucl Med* **30**:1636–1645.
80. Riva P, Franceschi G, Arista A and *et al.* (1997) Local application of radiolabeled monoclonal antibodies in the treatment of high grade malignant gliomas. A six-year clinical experience. *Cancer* **80**:2733–2742.
81. Bigner DD, Brown MT, Friedman AH and *et al.* (1998) Iodine-131-labeled antitenascin monoclonal antibody 81C6 treatment of patients with recurrent malignant gliomas: Phase I trial results. *J Clin Oncol* **16**:2202–2212.
82. Cokgor I, Akabani G, Kuan CT and *et al.* (2000) Phase trial results of iodine-131-labeled antitenascin monoclonal antibody 81C6 treatment of patients with newly diagnosed malignant gliomas. *J Clin Oncol* **18**:3862–3872.
83. Reardon DA, Akabani G, Colemen RE and *et al.* (2002) Phase II trial of murine $^{131}$I-labeled antitenascin monoclonal antibody 81C6 administered into surgically created resection cavities of patients with newly diagnosed malignant gliomas. *J Clin Oncol* **20**:1389–1397.
84. Hall WA and Fodstad O (1992) Immunotoxins and central nervous system neoplasia. *J Neurosurg* **76**:1–12.
85. Debinski W (2002) Local treatment of brain tumors with targeted chimera cytotoxic proteins. *Cancer Investig* **20**:801–809.
86. Cohen KA, Liu T, Bissonette R, Puri RK and Frankel AE (2003) $DAB_{389}$EGF fusion protein therapy of refractory glioblastoma multiforme. *Curr Pharm Biotechnol* **4**:39–49.
87. Pastan I, Chaudhary VK and Fitzgerald D (1992) Recombinant toxins as novel therapeutic agents. *Ann Rev Biochem* **61**:331–354.
88. Laske DW, Youle RJ and Oldfield EH (1997) Tumor regression with regional distribution of the targeted toxin TF-CRM107 in patients with malignant brain tumors. *Nature Med* **3**:1362–1368.
89. Weaver M and Laske DW (2003) Transferrin receptor ligand-targeted toxin conjugate (Tf-CRM107) for therapy of malignant gliomas. *J Neuro Oncol* **65**:3–13.
90. Sampson JH, Akabani G, Archer GE and *et al.* (2003) Progress report of a phase I study of the intracerebral microinfusion of a recombinant chimeric protein composed of transforming growth factor (TGF)-$\alpha$ and a mutated form of the

Pseudomonas exotoxin termed PE-38 (TP-38) for the treatment of malignant brain tumors. *J Neuro Oncol* **65**:27–35.

91. Weber F, Asher A, Bucholz R *et al.* (2003) Safety, tolerability, and tumor response of IL4-Pseudomonas exotoxin (NBI-3001) in patients with recurrent malignant glioma. *J Neuro Oncol* **64**:125–137.
92. Weingart J, Tatter S, Rosenfeld S *et al.* (2003) Intratumoral convection-enhanced delivery of IL13-PE38QQR cytotoxin for recurrent malignant glioma without planned resection: A phase I/II study. *Neuro-Oncol* **5**:357.
93. Prados M, Kunwar S, Lang FF *et al.* (2005) Final results of phase I/II studies of IL13-PE38QQR administered intratumorally (IT) and/or peritumorally (PT) via convection-enhanced delivery (CED) in patients undergoing tumor resection for recurrent malignant glioma. *Proc Am Soc Clin Oncol* **23**:115s.
94. Shin SU, Wu D, Ramanathan R, Pardridge WM and Morrison SL (1997) Functional and pharmacokinetic properties of antibody/avidin fusion proteins. *J Immunol* **158**:4797–4804.
95. Friden PM, Walus LR, Musso GF, Taylor MA, Malfroy B and Starzyk RM (1991) Anti-transferrin receptor antibody and antibody-drug conjugates cross the blood-brain barrier. *Proc Natl Acad Sci USA* **88**:4771–4775.
96. Pardridge WM, Buciak JL and Friden PM (1991) Selective transport of an anti-transferrin receptor antibody through the blood-brain barrier *in vivo*. *J Pharmacol Exp Ther* **259**:66–70.
97. Coloma MJ, Lee HJ, Kurihara A *et al.* (2000) Transport across the primate blood-brain barrier of a genetically engineered chimeric monoclonal antibody to the human insulin receptor. *Pharm Res* **17**:266–274.

# 15

# Protein and Peptide-Based Cancer Gene Therapy

*Sunil Chada and Rajagopal Ramesh*

## 1. Cancer Gene Therapy

The concept of gene therapy was initially envisioned for the treatment of genetic diseases, where the normal function of the cells will be restored by reintroducing the gene for which the cells are defective.[1,2] Since the initial documentation of the "proof of concept" for gene therapy in preclinical studies for the treatment of genetic diseases, the application of gene therapy technology has been extended for the treatment of a wide variety of human diseases that include cancer, acquired immunodeficiency syndrome (AIDS), arthritis, neurodegenerative disorders, peripheral vascular disease, and other acquired diseases.[3–8] For a more detailed and in-depth background on gene therapy, readers are advised to refer to previously published reports.[9–12] Gene therapy, thus, offers a new paradigm for treating a broad-spectrum of human diseases.

Although gene therapy has been tested and applied for the treatment of several human diseases, a major focus, however, has been in the treatment of cancer. This is because cancer is a disease with poor prognosis and remains a major killer worldwide.[13] Furthermore, despite significant advances made in the field of cancer treatment, the 5-year survival for patients diagnosed with epithelial cancer is poor.[13] In this article, we will discuss the various gene therapeutic approaches that have been tested or are being tested for treating cancer both in the laboratory and in the clinic, as well as the future prospects of this technology.

### 1.1. *Identifying the target gene/pathway for cancer therapy*

Cancer initiation and progression is the result of a series of insults or damage that a normal cell undergoes over time, during which several cellular genes are functionally activated or inactivated due to their deletions, amplifications and mutations in the DNA.[14,15] The consequence of such gene alterations can result in the loss of key protein expression that is required to maintain the normal cell function or in aberrant protein expression that is detrimental to the maintenance of the normal cell physiology. It should be noted that apart from the gene alterations described above, additional mechanisms such as epigenetic changes can also occur in cancer cells.[14,15] However, since epigenetic changes can be corrected by treating with small molecule inhibitors or chemicals. The major focus of cancer gene therapy has primarily been in restoring the normal gene functions that is lost due to deletion or mutations.

Among the several genes that are frequently activated and are detrimental to the normal cells are the oncogenes (c-myc, ras, c-fos) that promote rapid cell proliferation and transformation.[16,17] Genes that are inactivated or lost and are critical in maintaining the normal function of the cell include the tumor suppressor genes (TSGs) or the "gatekeepers".[18,19] Molecular genetic studies have revealed a large number of oncogenes and TSGs that have been deleted or mutated in a wide variety of human cancers (Table 1). In addition to oncogenes and tumor suppressor genes, alterations in a gene function that is part of a cell-signaling pathway can result in aberrant activation of the signaling pathway leading to continuous signaling for cancer cell proliferation and progression (Table 2). An example for altered gene function in a cell-signaling pathway is the amplification of the epidermal growth factor receptor in brain tumors.[20,21] Thus, a cancer cell can have single or multiple defects that are accumulated over time and can largely differ among different cancer types. The existence of multiple defects that is

**Table 1** Oncogenes and tumor suppressor genes identified in human cancers.*

| Oncogenes (activated) | | Tumor suppressor genes (inactivated/lost) | |
|---|---|---|---|
| c-myc | - broad spectrum of cancers | p53 | - broad spectrum of cancers |
| c-fos | - broad spectrum of cancers | BRCA1 | - breast, ovarian |
| K-ras | - pancreatic, lung, melanoma | PTEN | - colorectal, brain, prostate, lung, ovarian |
| Bcr-Abl | - leukemia | mda-7 | - broad spectrum of cancers |
| c-Src | - lung, breast, prostate | Fus1 | - lung, breast |
| | | Fhit | - lung, breast, bladder, lymphoma, colorectal |
| | | 101F6 | - lung |
| | | RassF1 | - lung, colorectal, endometrial, cervical, bladder, thyroid |
| | | p73 | - lymphoma, colorectal, cervical lung, leukemia, sarcoma |
| | | Rb | - broad spectrum of cancers |
| | | $P14^{Arf}$ | - melanoma, lymphoma, lung, breast |
| | | $p16^{waf1}$ | - lung, brain, ovarian, breast |
| | | p27 | - broad spectrum of cancers |
| | | TSC1/TSC2 | - hemangioma, brain, bladder, kidney |
| | | Maspin | - breast, prostate, thyroid |
| | | VHL | - broad spectrum of cancers |
| | | APC | - colorectal |
| | | SMAD-2 | - pancreatic, lung, colorectal |
| | | SMAD-4 | - pancreatic, lung, colorectal |
| | | Testin | - leukemia, lymphoma, breast, uterine |
| | | OPCML | - ovarian |
| | | PTP | - leukemias, lung, liver |
| | | RIZ | - liver, breast, lung, colon, melanoma bone, brain |

**Note*: The list shown above is an example of oncogenes that are activated and tumor suppressor genes that are mutated or lost in human cancer and does not encompass the complete list of genes.

**Table 2** Activated signaling pathways in cancer.*

| Pathway | Cancer |
|---|---|
| PI3 Kinase (PI3K) | Lung, prostate, colorectal, brain, ovarian |
| AKT Kinase | Lung, breast, colorectal, pancreas, prostate, melanoma |
| mTOR | Lung, breast, prostate, leukemia, lymphoma, melanoma |
| Src Kinase | Lung, breast, prostate, liver, brain |
| EGFR | Lung, brain, breast, ovarian |
| Her2/Neu | Lung, breast, ovarian |
| c-Kit | Leukemia, melanoma, lung, sarcoma, ovarian, gastrointestinal |
| PDGFR | Broad-spectrum of cancers |
| VEGFR | Broad-spectrum of cancers |
| B-RAF | Melanomas, pancreas, lung, thyroid, ovarian, brain |
| MAPK | Broad-spectrum of cancers |
| NFKB | Broad-spectrum of cancers |

**Note*: The list shown above is incomplete and is an example of various signaling pathways that are activated in human cancers.

often detected in a cancer cell provides an opportunity for targeting several of these defective genes. A potential caveat, however, is the need to select the appropriate and most effective target gene, which when reintroduced into the cancer cells, will revert the cancer cell into normal cell or make the cancer cells undergo cell death or "apoptosis". The various genes and pathways that have been tested for cancer therapy in preclinical studies and few of which have been tested in the clinic will be discussed below.

## 1.2. *Selection of gene therapy vector*

Apart from identifying the target gene for cancer therapy, another important factor for successful gene therapy is to determine the best gene delivery vehicle or "vector". Several gene delivery vectors (Table 3) have been developed and tested in the last 15–20 years, so as to achieve a very high efficiency of gene transfer and transgenic protein expression in the cancer cells with minimal to no effect on normal cells.[22,23] However, till date, the issue of developing an ideal gene delivery vector remains at the forefront of clinical applications of gene therapy. Controversy in the field of gene therapy still remains with scientific debates revolving whether the transfer of desired genes to appropriate cells and sufficient levels of expression for specific disease treatment have been achieved and how therapeutic

**Table 3** Gene delivery vectors for cancer gene therapy.

| Viral | Nonviral | Bacterial | Cellular |
|---|---|---|---|
| Retrovirus | Liposomes | Salmonella | Tumor cells |
| Adenovirus | PEI | E. Coli | Endothelial cells |
| Adeno-associated virus (AAV) | Naked plasmid DNA | Shigella | Mesenchymal stem cells |
| Lentivirus | Synthetic polymers | | |
| Vaccinia virus | | | |
| Sindbis virus | | | |
| Herpes Simplex virus | | | |
| Simian virus-40 (SV-40) | | | |

endpoints are best measured using this new class of drugs. It is anticipated that continued research in the area of gene transfer and the ability to achieve tissue-specific gene expression will hopefully resolve these issues in the majority of gene therapy protocols. A major outcome of the studies of the thousands of patients who have received gene therapy to date was that gene transfer was safe in the vast majority of them. The side effects observed have been very manageable in most cases with the notable exception of a patient death in one clinical trial.[24,25]

## 2. Protein-Based Cancer Gene Therapy

Protein-based cancer gene therapy works on the principle of introducing a therapeutic gene of interest into a cell which when efficiently transcribed and translated will result in the production of the appropriate protein and exert its antitumor effect in an autocrine and paracrine fashion. The advantage of protein-based cancer gene therapy unlike conventional protein-based therapy is that it is easy to produce and manufacture the gene therapy vector carrying the gene of interest, is less expensive, requires low doses of the vector to produce substantial amount of the protein intracellularly, and that protein expression can be specifically regulated at the target tumor site thereby minimizing the side effects. This technology has particularly been useful where protein production in purified form by conventional methods has posed technical challenges. The several protein-based cancer gene therapeutic approaches that have been tested till date are discussed below and are categorized based on the protein function of a gene and the molecular end-points that it produces when introduced into a cancer cell.

## 2.1. *Growth inhibition and apoptosis*

### 2.1.1. *Toxic proteins*

As the name indicates, the concept of introducing a gene into a cancer cell that encodes a protein that under appropriate conditions is toxic to the cells has been tested as a gene therapeutic approach in both preclinical and clinical setting. The gene that produces such toxic protein is commonly referred to as the "suicide gene" and the therapeutic approach using a suicide gene is called "suicide gene therapy".[26,27] In this gene therapy approach, cells expressing the suicide gene when exposed to a nontoxic prodrug convert the prodrug into an active toxic drug that kills the cells. The mechanism by which this cell killing occurs is that the inactive prodrug is converted into an active form of drug that acts as an inhibitor of DNA synthesis leading to cell death (Fig. 1). Thus only cells that express the protein are killed thereby achieving selectivity. The first suicide gene that was tested, as a potential cancer therapeutic was the herpes simplex virus thymidine kinase (HSV-TK) gene.[26–28] Tumor cells transduced with a retroviral expression vector, expressing HSV-TK when treated with a prodrug ganciclovir (GCV), showed effective cell killing.[26–28] Furthermore, Freeman *et al.*[29] showed that

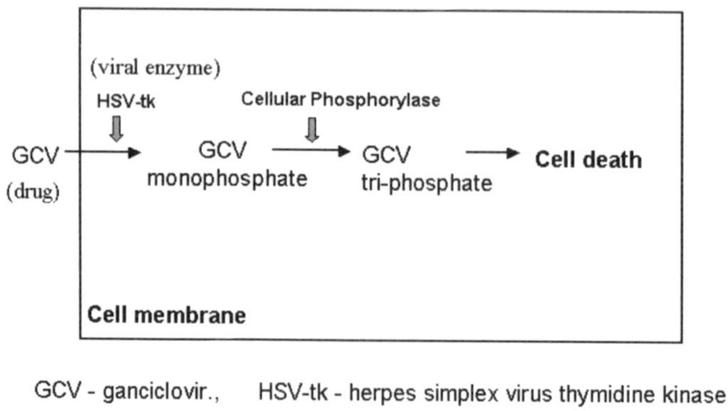


**Fig. 1** Schematic representation of the molecular mechanism for HSV-TK-based suicide gene therapy. The intracellular herpess simplex thymidine kinase converts the prodrug into an active drug by phosphorylation which when incorporated into the host DNA results inhibition of replication.

effective tumor cell killing occurred both *in vitro* and *in vivo* when only a fraction (10–25%) of tumor cells were transduced and expressed the HSV-TK protein. The underlying mechanism for this effective killing was shown to be due to a "bystander effect". Thus, the advantage of using suicide genes in cancer therapy was demonstrated where cells expressing the toxic gene exerted its toxic effect on adjacent cancer cells that do not express the toxic protein. Additionally, the "bystander effect" also overcame the limitation of retroviral vectors being able to transfect only actively dividing cells. Furthermore, induction of an immune response by the dying tumor cells was elicited that could protect mice from subsequent relapse.[30–32] Thus, retroviral-based suicide gene therapy was initiated and the concept of introducing toxic protein in tumor cells introduced.

Since the initial report of the HSV-TK-mediated bystander killing effect, a number of additional suicide genes have been identified and tested in the laboratory (Table 4). A few of the suicide genes such as the bacterial cytosine deaminase (CD) gene has also been shown to exert a bystander killing effect when CD-expressing cells are exposed to the prodrug 5-fluorocytosine (5-FC), a phenomenon akin to that observed with HSV-TK.[33,34] Similarly, studies using the nitroreducatse (NTR) gene showed enhanced cell killing when NTR expressing cells were exposed to CB1954.[35]

**Table 4** Suicide genes and prodrugs used for cancer gene therapy.

| Suicide gene | Prodrugs |
|---|---|
| HSV-TK | Ganciclovir (GCV)<br>Penciclovir (PCV)<br>Acyclovir (ACV)<br>Valacyclovir (VCV) |
| Cytosine deaminase (CD) | 5-Fluorocytosine (5-FC) |
| Purine nucleoside Phosphorylase (PNP) | 6-Methylpurine Deoxyriboside (MEP)<br>Fludarabine (FAMP) |
| NAH cytochrome-P450 (CYP) | Cyclophosphamide (CPA)<br>Ifosfamide (IFO) |
| Nitroreductase | Dinitroaziridinylbenzamide — CB1954<br>Dinitrobenzamider mustard — SN23862 |
| Thymidine phosphorylase (TP) | 5′-Deoxy-5-Fluorouridine (5′-DFUR) |
| Methionine-A,P-lyase (MET) | Selenomethionine (SeMET) |

Based on the preclinical results obtained from the studies using suicide gene-based cancer therapy approach, the first clinical trial of HSV-TK-based suicide gene therapy for human ovarian cancer was approved in 1991.[36] Since then, a number of clinical trials for the treatment of different types of cancers have been initiated and completed.[37–41] Most of these studies were Phase-I/II clinical trials that focused on determining the toxicity and maximum tolerated dose. Results from these early studies demonstrated that the treatment was well tolerated with no significant toxicity or adverse events. However, since the gene delivery vector system used in these studies were applicable only for the treatment of solid localized tumors but not for metastatic cancer, treatment of patients diagnosed with metastatic cancer was not feasible using this approach.

With the recent advances made in the area of gene delivery vectors, increased and selective expression of the suicide genes in tumor cells have been demonstrated using a variety of vectors. Recombinant adeno-associated virus (rAAV) expressing HSV-TK under the control of E2F alpha promoter was shown to inhibit human sarcoma growth *in vitro*.[42] Similarly, intratumoral administration of an attenuated salmanoella expression vector expressing the cytosine deaminase gene (TAPET-CD) into refractory cancer patients showed no toxicity although this study had a small number of patients.[43] Thus, the use of novel gene delivery vectors to selectively express toxic genes in cancer cells and systemically treat metastatic cancers appears to hold promise and become a reality in the near future.

In addition to suicide genes, studies have also been conducted where bacterial toxins such as diphtheria toxin A when specifically expressed in tumor cells can induce tumor cell death.[44] In this study using a nonviral liposomal vector, transfer of the diphtheria toxin A in prostate cancer cells showed cytopathic effects and was indicated to be useful for systemic treatment of patients with metastatic prostate cancer.[44] Studies investigating the use of proapoptotic genes that are responsible for inducing apoptosis have also been examined as a cancer gene therapeutic approach. Overexpression of proapoptotic proteins such as BAX, BAK, BIK, Fas, FasL, TRAIL, and Caspase-8 has shown effective tumor growth inhibition in preclinical studies.[45–49] Although most of these studies have been limited to treatment of subcutaneous tumors, it is anticipated that some of these proteins will be useful for systemic treatment of metastatic cancer. The rate-limiting step in using these proapoptotic genes has been that the proteins also exert some cytotoxicity in normal cells. Thus, design of novel vectors that can specifically regulate the expression of these proteins can minimize the cytotoxic effects on normal cells. One approach for tumor-specific protein expression is the regulation of target proteins under the control of the human telomerase reverse transcriptase (hTERT) promoter.[50] Preclinical studies demonstrate that hTERT-mediated protein expression selectively occurs in cancer cells that express high levels of telomerase enzyme resulting in tumor suppression.[51–53] It is to be noted that although hTERT-mediated gene expression levels are lower than that observed when using the conventional cytomegalovirus (CMV) promoter, the tumor cell specificity achieved is higher for hTERT promoter than for CMV promoter, thereby reducing the cytotoxic effects on normal cells. Similarly, lentiviral-mediated expression of diphtheria toxin A (DTA) under the control of prostate-specific antigen promoter resulted in eradication of prostate cancer cells both *in vitro* and *in vivo*.[54] Additional tumor-specific promoters are also being currently examined in several laboratories and are yet to be seen which of these promoters will ultimately be used in the clinic to achieve tumor-selective expression of toxic proteins. Whatever the gene or promoter to be used, it is now clear from these preclinical studies that expression of toxic proteins via gene therapy is feasible for cancer treatment.

### 2.1.2. *Tumor suppressor proteins*

The TSGs a.k.a. "gatekeepers" help in maintaining the normal function of the cells by preventing accumulation of DNA damage in the cells. Cells that are extensively damaged and cannot be repaired are quickly eliminated through a process called "apoptosis". A number of tumor suppressor genes have been identified in the recent past and with several more yet be identified.[55–58] However, inactivation of TSGs due to genetic (mutations, deletions) or epigenetic (hypermethylation) mechanisms results in the loss of the tumor suppressor function leading to uncontrolled cell proliferation and culminating in cancer. It appears that the majority of cancers have mutation or alterations in one or more of the TSGs. Thus, restoration of the tumor suppressor function by introducing the wild-type TSG into a cancer cell should result in inhibiting tumor growth and is the basis for tumor suppressor cancer gene therapy. Due to the large number of TSGs identified till date with several of these genes tested for cancer gene therapy, it will not be possible for the authors to discuss all of TSGs based gene therapy studies. However, the authors will attempt to discuss a few gene therapy studies that have been of significance and have been tested in the clinic.

Among the several TSGs that have been tested as a therapeutic, studies using the p53 TSG have dominated the cancer gene therapy field. This is because p53 is mutated in more than 50% of human cancers making it more applicable to a wide-variety of cancers.[59] Early studies using retroviral vector showed restoration of wild-type 53 (wt-p53) to lung cancer cells resulted in growth inhibition and apoptosis.[60–62] Subsequent studies by several investigators using different gene delivery vectors showed that restoration of wt-p53 expression in a broad-spectrum of cancer cells results in tumor growth inhibition both *in vitro* and *in vivo*.[63,64] Furthermore, Xu *et al.*[62] reported a p53-mediated bystander effect *in vivo*. Additionally p53-based gene therapy when combined with chemotherapy or radiotherapy showed enhanced tumor killing compared to monotherapy suggesting combination therapy is best suited for cancer treatment.[65–67] Based on these results, the first phase-I gene therapy-based clinical trial to treat non small cell lung cancer using the p53 tumor suppressor gene was carried out using a retroviral vector.[68] In this trial, a small number of lung cancer patients were treated with the p53-vector by directly injecting into single tumor lesions. Tumor regression or tumor stabilization was demonstrated in a small number of patients who were treated with p53 gene therapy. Additionally, no significant toxicity or adverse events were reported indicating that the treatment was well tolerated. The results from this trial established the "proof of concept" and the potential success for cancer gene therapy. Based on the initial success of p53-based gene therapy and with the advent of adenoviral vector, additional phase I/II trials were initiated for treatment of nonsmall cell lung cancer, gliomas, ovarian, head and neck, bladder, and breast cancer.[65–68] In these trials, an adenoviral vector expressing the p53 gene (Ad-p53) was injected intratumorally either as monotherapy or in conjunction with radiation or chemotherapy.[69–73] The results from these trials were mixed in that tumor response was reported in nonsmall cell lung cancer, head and neck cancer, gliomas, and breast cancer patients, but not in ovarian and bladder cancer patients. The difference in the clinical response to p53 among the different cancer types is not clear and remains to be investigated. Several possibilities exist that include additional alterations in the tumors, difference in transgene expression levels within a tumor type, treatment schedule. Despite these limitations it is evident that in all of these studies treatment was largely well tolerated with minimal toxicity. It is anticipated with further design of the clinical trial and with the incorporation of advanced molecular technologies, p53-based gene therapy will be useful for systemic cancer therapy. Additionally with the recent demonstration of efficient p53 gene delivery to disseminated lung tumors that resulted in inhibition of experimental metastases following systemic injection of DOTAP:cholesterol (DOTAP:Chol) liposome containing p53 DNA raises the hope that systemic gene therapy for the treatment of disseminated cancer will soon be available for clinical testing.[74]

Apart from p53 other TSGs have also been tested in preclinical studies. Expression of wild-type p16 gene, frequently inactivated in cancer cells by methylation, induced cell cycle arrest and apoptosis in tumor but not in normal cells.[75–76] Similarly, introduction of cell cycle inhibitors such as p21 or p27 has been shown to inhibit tumor growth.[77,78] Recent studies by us and others have shown that expression of novel TSGs (Fus1, Fhit, 101F6, and RASSF1) located on the 3p-chromosome and frequently lost in lung cancer, results in tumor growth inhibition and apoptosis of lung tumor cells both *in vitro* and *in vivo*.[74,79–81] In these efficacy studies, intratumoral or systemic treatment of lung tumor-bearing mice with DOTAP:Cholesterol liposome carrying Fus1, Fhit, or 101F6 gene demonstrated inhibition of lung tumor growth and increased animal survival (Fig. 2). Based on these results, we have initiated a phase-I clinical trial for systemic treatment of non small lung cancer patients with DOTAP:Chol Fus1 complex at M.D. Anderson Cancer Center, Houston, TX, USA. Patient accrual is currently ongoing and it is anticipated that the results of this trial will be available in the next couple of years.

Studies from our laboratory and others have also demonstrated effective tumor growth inhibition following adenovirus-mediated PTEN (Ad-PTEN) gene expression.[82–85] More recently, we and others have shown that overexpression of novel tumor suppressor/cytokine gene called melanoma differentiation associated-7 (mda-7) gene/interleukin-24 (IL-24) results in killing of a broad-spectrum of human cancer cells both *in vitro* and *in vivo*.[86,87] Furthermore, systemic delivery of mda-7 encapsulated in nanoparticles showed a marked reduction in experimental lung metastasis that was associated with reduced vascularization.[88] The advantage of mda-7-based gene therapy is that mda-7 has multiple

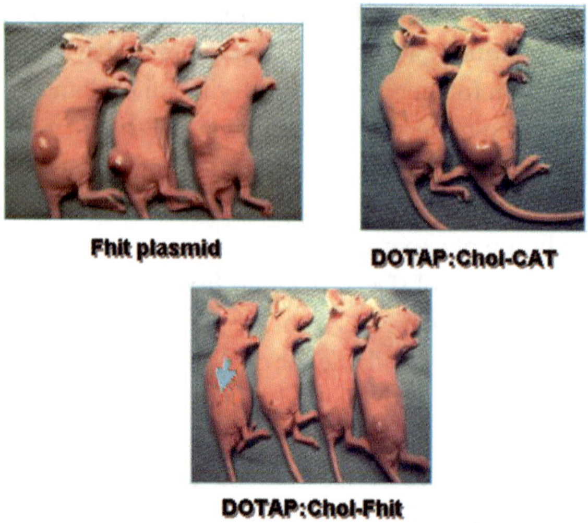


**Fig. 2** Therapeutic effect of DOTAP:cholesterol (DOTAP:Chol)-Fhit liposome complex on subcutaneous lung tumor. Human H1299 lung tumor xenografts established in nude mice were untreated or treated with DOTAP:Chol-CAT complex or DOTAP:Chol-Fhit complex daily for six doses (50 μg/dose). A significant inhibition of tumor growth was observed in mice treated with DOTAP:Chol-Fhit complex compared to tumor growth in mice that were untreated or treated with DOTAP:Chol-CAT complex.

functions that include proapoptotic,[89–91] antiangiogenic[92,93] antimetastatic (Fig. 3)[94] and proimmune[95] properties. Additionally, the MDA-7 protein is secreted and the secreted protein can exert antitumor and antiangiogenic activity.[93,96] Thus, mda-7 can exert its tumor suppressive activity both in a local autocrine fashion and in a distal paracrine fashion. Furthermore, Mda-7-based gene therapy when combined with chemotherapy (Fig. 4; unpublished data), radiation therapy,[97] biologic therapy,[98] or nonsteroidal anti-inflammatory drugs (NSAIDS)[99] showed enhanced tumor growth inhibition and cell killing.

Based on the unique properties of mda-7, it is envisioned that mda-7-based gene therapy will be more appropriate for treatment of metastatic cancers. As a first step toward achieving this goal, a Phase-I clinical trial was recently initiated and completed.[100,101] In this trial, 22 patients were treated with Ad-mda7 (INGN241) intratumorally in a dose-escalating fashion to determine the maximum tolerated dose (MTD). Results from this trial showed that the treatment was well tolerated by the patients

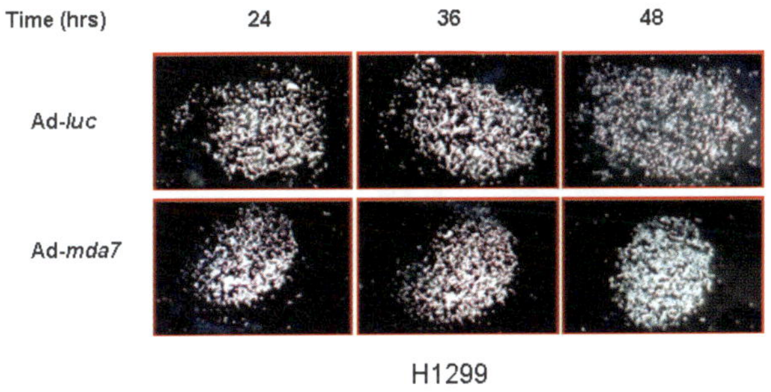


**Fig. 3** Inhibitory effects of mda-7 on tumor cell migration. Human lung tumor (A549) cells were treated with Ad-luc or Ad-mda7 and plated on gelatin coated glass slides and observed for migration at various time points. A significant inhibition of cell migration was observed in Ad-mda7 treated cells compared to Ad-luc treated cells.

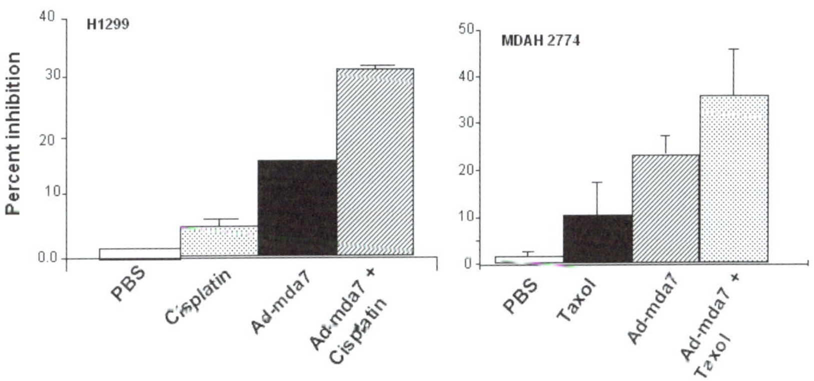


**Fig. 4** Enhanced growth inhibitory effects of Ad-mda7 in combination with chemotherapy. Human lung (H1299) and ovarian (MDAH 2774) cells were treated with Ad-mda7. Three-hours after treatment cells were exposed to cisplatin (H1299) or taxol (MDAH2774). Cells were analyzed for growth inhibitory effects at 72 hr after treatment. Cells treated with Ad-mda7 alone, cisplatin alone, and taxol alone served as controls. A significant inhibition of cell proliferation was observed in cells that were treated with Ad-mda7 plus cisplatin or taxol compared to other treatment groups. Error bars denote standard error.

with minimal toxicity. Surprisingly, in a small number of patients clinical response was observed that correlated with transgene expression and tumor cell apoptosis. Additionally, apart from an inflammatory response as indicated by the production of proinflammatory cytokines in a subset of patients' increase in $CD8^+$ T-cells were observed indicating activation of host-immune response. All of these observations indicate that the *in vivo* tumor response following Ad-mda7 treatment is due to a combination of the antitumor and proimmune activities of mda-7. In this study, the correlation between tumor response and CD31 positive staining, a marker for angiogenesis, however, was not examined and hence the antiangiogenic activity of mda-7 *in vivo* remains to be determined.

Based on these encouraging results, a phase-II clinical trial for treatment of intransit melanoma with Ad-mda7 is being initiated at M.D. Anderson Cancer Center, Houston, TX, USA. The results from this trial are anticipated in the next few years. However, the results of these studies will be from intratumoral injections into individual tumors. Thus, as with p53-gene therapy, future studies that are focused on systemic mda-7-based gene therapy are warranted for the treatment of disseminated tumors.

### 2.1.3. *Antiangiogenic proteins*

The application of gene therapy targeting tumor angiogenesis has also been tested in various tumor models. The rationale for this approach is that for tumors to grow and metastasize, there is a continuous need of blood supply provided by the tumor vasculature. In fact, a direct correlation between tumor vascularization and tumor metastases exists.[102] Furthermore, degree of tumor vascularization as indicated by the CD31 positive endothelial cells is often inversely associated with prognosis.[103] Thus, inhibition of tumor endothelial cell proliferation should result in tumor growth inhibition and metastases. A number of endogenous proteins (angiostatin, endostatin) have been shown to possess antiangiogenic activity.[104–107] Treatment of xenograft tumors established in mice with purified antiangiogenic proteins has been shown to suppress tumor growth by inducing growth arrest and cell death of endothelial cells.[104–107] Although these studies have established the concept of using antiangiogenic proteins for cancer therapy, the difficulty often faced both in an academic and in an industrial setting is the availability of purified functional protein in large quantities. Furthermore, the cost of producing purified protein is very expensive and as a result the enthusiasm for protein-based therapy is often dampened. To circumvent these problems, expression of antiangiogenic proteins by gene therapy has been investigated in several laboratories. Expression of angiostatin, endostatin or a combination of both proteins using gene delivery vectors showed tumor growth inhibition.[108–111] Using angiostatin plasmid DNA encapsulated in a cationic liposome, we have tested the therapeutic effects of angiostatin on subcutaneous lung tumor xenografts established in nude mice. Intratumoral administration of the liposome–angiostatin complex daily for six doses (50 μg DNA) showed delay in tumor growth compared to control mice that were treated with control plasmid DNA (Fig. 5; unpublished data). Similarly, inhibition of glioma cell growth expressing exogenous pigment epithelium-derived factor (PEDF) was demonstrated *in vivo*.[112] Using mda7 gene, we demonstrated inhibition of tumor growth involved antiangiogenic mechanisms.[93] In this study, we showed that secreted mda-7 protein has direct antiangiogenic activity on tumor endothelial cells both *in vitro* and *in vivo* (Fig. 6). These findings raise the possibility of using mda-7 as an antiangiogenic drug and expression of mda-7 protein via gene therapy offers a convenient way of expressing the protein intracellularly at concentrations that is effective in producing an antitumor effect.

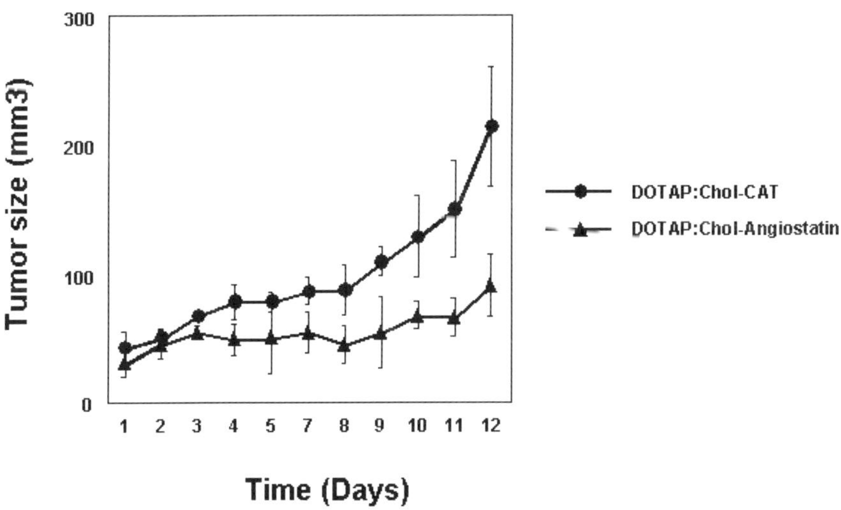


**Fig. 5** Therapeutic effect of DOTAP:Cholesterol (DOTAP:Chol)-angiostatin liposome complex on subcutaneous lung tumor. Human H1299 lung tumor xenografts established in nude mice were treated with DOTAP:Chol-CAT complex or DOTAP:Chol-Angiostatin complex daily for six doses (50 μg/dose). A significant inhibition of tumor growth was observed in mice treated with DOTAP:Chol-Angiostatin complex compared to tumor growth in mice that were treated with DOTAP:Chol-CAT complex. Error bars denote standard error.

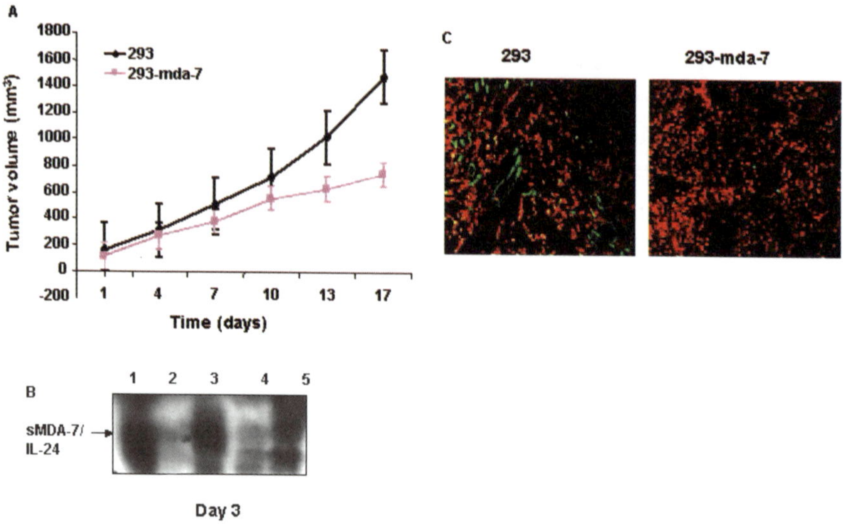


**Fig. 6** Secreted mda-7 protein inhibits tumor growth by inhibiting angiogenesis. Human embryonic kidney 293 cells were stably expressing mda-7 protein (293-mda7) were mixed with A549 lung tumor cells and implanted subcutaneously into nude mice. Mice that were implanted with a mixture of A549 and parental 293 cells served as control. Tumor growth was measured at regular intervals and expressed as the average mean of group. (**A**), A significant inhibition of tumor growth was observed in mice implanted with 293-mda7 cells compared to mice implanted with parental 293 cells. (**B**), MDA-7 protein was detected by western blotting in the serum of mice implanted with A549 and 293-mda7 cells (lanes 1,3,5) but not in the serum of mice implanted with A549 and parental 293 cells (lanes 2 and 4). (**C**), tumor tissues from mice implanted with A549 and 293 or 293-mda7 cells were subjected to double immunofluroscence staining for tumor cell (red) and endothelial cell (green) apoptosis. Increased tumor cell and endothelial cell apoptosis was observed in tumor tissues obtained from mice implanted with A549 plus 293-mda7 compared to those from mice implanted with A549 plus 293 cells. Magnification × 100 (Fig. reproduced with permission from *Cancer Res* (2003) Ramesh *et al.*, **63**: 5105–5113).

Since inhibition of tumor growth via antiangiogenic mechanisms is a slow process and requires sustained expression of the antiangiogenic protein, the use of adeno-associated virus (AAV) and lentivirus-based gene therapy has been investigated. Both AAV and lentivirus can integrate into the host genome resulting in continuous protein production. Preclinical studies have shown expression of antiangiogenic proteins using AAV or lentiviral vectors results in tumor growth suppression and prolonged animal survival.[113–116] Although all of these studies are exciting, it is expected that antiangiogenic-based gene therapy will need further investigation and validation prior to testing in the clinic.

## 2.2. *Enhancement of the immune response*

In addition to tumor-suppressor-based gene therapy, studies have also been carried out investigating cytokine-based gene therapy or immunotherapy. The rationale for cytokine-based cancer gene therapy is that cancer cells often escape the host-immune system by altering their structural or functional proteins at the cell surface. Alternatively, cancer cells produce cytokines and growth factors that are immunosuppressive thereby inactivating the host immune cells such as the tumor infiltrating lymphocytes (TILs). Thus, by harnessing the power of the patients' immune system by reintroducing cytokine genes, not only can one inhibit primary tumors but also distal metastatic tumors.[117] The various approaches that have been tested to elicit a host immune response against tumors cells include expression of cytokines, costimulatory molecules, tumor-associated antigens, and major histocompatability antigens by tumor cells. Preclinical experiments have demonstrated that each of the approaches mentioned above have been effective in eradicating established tumors in mice.[118–122] Recent studies from our laboratory using mda-7/IL-24 gene showed that treatment of immunocompetent mice bearing sygeneic tumors with Ad-mda7 resulted in tumor regression. However, treatment of xenograft human lung tumors established in nude mice with Ad-mda7 resulted in tumor suppression but not tumor regression indicating a role for host-immune cells in eradicating Ad-mda7 treated tumor cells.[123] When the tumor-free syngeneic mice were challenged with parental tumor cells, no tumor growth was observed compared with tumor growth in naive control mice. Furthermore, analysis of the serum cytokine levels showed increased tumor-necrosis factor-alpha (TNF-$\alpha$), granulocyte-macrophage colony-stimulating factor (GM-CSF) and interferon-alpha (IFN-$\alpha$), cytokines associated with TH1-response in Ad-mda-7 treated mice compared to control mice. Analysis of splenocytes indicated increased $CD8^+$ cells in Ad-mda7 treated mice compared to control mice. All of these results indicated activation of the host-immune response in Ad-mda7 treated mice. However, it is not known whether the immune response leading to Ad-mda7-induced antitumor activity is mediated by the secreted mda-7 protein, due to efficient tumor cell antigen presentation or due to a combination of both. Previous study by Caudell *et al.*[124], using purified human mda-7 protein, showed that treatment of human peripheral mononuclear cells (PBMC's) with mda-7 protein results in production of proinflammatory cytokines *in vitro*. These results concur with our *in vivo* findings and suggest that secreted mda-7 protein can directly activate the immune system. However, additional studies are warranted to support these claims and are currently being tested in our laboratory.

Studies have also been conducted where immunotherapy has been tested in combination with TSG therapy. For example, treatment of tumors established in nude mice with Ad-p53 and IL-12 resulted in enhanced tumor growth inhibition and regression compared to tumor growth in mice that were treated with either Ad-p53 or IL-12 alone.[125] Similarly, systemic treatment of syngeneic hepatoma-bearing mice with a nonviral gene delivery vector containing $p21^{waf-1}$ in combination with GM-CSF resulted in effective tumor suppression and increased animal survival.[126] These results demonstrate that combination therapy is another method of enhancing antitumor activity and host-immune activity that results in effective tumor suppression. This strategy is still being tested in preclinical studies and is anticipated that in the near future this combination treatment strategy will be tested in early phase-I clinical trials.

More recently, the role of dendritic cells in host-immune response has been studied. It has been shown that dendritic cells are efficient antigen presenting cells (APCs) that are often inactivated in the tumor microenvironment or unable to recognize the tumor antigen due to alteration at the tumor antigen or epitope. As a result, these dendritic cells are unable to efficiently present the tumor antigens to the T-cells, thereby resulting in tumor escape. To circumvent this problem, studies have been carried out where dendritic cells have been transduced with a tumor antigen *ex vivo* via gene therapy and reintroduced into the host. Reintroduction of the *ex vivo* modified dendritic cells has been shown to stimulate the host immune response resulting in tumor suppression.[127] Similarly, pulsing of dendritic cells with p53 peptides has been shown to be effective inducing an immune response.[128] Vaccination of patients with advanced cancer with dendritic cells pulsed with MUC1-derived peptides resulted in induction of immunological response with a subset of patients showing regression of metastases and a few showing objective response.[129] Furthermore, the treatment was well tolerated with no significant toxicity. These results show that enhancement of immune response in cancer patients can result in antitumor activity against primary and metastatic cancers. Furthermore, the occurrence of relapse after cytokine gene therapy is low due to the ability of the stimulated host immune system to identify and eliminate any cancer cells that may persist. Based on the several preclinical studies conducted till date, a number of phase-I trials based on immunotherapy for cancer have been initiated and a few of them completed. Preliminary reports have indicated promising results.[130–132] However, it is yet to be seen which of the immunotherapy strategies, being currently tested in phase-I clinical trials, will hold through subsequent phase II and III trials for translation into the clinic for treatment of cancer.

## 3. Peptide-Based Cancer Gene Therapy

Peptide-based cancer gene therapy unlike protein-based cancer therapy has primarily focused on tumor targeting and increasing the transduction efficiency of gene delivery vectors to achieve very high expression levels of the therapeutic protein in tumor cells. However, peptide-based cancer gene therapy has also been investigated for promoting tumor cell death or enhancing the immune response against the growing tumor cells. The peptide-based cancer gene therapy strategies used as a means for cancer targeting and enhanced protein expression will be discussed below.

### 3.1. *Tumor targeting and gene transduction*

For effective treatment of cancer by gene therapy, it is essential that the gene delivery vector reach the tumor site efficiently and specifically and transduce the tumor cells with very high efficiency. In addition to transduction, the gene once inside the cancer cell should be able to move from the cytoplasm to the nucleus for messenger RNA transcription. Although several gene delivery vectors have been developed and shown to deliver therapeutic genes to tumor cells and exert antitumor activity *in vitro* and *in vivo*, the efficiency of transduction varies and is often poor, depending on the vector type used. Furthermore, transduction of normal cells adjacent to the tumor and at distant sites has also been observed with various gene vectors raising the potential for systemic toxic effects. For example, adenoviral vectors express the transgene at very high levels both *in vitro* and *in vivo* and are the vector of choice for treatment of localized solid tumors. However, systemic administration of adenoviral vector for cancer treatment is not feasible as they induce a strong immune response as well as have a very affinity for the liver thereby causing toxicity. In contrast, nonviral vectors are less immune and are less toxic but the transgene expression is poor. Thus, a major hurdle in cancer gene therapy is the inability to deliver the gene-vector specifically to the tumor and obtain high expression levels of the therapeutic protein within the tumor microenvironment.

To overcome some of the existing limitations, several investigators have attempted to modify the vector or the plasmid DNA, using peptides to achieve tumor targeted gene delivery and enhanced gene expression. Oh *et al.*[133] showed enhanced gene delivery to cancer cells when a small transferrin peptide (Tf peptide) that binds to the human transferrin receptor (hTR) was linked to the C-terminus of the adenoviral fiber protein. The ability of the Tf peptide-modified adenoviral vector to increase gene delivery in cancer cells is due to the increased transferrin receptor expression in cancer cells compared to normal cells. Transferrin receptor expression is overexpressed at least 100-fold in several cancers including cancer of the liver, pancreas, breast, and prostate.[134] Attachment of RGD peptide motif to the fiber knob of an adenoviral vector expressing the HSV-tk suicide gene was shown to efficiently infect tumor cells that are deficient for the coxsackievirus and adenovirus receptor (CAR) and induce enhanced cell death.[135] Attachment of RGD peptide to the fiber knob of adenovirus was shown to infect CAR-negative tumor cells as efficiently as CAR-positive tumor cells.[136] RGD containing peptide binds to alpha (v) beta3-integrin and facilitate viral entry. Furthermore, in the same study, pegylation of the RGD-modified adenoviral vector was shown to evade the antibody-mediated humoral response. Similarly, RGD-peptide modified adenovirus vector expressing a transgene under the control of the fms-like tyrosine kinase receptor (FLT-1) was shown to specifically target tumor endothelial cells demonstrating the feasibility of the modified adenoviral vector to target tumor vasculature that express very high levels of alpha (v) beta3-integrin.[137] Apart from adenoviral vector, RGD-peptide when attached to the surface of liposomes and injected *in vivo* was shown to target activated platelets and suggested to be a potential vascular drug or gene delivery system.[138] *In vitro* studies from our laboratory showed that cells treated with

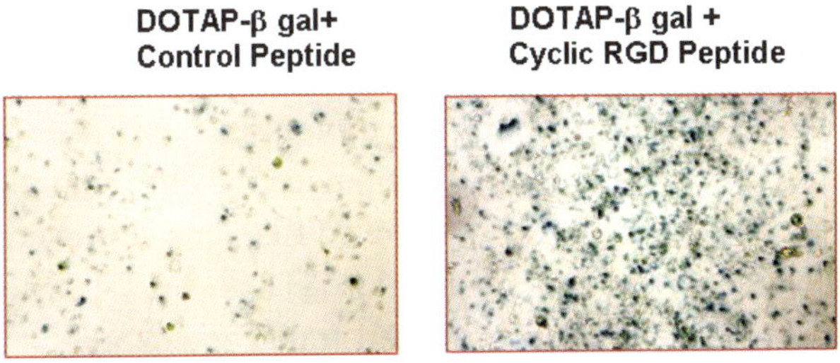


**Fig. 7** Cyclic RGD peptide enhances transfection. Human lung tumor (H226Br) cells were transfected with a mixture of cyclic RGD peptide and DOTAP:Chol-$\beta$-gal complex or with a mixture of cyclic control peptide and DOTAP:Chol-$\beta$-gal complex *in vitro*. Cells were analyzed for beta-galactosidase expression at 24 hr after transfection. Increased transgene expression was observed in cells that were treated with cyclic RGD peptide compared to cells that were treated with control peptide. Magnification × 100.

mixture of cyclic-RGD peptide and liposome enhanced transgene expression (Fig. 7; unpublished data). Another peptide that has been tested for increasing transduction efficiency is the human immunodeficiency virus transactivator of transcription (Tat) peptide. Pretreatment of adenovirus with a solution containing Tat peptide prior to transfection resulted in increased transduction of tumor cells and increased gene expression.[139] Similarly, incorporation of Tat peptide with solid lipid nanoparticle resulted in enhanced transduction and gene expression.[140] Additional peptides that have been attached to viral and nonviral vectors include peptide sequences from the Drosophila homeporotein antennapedia transcription protein (Antp),[139] membrane penetrating peptide (MPP),[141] modified branched peptides with a histidine-rich tail,[142] and penetratin, a 16-residue water soluble peptide.[143]

As indicated above in addition to the problems of poor vector transduction, another problem often encountered in cancer gene therapy is the inability to achieve desirable levels of intracellular protein expression. This is due to several reasons among which are the entrapment of the DNA in endosomes/lysosomes, degradation of DNA in the cytoplasm and poor movement of the DNA from the cytoplasm to the nucleus due to lack of nuclear localization signal (NLS). As a result, eradication of tumor cells is not complete, leading to the possibility of relapse. To overcome these limitations, studies have been carried out using small polybasic peptides derived from the transduction domains of certain proteins. The best characterized are the drosophila antennapedia protein (Antp), the herpes simplex virus structural protein VP22 and the HIV Tat peptide.[144] Expression of VP22 and HSV-TK as VP22-HSV-TK fusion protein in tumor cells was shown to increase transgene expression and enhance the antitumor activity both *in vitro* and *in vivo*.[145] Kong *et al.*[146] showed that lentiviral mediated delivery of MUC1 antibody targeted VP22-HSV-TK in MUC1 positive human ovarian carcinoma resulted in increased transgene expression leading to efficient tumor cell killing and increased animal survival. Induction of tumor cell apoptosis was demonstrated in p53 mutant cancer cells when the cells were transduced with an adenoviral vector expressing VP22-p53 as a fusion protein.[147] Use of VP22 has also been tested for enhancing the immune response. Alphavirus-mediated expression of VP22 with the human papilloma virus (HPV-16) E7 protein (VP22-E7) showed enhanced immune response leading to potent antitumor response.[148]

Studies have also shown that the use of Tat in conjunction with a therapeutic gene results in enhanced antitumor activity. Tasciotti *et al.*[149] showed that the fusion of Tat with HSV-TK resulted in increased expression and translocation of the thymidine kinase enzyme in cells which when exposed to the prodrug acyclovir were highly sensitive and underwent apoptosis. In another study, transduction of tumor cells with mdm-2-binding domain derived from $p14^{ARF}$ coupled to Tat showed increased cell killing compared with cells transduced with $p14^{ARF}$ alone or transduced with Tat alone.[150] More recently, Fang *et al.*[151] using recombinant adeno-associated virus (AAC) have shown that coupling of a short self-processing peptide (2A) derived from the foot-and-mouth disease virus (FAMDV) with a monoclonal antibody expression cassette results in stable expression of the antibody reaching therapeutic levels *in vivo*. In this study, the authors were able to express a mouse anti-VEGFR2 antibody in tumor-bearing mice that resulted in tumor suppression. In a separate study, Hu *et al.*[152] showed expression of 24-amino-acid N-terminal end of the p55 regulatory subunit of PI3K in gastric carcinoma cells resulted in cell-cycle arrest. The N-terminal amino acid of the p55 regulatory unit of PI3K is a retinoblastoma (Rb)-binding domain and affects Rb function and or Rb-E2F interaction. Additional potential peptides that have been proposed for use are the Zinc finger peptides.[153] The results from these studies indicate the potential use of peptides either alone or in conjunction with a therapeutic protein as a novel gene therapy strategy for cancer. However further preclinical testing and validation is needed prior to translating into the clinic.

## 4. Conclusions

With the completion of the human genome project, several novel genes have been identified that are frequently lost or altered in human cancer cells, thereby providing a panel of genes that can be used in cancer gene therapy.[154] Furthermore, it has allowed us to better understand the molecular biology of the disease. Additionally, the continuous ongoing effort to develop an efficient gene delivery vector has led to the development of several novel viral and nonviral targeted vectors that can specifically target the cancer cells. More recently, novel technologies such as nanoparticle-based cancer gene therapy have been initiated and appeared to hold promise.[155] Thus, it is anticipated that in the next few years, the existing challenges in cancer gene therapy will be overcome, resulting in a therapy that will be tailored to treat each cancer patient based on the genetic defects identified in that cancer type. With the large number of gene-based clinical trials conducted till date reporting no treatment-related toxicity, it is evident that cancer gene therapy holds much promise in the future.

## References


1. Anderson WF (1984) Prospects for human gene therapy. *Science* **226**:401–409.
2. (1990) Clinical protocol: The ADA human gene therapy clinical protocol. *Hum Gene Ther* **1**:327–362.
3. Bandara G, Robbins PD, Georgescu HI, Mueller GM, Glorioso JC and Evans CH (1992) Gene transfer to synoviocytes: Prospects for gene treatment of arthritis. *DNA Cell Biol* **11**:227–231.
4. Bellon G, Michel-Calemard L, Thouvenot D, Jagneaux V, Poitvein F, Malcus C, Accart N, Layani MP, Aymard M, Bernon H, Bienvenu J, Courtney M, Doring G, Gilly B, Gilly R, Lamy D, Levrey H, Morel Y, Paulin C, Perraud F, Rodillon L, Sene C, So S, Touraine-Moulin F, Schatz C and Pavirani A (1997) Aerosol administration of a recombinant adenovirus expressing CFTR to cystic fibrosis patients: A phase I clinical trial. *Hum Gene Ther* **8**:15–25.
5. Bohnlein E (1997) Gene therapy for HIV-1 disease in Concepts in Gene Therapy. Strauss M and Barranger JA (eds.) *Walter de Gruyter, New York*, pp. 431–465.

6. Bordignon C, Mavilio F, Ferrari G, Sevida P, Ugazio AG, Notarangelo LD, Gilboa E, Rossini S, O'Reilly RJ and Smith CA (1993) Transfer of the ADA gene into bone marrow cells and peripheral blood lymphocytes for the treatment of patients affected by ADA-deficient SCID. *Hum Gene Ther* **4**:513–520.
7. Brenner MK (1994) Genetic marking and manipulation of hematopoietic progenitor cells using retoviral vectors. *Immunomethods* **5**:204–210
8. Bunnell BA and Morgan RA (1998) Gene therapy for infectious diseases. *Clin Microbiol Rev* **11**:42–56.
9. Anderson WF (1995) Gene therapy. *Sci Am* **273**:124–128.
10. Friedman T (1989) Progress towards human gene therapy. *Science* **244**: 1275–1281.
11. Mulligan RC (1993) The basic science of gene therapy. *Science* **260**:926–932.
12. Roth JA and Christiano RJ (1997) Gene therapy for cancer: What have we done and where are we going? *J Natl Cancer Inst* **89**:21–39.
13. Greenlee R, Murray T, Bolden S and Wingo PA (2000) Cancer statistics CA. *Cancer J Clin* **50**:7–33.
14. Khan J, Wei JS, Ringner M, Saal LH, Ladanyi M, Westermann F, Berthold F, Schwab M, Antonescu CR, Peterson C and Meltzer PS (2001) Classification and diagnostic prediction of cancers using gene expression profiling and artificial neural networks. *Nat Med* **7**:673–679.
15. Esteller M (2005) Aberrant DNA methylation as a cancer-inducing mechanism. *Annu Rev Pharmacol Toxicol* **45**:629–656.
16. Shachaf CM, Kopelman AM, Aranitis C, Karisson CA, Beer S, Mandi S, Bachmann MH, Borowsky AD, Ruebner B, Cardiff RD, Yang Q, Bishop JM, Contag CH and Felsher DW (2004) MYC inactivation uncovers pluripotent differentiation and tumour dormancy in hepatocellular cancer. *Nature* **431**: 1112–1117.
17. Kenny PA and Bissell MJ (2003) Tumor reversion: Correction of malignant behavior by microenvironmental cues. *Int J Cancer* **107**:688–695.
18. Haber D and Harlow E (1997) Tumour-suppressor genes: Evolving definitions in the genomic age. *Nat Genet* **16**:320–322.
19. Bertelsen AH, Beaudry GA, Stoller TJ, Trotta PP and Sherman MI (1995) Tumor suppressor genes: Prospects for cancer therapies. *Biotechnology* **13**:127–131.
20. Ohgaki H (2005) Genetic pathways to glioblastomas. *Neuropathology* **25**:1–7.
21. Barber TD, Vogelstein B, Kinzler KW and Velculescu VE (2004) Somatic mutations of EGFR in colorectal cancers and glioblastomas. *N Engl J Med* **351**:2883.
22. Gardlik R, Palffy R, Hodosy J, Lukacs J, Turna J and Celec P (2005) Vectors and delivery systems in gene therapy. *Med Sci Monit* **11**:RA110–121 [Epub ahead of print].

23. Verma IM and Weitzman MD (2004) Gene therapy: Twenty-first century medicine. *Annu Rev Biochem* **74**:711–738.
24. Check E (2005) Sanctions agreed over teenager's gene-therapy death. *Nature* **433**:674.
25. Boyce N (1999) Who needs to know? Confusion reigns over the reporting of gene therapy deaths. *New Sci* **164**:15.
26. Moolten FL (1986) Tumor chemosensitivity conferred by inserted herpes thymidine kinase gene: Paradigm for a prospective cancer control strategy. *Cancer Res* **46**:5276–5281.
27. Whartenby KA, Marrogi AJ, Ramesh R, Freeman JL and Freeman SM (1995) The biology of cancer gene therapy. *Lab Invest* **72**:131–141.
28. Freeman SM, Wharternby KA, Koeplin DS, Moolten FL, Abboud CN and Abraham GN (1992) Tumor regression when a fraction of the tumor mass contains the HSV-TK gene. *J Cell Biochem* **16F**:47.
29. Freeman SM, Abboud CN, Whartenby KA and Abraham CN (1993) The bystander effect: Tumor regression when a fraction of the tumor mass is genetically modified. *Cancer Res* **53**:5274–5284.
30. Ramesh R, Munshi A, Marrogi AJ, Abboud CN and Freeman SM (1996) Expression of costimulatory molecules: B7 and ICAM upregulation after treatment with a suicide gene. *Cancer Gene Ther* **3**:373–384.
31. Ramesh R, Marrogi AJ, Munshi A, Abboud CN and Freeman SM (1996) *In vivo* analysis of the "bystander effect": A cytokine cascade. *Exp Hematol* **24**: 829–838.
32. Freeman SM, Ramesh R and Marrogi AJ (1997) Immune system in suicide gene therapy. *Lancet* **349**:2–3.
33. Mullan CA, Kilstrup M and Blaese RM (1992) Transfer of the bacterial gene for cytosine deaminase to mammalian cells confers lethal senstivity to 5-fluorocytosine: A negative selection system. *Proc Natl Acad Sci USA* **89**:33–37.
34. Huber BE, Austin EA, Good SS, Knick VC, Tibbels S and Richards C (1993) *In vivo* antitumor activity of 5-fluorocytosine on human colorectal carcinoma cells genetically modified to express cytosine deaminase. *Cancer Res* **53**: 4619–4626.
35. Djeha HA, Todryk SM, Pelech S, Wrighton CJ, Irvine AS, Mountain A and Lipinski KS (2005) Antitumor immune responses mediated by adenoviral GDEPT using nitroreductase/CB1954 is enhanced by high-level coexpression of heat shock protein 70. *Cancer Gene Ther* **12**:560–571.
36. Freeman SM, McCune C, Angel C, Abraham GN and Abboud CN (1992) Treatment of ovarian cancer using HSV-TK gene modified vaccine-regulatory issues. *Hum Gene Ther* **3**:342–349.

37. Eck SL, Alavi JB, Alavi A, Davis A, Hackney D, Judy K, Mollman J, Phillips PC, Wheeldon EB and Wilson JM (1996) Treatment of advanced CNS malignancies with the recombinant adenovirus H5.010RSVTK: A phase I trial. *Hum Gene Ther* **7**:1465–1482.
38. Alvarez RD and Curiel DT (1997) A phase I study of recombinant adenoviral vector-mediated intraperitoneal delivery of herpes simplex virus thymidine kinase gene and intravenous ganciclovir for previously treated ovarian cancer and extraovarian cancer patients. *Hum Gene Ther* **8**:597–613.
39. Schwarzenberger P, Harrison L, Weinacker A, Marrogi AJ, Byrne P, Ramesh R, Theodossiou C, Gaumer R, Freeman SM and Kolls J (1998) Clinical Protocol: gene therapy for malignant mesothelioma: A novel approach for an incurable cancer with increased incidencein Louisiana. *Hum Gene Ther* **9**:2641–2649.
40. Klatzmann D, Valery CA, Bensimon G, Marro B, Boyer O, Mokhtari K, Diquet B, Salzmann JL and Philippon J (1998) A phase I/II study of herpes simplex virus type I thymidine kinase "suicide" gene therapy for recurrent glioblastoma. Study group on gene therapy for glioblastoma. *Hum Gene Ther* **9**:2595–2604.
41. Floeth FW, Shand N, Bojar H, Prisack HB, Felsberg J, Neuen-Jacob E, Aulich A, Burger KJ, Bock WJ and Weber F (2001) Local inflammation and devascularization-in vivo mechanisms of the "bystander effect" in VPC-mediated HSV-TK/GCV gene therapy for human malignant glioma. *Cancer Gene Ther* **8**:843–851.
42. Vedlwijk MR, Berlinghoff S, Laufs S, Hengge UR, Zeller WJ, Wenz F and Fruehauf S (2004) Suicide gene therapy of sarcoma cell lines using recombinant adeno-associated virus 2 vectors. *Cancer Gene Ther* **11**:577–584.
43. Nemunaitis J, Cunningham C, Senzer N, Kuhn J, Cramm J, Litz C, Cavagnolo R, Cahill A, Clairmont C and Sznol M (2003) Pilot trial of genetically modified, attenuated salmonella expressing the E.coli cytosine deaminase gene in refractory cancer patients. *Cancer Gene Ther* **10**:737–744.
44. Pang S (2000) Targeting and eradicating cancer cells by a prostate-specific vector carrying the diphtheria toxin A gene. *Cancer Gene Ther* **7**:991–996.
45. Kagawa S, He C, Gu J, Koch P, Rha SJ, Roth JA, Curley SA, Stephens LC and Fang B (2001) Antitumor activity and bystander effects of the tumor necrosis factor-related apoptosis-inducing ligand (TRAIL) gene. *Cancer Res* **61**: 3330–3338.
46. Pataer A, Smythe WR, Yu R, Fang B, McDonnell T, Roth JA and Swisher SG (2001) Adenovirus-mediated Bak gene transfer induces apoptosis in mesothelioma cell lines. *J Thorac Cardiovasc Surg* **121**:61–67.

47. Kagawa S, Pearson SA, Ji L, Xu K, McDonnell TJ, Swisher SG, Roth JA and Fang B (2000) A binary adenoviral vector system for expressing high levels of the proapoptotic gene bax. *Gene Ther* **7**:75–79.
48. Norris JS, Hyer ML, Voelkel-Johnson C, Lowe SL, Rubinchik S and Dong JY (2001) The use of Fas Ligand, TRAIL and Bax in gene therapy of prostate cancer. *Curr Gene Ther* **1**:123–136.
49. Tong Y, Yang Q, Vater C, Venkatesh LK, Custeau D, Chittenden T, Chinnadurai G and Gourdeau H (2001) The pro-apoptotic protein, Bik, exhibits potent antitumor activity that is dependent on its BH3 domain. *Mol Cancer Ther* **1**:95–102.
50. Janknecht R (2004) On the road to immortality: hTERT upregulation in cancer cells. *FEBS Lett* **564**:9–13.
51. Boyd M, Mairs RJ, Keith WN, Ross SC, Welsh P, Akabani G, Owens J, Vaidyanathan G, Carruthers R, Dorrens J and Zalutsky MR (2004) An efficient targeted radiotherapy/gene therapy strategy utilizing human telomerase promoters and radioastatine and harnessing radiation-mediated bystander effects. *J Gene Med* **6**:937–947.
52. Jacob D, Bahra M, Schumacher G, Neuhaus P and Fang B (2004) Gene therapy in colon cancer cells with a fiber-modified adenovector expressing the TRAIL gene driven by the hTERT promoter. *Anticancer Res* **24**:3075–3079.
53. Jacob D, Schumacher G, Bahra M, Davis J, Zhu HB, Zhang LD, Teraishi F, Neuhaus P and Fang BL (2005) Fiber-modified adenoviral vector expressing the tumor necrosis factor-related apoptosis-inducing ligand gene from the human telomerase reverse transcriptase promoter induces apoptosis in human hepatocellular carcinoma cells. *World J Gastroenterol* **11**:2552–2556.
54. Yu D, Chen D, Chiu C, Razmazma B, Chow YH and Pang S (2001) Prostate-specific targeting using PSA promoter-based lentiviral vectors. *Cancer Gene Ther* **8**:628–635.
55. Greenblatt MS, Bennett WP, Hollstein M and Harris CC (1994) Mutations in the p53 tumor suppressor gene: Clues to cancer etiology and molecular pathogenesis. *Cancer Res* **54**:4855–4978.
56. Hartwell LH and Kastan MB (1994) Cell cycle control and cancer. *Science* **266**:1821–1828.
57. Harper JW, Adami GR, Wei N, Keyomarsi K and Elledge SJ (1993) The p21 Cdk-interacting protein Cip1 is a potent inhibitor of G1 cyclin-dependent kinases. *Cell* **75**:805–816.
58. Weinberg RA (1995) The retinoblastoma protein and cell cycle control. *Cell* **81**:603–608.

59. Perry ME and Levine AJ (1993) Tumor-suppressor p53 and the cell cycle. *Curr Opin Genet Dev* **3**:50–54.
60. Cai DW, Mukhopadhyay T, Liu YJ, Fujiwara T and Roth JA (1993) Stable expression of the wild-type p53 gene in human lung cancer cells after retrovirus-mediated gene transfer. *Hum Gene Ther* **4**:617–624.
61. Fujiwara T, Cai DW, Georges RN, Mukhopadhyay T, Grimm EA and Roth JA (1994) Therapeutic effect of a retroviral wild-type p53 expression vector in an orthotopic lung cancer model. *J Natl Cancer Inst* **86**:1458–1462.
62. Xu M, Kumar D, Srinivas S, Detolla LJ, Yu SF, Stass SA and Mixson AJ (1997) Parenteral gene therapy with p53 inhibits human breast tumors *in vivo* through a bystander mechanism without evidence of toxicity. *Hum Gene Ther* **8**:177–185.
63. Nakase M, Inui M, Okumura K, Kamei T, Nakamura S and Tagawa T (2005) p53 gene therapy of human osteosarcoma using a transferrin-modified cationic liposome. *Mol Cancer Ther* **4**:625–631.
64. Lane D (2004) Curing cancer with p53. *N Engl J Med* **350**:2711–2712.
65. Spitz FR, Nguyen D, Skibber JM, Meyn RE, Cristiano RJ and Roth JA (1996) Adenoviral-mediated wild-type p53 gene expression sensitizes colorectal cancer cells to ionizing radiation. *Clin Cancer Res* **2**:1665–1671.
66. Colletier PJ, Ashoori F, Cowen D, Meyn RE, Tofilon P, Meistrich ME and Pollack A (2000) Adenoviral-mediated p53 transgene expression sensitizes both wild-type and null p53 prostate cancer cells *in vitro* to radiation. *Int J Radiat Oncol Biol Phys* **48**:1507–1512.
67. Quist SR, Wang-Gohrke S, Kohler T, Kreienberg R and Runnebaum IB (2004) Cooperative effect of adenoviral p53 gene therapy and standard chemotherapy in ovarian cancer cells independent of the endogenous p53 status. *Cancer Gene Ther* **11**:547–554.
68. Roth JA, Nguyen D, Lawrence DD, Kemp BL, Carrasco CH, Ferson DZ, Hong WK, Komaki R, Lee JJ, Nesbitt JC, Pisters KM, Putnam JB, Schea R, Shin DM, Walsh GL, Dolormente MM, Han CI, Martin FD, Yen N, Xu K, Stephens LC, McDonell TJ, Mukopadhyay T and Cai D (1996) Retrovirus-mediated wild type p53 gene transfer to tumors of patients with lung cancer. *Nat Med* **2**:974–975.
69. Swisher S, Roth JA, Nemunaitis J, Lawrence DD, Kemp BL, Carrasco CH, Connors DG, El-Naggar AK, Fossela F, Glisson BS, Hong WK, Kurie JM, Lee JS, Mack M, Merritt JA, Nguyen DM, Nesbit JC, Perez-Soler R, Pisters KM, Putnam JB Jr, Richli WR, Savin M and Waugh MK (1999) Adenovirus-mediated p53 gene transfer in advanced non-small-cell lung cancer. *J Natl Cancer Inst* **91**:763–771.

70. Buller RE, Runnebaum IB, Karlan BY, Horowitz JA, Shahin M, Buekers T, Petrauskas S, Kreienberg R, Slamon D and Pegram M (2002) A phase I/II trial of rAd/p53 (SCH 58500) gene replacement in recurrent ovarian cancer. *Cancer Gene Ther* **9**:553–566.
71. Kuball J, Wen SF, Leissner J, Atkins D, Meinhardt P, Quijano E, Engler H, Hutchins B, Maneval DC, Grace MJ, Fritz MA, Storkel S, Thuroff JW, Huber C and Schuler M (2002) Successful adenovirus-mediated wild-type p53 gene transfer in patients with bladder cancer by intravesical vector instillation. *J Clin Oncol* **20**:957–965.
72. Lang FF, Bruner JM, Fuller GN, Aldape K, Prados MD, Chang S, Berger MS, McDermott MW, Kunwar SM, Junck LR, Chandler W, Zwiebel JA, Kaplan RS and Yung WK (2003) Phase I trial of adenovirus-mediated p53 gene therapy for recurrent glioma: Biological and clinical results. *J Clin Oncol* **21**: 2508–2518.
73. Schuler M, Herrmann R, De Greve JL, Stewart AK, Gatzemeier U, Stewart DJ, Laufman L, Gralla R, Kuball J, Buhl R, Heussel CP, Kommoss F, Perruchoud AP, Shepherd FA, Fritz MA, Horowitz JA, Huber C and Rochlitz C (2001) Adenovirus-mediated wild-type p53 gene transfer in patients receiving chemotherapy for advanced non-small-cell lung cancer: Results of a multicenter phase II study. *J Clin Oncol* **19**:1750–1758.
74. Ramesh R, Saeki T, Templeton NS, Ji L, Stephens LC, Ito I, Wilson DR, Wu Z, Branch CD, Minna JD and Roth JA (2001) Successful treatment of primary and disseminated human lung cancers by systemic delivery of tumor suppressor genes using an improved liposome vector. *Mol Ther* **3**:1–14.
75. Lee AW, Li JH, Shi W, Li A, Ng E, Liu TJ, Klamut HJ and Liu FF (2003) p16 gene therapy: a potentially efficacious modality for nasopharyngeal carcinoma. *Mol Cancer Ther* **2**:961–969.
76. Kawakami Y, Hama S, Hiura M, Nogawa T, Chiba T, Yokoyama T, Takashima S, Tajiri H, Eguchi K, Nagai N, Shigemasa K, Ohama K, Kurisu K and Heike Y (2001) Adenovirus-mediated p16 gene transfer changes the sensitivity to taxanes and Vinca alkaloids of human ovarian cancer cells. *Anticancer Res* **21**:2537–2545.
77. Yang ZY, Perkins ND, Ohno T, Nabel EG and Nabel GJ (1995) The p21 cyclin-dependent kinase inhibitor suppresses tumorigenicity *in vivo*. *Nat Med* **1**:1395–1403.
78. Sasaki T, Katayose Y, Suzuki M, Yamamoto K, Shiraso S, Mizuma M, Unno M, Takeuchi H, Lee CT and Matsuno S (2004) Adenovirus expressing mutant p27kip1 enhanced apoptosis against cholangiocarcinoma than adenovirus-p27kip1 wild type. *Hepatogastroenterology* **5**:68–75.

79. Ito I, Ji L, Tanaka F, Saito Y, Gopalan B, Branch CD, Xu K, Atkinson EN, Bekele BN, Stephens LC, Minna JD, Roth JA and Ramesh R (2004) Liposomal vector mediated delivery of the 3p FUS1 gene demonstrates potent antitumor activity against human lung cancer *in vivo*. *Cancer Gene Ther* **11**:733–739.
80. Gopalan B, Ito I, Branch CD, Stephens C, Roth JA and Ramesh R (2004) Nanoparticle based systemic gene therapy for lung cancer: molecular mechanisms and strategies to suppress nanoparticle-mediated inflammatory response. *Technol Cancer Res Treat* **3**:647–657.
81. Ji L, Nishizaki M, Gao B, Burbee D, Kondo M, Kamibayashi C, Xu K, Yen N, Atkinson EN, Fang B, Lerman MI, Roth JA and Minna JD (2002) Expression of several genes in the human chromosome 3p21.3 homozygous deletion region by an adenovirus vector results in tumor suppressor activities *in vitro* and *in vivo*. *Cancer Res* **62**:2715–2720.
82. Saito Y, Gopalan B, Mhashilkar AM, Roth JA, Chada S, Zumstein L and Ramesh R (2003) Adenovirus-mediated PTEN treatment combined with caffeine produces a synergistic therapeutic effect in colorectal cancer cells. *Cancer Gene Ther* **10**:803–813.
83. Saito Y, Swanson X, Mhashilkar AM, Oida Y, Schrock R, Branch CD, Chada S, Zumstein L and Ramesh R (2003) Adenovirus-mediated transfer of the PTEN gene inhibits human colorectal cancer growth *in vitro* and *in vivo*. *Gene Ther* **10**:1961–1969.
84. Davies MA, Lu Y, Sano T, Fang X, Tang P, LaPushin R, Koul D, Bookstein R, Stokoe D, Yung WK, Mills GB and Steck PA (1998) Adenoviral transgene expression of MMAC/PTEN in human glioma cells inhibits Akt activation and induces anoikis. *Cancer Res* **58**:5285–5290.
85. Seki M, Iwakawa J, Cheng H and Cheng PW (2002) p53 and PTEN/MMAC1/TEP1 gene therapy of human prostate PC-3 carcinoma xenograft, using transferrin-facilitated lipofection gene delivery strategy. *Hum Gene Ther* **13**:761–773.
86. Chada S, Sutton RB, Ekmekcioglu S, Ellerhorst J, Mumm JB, Leitner WW, Yang HY, Sahin AA, Hunt KK, Fuson KL, Poindexter N, Roth JA, Ramesh R, Grimm EA and Mhashilkar AM (2004) MDA-7/IL-24 is a unique cytokine–tumor suppressor in the IL-10 family. *Int Immunopharmacol* **4**:649–667.
87. Fisher PB, Gopalkrishnan RV, Chada S, Ramesh R, Grimm EA, Rosenfeld MR, Curiel DT and Dent P (2003) mda-7/IL-24, a novel cancer selective apoptosis inducing cytokine gene: From the laboratory into the clinic. *Cancer Biol Ther* **2**:S23–S37.
88. Ramesh R, Ito I, Saito Y, Wu Z, Mhashikar AM, Wilson DR, Branch CD, Roth JA and Chada S (2004) Local and systemic inhibition of lung tumor growth after nanoparticle-mediated mda-7/IL-24 gene delivery. *DNA Cell Biol* **23**:850–857.

89. Saeki T, Mhashilkar A, Chada S, Branch C, Roth JA and Ramesh R (2000) Tumor-suppressive effects by adenovirus-mediated mda-7 gene transfer in non-small cell lung cancer cell *in vitro*. *Gene Ther* **7**:2051–2057.
90. Mhashilkar AM, Schrock RD, Hindi M, Liao J, Sieger K, Kourouma F, Zou-Yang XH, Onishi E, Takh O, Vedvick TS, Fanger G, Stewart L, Watson GJ, Snary D, Fisher PB, Saeki T, Roth JA, Ramesh R and Chada S (2001) Melanoma differentiation associated gene-7 (mda-7): A novel anti-tumor gene for cancer gene therapy. *Mol Med* **7**:271–282.
91. Yacoub A, Mitchell C, Brannon J, Rosenberg E, Qiao L, McKinstry R, Linehan WM, Su ZS, Sarkar D, Lebedeva IV, Valerie K, Gopalkrishnan RV, Grant S, Fisher PB and Dent P (2003) MDA-7 (interleukin-24) inhibits the proliferation of renal carcinoma cells and interacts with free radicals to promote cell death and loss of reproductive capacity. *Mol Cancer Ther* **2**:623–632.
92. Saeki T, Mhashilkar A, Swanson X, Zou-Yang XH, Sieger K, Kawabe S, Branch CD, Zumstein L, Meyn RE, Roth JA, Chada S and Ramesh R (2002) Inhibition of human lung cancer growth following adenovirus-mediated mda-7 gene expression *in vivo*. *Oncogene* **21**:4558–4566.
93. Ramesh R, Mhashilkar AM, Tanaka F, Saito Y, Branch CD, Sieger K, Mumm JB, Stewart AL, Boquoi A, Dumoutier L, Grimm EA, Renauld JC, Kotenko S and Chada S (2003) Melanoma differentiation-associated gene 7/interleukin (IL)-24 is a novel ligand that regulates angiogenesis via the IL-22 receptor. *Cancer Res* **63**:5105–5113.
94. Ramesh R, Ito I, Gopalan B, Saito Y, Mhashilkar AM and Chada S (2004) Ectopic production of MDA-7/IL-24 inhibits invasion and migration of human lung cancer cells. *Mol Ther* **9**:510–518.
95. Gopalkrishnan RV, Sauane M and Fisher PB (2004) Cytokine and tumor cell apoptosis inducing activity of mda-7/IL-24. *Int Immunopharmacol* **4**: 635–647.
96. Chada S, Mhashilkar AM, Ramesh R, Mumm JB, Sutton RB, Bocangel D, Zheng M, Grimm EA and Ekmekcioglu S (2004) Bystander activity of Ad-mda7: Human MDA-7 protein kills melanoma cells via an IL-20 receptor-dependent but STAT3-independent mechanism. *Mol Ther* **10**:1085–1095.
97. Oida Y, Gopalan B, Miyahara R, Inoue S, Branch CD, Mhashilkar AM, Lin E, Bekele BN, Roth JA, Chada S and Ramesh R (2005) Sulindac enhances adenoviral vector expressing mda-7/IL-24-mediated apoptosis in human lung cancer. *Mol Cancer Ther* **4**:291–304.
98. McKenzie T, Liu Y, Fanale M, Swisher SG, Chada S and Hunt KK (2004) Combination therapy of Ad-mda7 and trastuzumab increases cell death in Her-2/neu-overexpressing breast cancer cells. *Surgery* **136**:437–442.

99. Nishikawa T, Ramesh R, Munshi A, Chada S and Meyn RE (2004) Adenovirus-mediated mda-7 (IL24) gene therapy suppresses angiogenesis and sensitizes NSCLC xenograft tumors to radiation. *Mol Ther* **9**:818–828.
100. Tong AW, Nemunaitis J, Su D, Zhang Y, Cunningham C, Senzer N, Netto G, Rich D, Mhashilkar A, Parker K, Coffee K, Ramesh R, Ekmekcioglu S, Grimm EA, van Wart Hood J, Merritt J and Chada S (2005) Intratumoral injection of INGN 241, a nonreplicating adenovector expressing the melanoma-differentiation associated gene-7 (mda-7/IL24): Biologic outcome in advanced cancer patients. *Mol Ther* **11**:160–172.
101. Cunningham CC, Chada S, Merritt JA, Tong A, Senzer N, Zhang Y, Mhashilkar A, Parker K, Vukelja S, Richards D, Hood J, Coffee K and Nemunaitis J (2005) Clinical and local biological effects of an intratumoral injection of mda-7 (IL24; INGN 241) in patients with advanced carcinoma: A phase I study. *Mol Ther* **11**:149–159.
102. Carmeliet P and Jain RK (2000) Angiogenesis in cancer and other diseases. *Nature* **407**:249–257.
103. Weidner N (1993) Tumor angiogenesis: Review of current applications in tumor prognostication. *Semin Diagn Pathol* **10**:302–313.
104. O'Reilly MS, Holmgren L, Shing Y, Chen C, Rosenthal RA, Moses M, Lane WS, Cao Y, Sage EH and Folkman J (1994) Angiostatin: A novel angiogenesis inhibitor that mediates the suppression of metastases by a Lewis lung carcinoma. *Cell* **79**:315–328.
105. O'Reilly MS, Boehm T, Shing Y, Fukai N, Vasios G, Lane WS, Flynn E, Birkhead JR, Olsen BR and Folkman J (1997) Endostatin: An endogenous inhibitor of angiogenesis and tumor growth. *Cell* **88**:277–285.
106. Maeshima Y, Sudhakar A, Lively JC, Ueki K, Kharbanda S, Kahn CR, Sonenberg N, Hynes RO and Kalluri R (2002) Tumstatin, an endothelial cell-specific inhibitor of protein synthesis. *Science* **295**:140–143.
107. Rege TA, Fears CY and Gladson CL (2005) Endogenous inhibitors of angiogenesis in malignant gliomas: nature's antiangiogenic therapy. *Neuro-oncol* **7**:106–121.
108. Morini M, Albini A, Lorusso G, Moelling K, Lu B, Cilli M, Ferrini S and Noonan DM (2004) Prevention of angiogenesis by naked DNA IL-12 gene transfer: angioprevention by immunogene therapy. *Gene Ther* **11**:284–291.
109. Blezinger P, Yin G, Xie L, Wang J, Matar M, Bishop JS and Min W (1999) Intravenous delivery of an endostatin gene complexed in cationic lipid inhibits systemic angiogenesis and tumor growth in murine models. *Angiogenesis* **3**: 205–210.
110. Dkhissi F, Lu H, Soria C, Opolon P, Griscelli F, Liu H, Khattar P, Mishal Z, Perricaudet M and Li H (2003) Endostatin exhibits a direct antitumor effect

in addition to its antiangiogenic activity in colon cancer cells. *Hum Gene Ther* **14**:997–1008.

111. Wang L, Schmitz V, Perez-Mediavilla A, Izal I, Prieto J and Qian C (2003) Suppression of angiogenesis and tumor growth by adenoviral-mediated gene transfer of pigment epithelium-derived factor. *Mol Ther* **8**:72–79.
112. Guan M, Pang CP, Yam HF, Cheung KF, Liu WW and Lu Y (2004) Inhibition of glioma invasion by overexpression of pigment epithelium-derived factor. *Cancer Gene Ther* **11**:325–332.
113. Noro T, Miyake K, Suzuki-Miyake N, Igarashi T, Uchida E, Misawa T, Yamazaki Y and Shimada T (2004) Adeno-associated viral vector-mediated expression of endostatin inhibits tumor growth and metastasis in an orthotropic pancreatic cancer model in hamsters. *Cancer Res* **64**:7486–7490.
114. Ponnazhagan S, Mahendra G, Kumar S, Shaw DR, Stockard CR, Grizzle WE and Meleth S (2004) Adeno-associated virus 2-mediated antiangiogenic cancer gene therapy: Long-term efficacy of a vector encoding angiostatin and endostatin over vectors encoding a single factor. *Cancer Res* **64**: 1781–1787.
115. Zacchigna S, Zentilin L, Morini M, Dell'Eva R, Noonan DM, Albini A and Giacca M (2004) AAV-mediated gene transfer of tissue inhibitor of metalloproteinases-1 inhibits vascular tumor growth and angiogenesis *in vivo*. *Cancer Gene Ther* **11**:73–80
116. Pfeifer A, Kessler T, Silletti S, Cheresh DA and Verma IM (2000) Suppression of angiogenesis by lentiviral delivery of PEX, a noncatalytic fragment of matrix metalloproteinase 2. *Proc Natl Acad Sci USA* **97**:12227–12232.
117. Ochsenebin AF (2002) Principles of tumor immunosurveillance and implications for immunotherapy. *Cancer Gene Ther* **9**:1043–1055.
118. Hodge JW, Abrams S, Schlom J and Kantor JA (1994) Induction of antitumor immunity by recombinant vaccinia viruses expressing B7-1 or B7-2 costimulatory molecules. *Cancer Res* **54**:5552–5555.
119. McLaughlin JP, Abrams S, Kantor J, Dobrzanski MJ, Greenbaum J, Schlom J and Greiner JW (1997) Immunization with a syngeneic tumor infected with recombinant vaccinia virus expressing granulocyte-macrophage colony-stimulating factor (GM-CSF) induces tumor regression and long-lasting systemic immunity. *J Immunother* **20**:449–459.
120. Melani C, Sule-Suso J, Arienti F, MacCalli C, Passerini F, Colombo M and Parmiani G (1995) A human melanoma cell line transduced with interleukin-4 gene by a retroviral vector releases biologically active IL-4 and maintains the original tumor antigenic phenotype. *Hum Gene Ther* **6**:1427–1436.
121. Toda M, Martuza RL, Kojima H and Rabkin SD (1998) *In situ* cancer vaccination: An IL-12 defective vector/replication-competent herpes simplex

virus combination induces local and systemic antitumor activity. *J Immunol* 4457–4464.

122. Wallack MK and Michaelides M (1984) Serologic response to human melanoma lines from patients with melanoma undergoing treatment with vaccinia melanoma oncolysates. *Surgery* **96**:791–800.
123. Miyahara R, Banerjee S, Kawano K, Efferson C, Tsuda N, Miyahara Y, Ioannides CG, Chada S and Ramesh R (2005) Induction of anticancer immunity by mda-7/IL-24. *Proc Am Assoc Cancer Res* **44**:1410.
124. Caudell EG, Mumm JB, Poindexter N, Ekmekcioglu S, Mhashilkar AM, Yang XH, Retter MW, Hill P, Chada S and Grimm EA (2002) The protein product of the tumor suppressor gene, melanoma differentiation-associated gene 7, exhibits immunostimulatory activity and is designated IL-24. *J Immunol* **168**:6041–6046.
125. Chen B, Timiryasova TM, Andres ML, Kajioka EH, Dutta-Roy R, Gridley DS and Fodor I (2000) Evaluation of combined vaccinia virus-mediated antitumor gene therapy with p53, IL-2, and IL-12 in a glioma model. *Cancer Gene Ther* **7**:1437–1447.
126. Liu X, Tian PK, Ju DW, Zhang MH, Yao M, Cao XT and Gu JR (2003) Systemic genetic transfer of p21WAF-1 and GM-CSF utilizing of a novel oligopeptide-based EGF receptor targeting polyplex. *Cancer Gene Ther* **10**:529–539.
127. Tan PH, Beutelspacher SC, Xue SA, Wang YH, Mitchell P, McAlister JC, Larkin DF, McClure MO, Stauss HJ, Ritter MA, Lombardi G and George AJ (2005) Modulation of human dendritic-cell function following transduction with viral vectors: Implications for gene therapy. *Blood* **105**:3824–3832.
128. Gabrilovitch D, Cunningham HT and Carbone DP (1996) IL-12 and mutant P53 peptide-pulsed dendritic cells for the specific immunotherapy of cancer. *J Immunother Emphasis Tumor Immunol* **19**:414–418.
129. Wierecky J, Mueller M and Brossart P (2005) Dendritic cell-based cancer immunotherapy targeting MUC-1. *Cancer Immunol Immunother* Apr 28; [Epub ahead of print].
130. (1990) Clinical protocol: TNF/TIL human gene therapy clinical protocol. *Hum Gene Ther* **1**:443.
131. Moller P, Sun Y, Dorbic T, Alijagic S, Makki A, Jurgovsky, Schroff M, Henz BM, Wittig B and Schadendorf D (1998) Vaccination with IL-7 gene–modified autologous melanoma cells can enhance the anti-melanoma lytic activity in peripheral blood of patients with a good clinical performance status: a clinical phase I study. *Br J Cancer* **77**:1907–1916.
132. Sun Y, Jurgovsky K, Moller P, Alijagic S, Dorbic T, Georgieva, Wittig B and Schadendorf D (1998) Vaccination with IL-12 gene-modified autologous

melanoma cells: preclinical results and a first clinical phase I study. *Gene Ther* **5**:481–490.

133. Oh KS, Engler JA and Joung I (2005) Enhancement of gene delivery to cancer cells by a retargeted adenovirus. *J Microbiol* **43**:179–182.
134. Li H, Sun H and Qian ZM (2002) The role of the transferrin-transferrin receptor system in drug delivery and targeting. *Trends Pharmacol Sci* **23**:206–209.
135. Mizuguchi H and Hayakawa T (2002) Enhanced antitumor effect and reduced vector dissemination with fiber-modified adenovirus vectors expressing herpes simplex virus thymidine kinase. *Cancer Gene Ther* **9**:236–242.
136. Eto Y, Gao JQ, Sekiguchi F, Kurachi S, Katayama K, Maeda M, Kawasaki K, Mizuguchi H, Hayakawa T, Tsutsumi Y, Mayumi T and Nakagawa S (2005) PEGylated adenovirus vectors containing RGD peptides on the tip of PEG show high transduction efficiency and antibody evasion ability. *J Gene Med* **7**:604–612.
137. Work LM, Ritchie N, Nicklin SA, Reynolds PN and Baker AH (2004) Dual targeting of gene delivery by genetic modification of adenovirus serotype 5 fibers and cell-selective transcriptional control. *Gene Ther* **11**:1296–1300.
138. Gupta AS, Huang G, Lestini BJ, Sagnella S, Kottke-Marchant K and Marchant RE (2005) RGD-modified liposomes targeted to activated platelets as a potential vascular drug delivery system. *Thromb Haemost* **93**:106–114.
139. Gratton JP, Yu J, Griffith JW, Babbitt RW, Scotland RS, Hickey R, Giordano FJ and Sessa WC (2003) Cell-permeable peptides improve cellular uptake and therapeutic gene delivery of replication-deficient viruses in cells and *in vivo*. *Nat Med* **9**:357–362.
140. Rudolph C, Schillinger U, Ortiz A, Tabatt K, Plank C, Muller RH and Rosenecker J (2004) Application of novel solid lipid nanoparticle (SLN)-gene vector formulations based on a dimeric HIV-1 TAT-peptide *in vitro* and *in vivo*. *Pharm Res* **21**:1662–1669.
141. Xie J, Yu BF, Xu J, Zhang YH, Cheng NL, Niu B, Hu XN, Xiang Q and Zhang ZG (2005) Protein transduction domain of membrane penetrating peptide can efficiently deliver DNA and protein into mouse liver for gene therapy. *Hepatobiliary Pancreat Dis Int* **4**:90–93.
142. Leng Q and Mixson AJ (2005) Modified branched peptides with a histidine-rich tail enhance in vitro gene transfection. *Nucleic Acids Res* **33**:e40.
143. Christiaens B, Dubruel P, Grooten J, Goethals M, Vandekerckhove J, Schacht E and Rosseneu M (2005) Enhancement of polymethacrylate-mediated gene delivery by Penetratin. *Eur J Pharm Sci* **24**:525–537.
144. Wadia JS and Dowdy SF (2005) Transmembrane delivery of protein and peptide drugs by TAT-mediated transduction in the treatment of cancer. *Adv Drug Deliv Rev* **57**:579–596.

145. Qiu Z, Harms JS, Zhu J and Splitter GA (2004) Bovine herpesvirus tegument protein VP22 enhances thymidine kinase/ganciclovir suicide gene therapy for neuroblastomas compared to herpes simplex virus VP22. *J Virol* **78**:4224–4233.
146. Kong B, Wang W, Liu C, Song L, Ma D, Qu X, Jiang J, Yang X, Zhang Y, Wang B, Wei MQ and Yang Q (2003) Efficacy of lentivirus-mediated and MUC1 antibody-targeted VP22-TK/GCV suicide gene therapy for ovarian cancer. *In Vivo* **17**:153–156.
147. Roy I, Holle L, Song W, Holle E, Wagner T and Yu X (2002) Efficient translocation and apoptosis induction by adenovirus encoded VP22-p53 fusion protein in human tumor cells in vitro. *Anticancer Res* **22**:3185–3189.
148. Cheng WF, Hung CF, Hsu KF, Chai CY, He L, Polo JM, Slater LA, Ling M and Wu T (2002) Cancer immunotherapy using Sindbis virus replicon particles encoding a VP22-antigen fusion. *Hum Gene Ther* **13**:553–568.
149. Tasciotti E, Zoppe M and Giacca M (2003) Transcellular transfer of active HSV-1 thymidine kinase mediated by an 11-amino-acid peptide from HIV-1 Tat. *Cancer Gene Ther* **10**:64–74.
150. Jarajapu YP, Baltunis J, Knot HJ and Sullivan SM (2005) Biological evaluation of penetration domain and killing domain peptides. *J Gene Med* **7**:908–917.
151. Fang J, Qian JJ, Yi S, Harding TC, Tu GH, Vanroey M and Jooss K (2005) Stable antibody expression at therapeutic levels using the 2A peptide. *Nat Biotechnol* **23**:584–590. Epub 2005 Apr 17.
152. Hu J, Liu S, Wang J, Luo X, Gao X, Xia X, Feng Y, Tao D, Wang G, Li X, Zhao J, Ding H, Reed E, Li QQ and Gong J (2005) Overexpression of the N-terminal end of the p55gamma regulatory subunit of phosphatidylinositol 3-kinase blocks cell cycle progression in gastric carcinoma cells. *Int J Oncol* **26**: 1321–1327.
153. Corbi N, Libri V, Onori A and Passananti C (2004) Synthetic zinc finger peptides: Old and novel applications. *Biochem Cell Biol* **82**:428–436.
154. Venter JC, Adams MD and Myers EW *et al.* (2001) The sequence of the human genome. *Science* **29**:1304–1351. Erratum in: *Science* 2001 **292**:1838.
155. Kubik T, Bogunia-Kubik K and Sugaisaka M (2005) Nanotechnology on duty in medical applications. *Curr Pharm Biotechnol* **6**:17–33.

# Index